Introduction to
Electronic Defense Systems

For a complete listing of the *Artech House Radar Library*,
turn to the back of this book . . .

Introduction to
Electronic Defense Systems

Filippo Neri

Artech House
Boston • London

Library of Congress Cataloging-in-Publication Data
Neri, Filippo.
 [Introduzione ai sistemi di difesa elettronica. English.]
 Introduction to electronic defense systems / Filippo Neri.
  p.   cm.
 Translation of: Introduzione ai sistemi di difesa elettronica.
 Includes bibliographical references and index.
 ISBN 0-89006-553-5
  1. Electronics in military engineering. I. Title.
 UG485.N4713 1991   91-28823
 623'.043—dc20   CIP

© 1991 Artech House, Inc.
685 Canton Street
Norwood, MA 02062

International Standard Book Number: 0-89006-553-5
Library of Congress Catalog Card Number: 91-28823

10 9 8 7 6 5 4 3 2 1

Foreword

Winston Churchill called it the "wizard war," the Soviets call it "electronic combat," the West calls it "electronic warfare" (EW), but from stealth technology to decoys, from *electronic support measures* (ESM) to off-board jamming, Filippo Neri calls it "electronic defense."

For most platforms electronic defense is not a stand-alone system but a means to improve survivability for an isolated target or for an entire battle group. Electronic warfare, in all its varieties, was used extensively in the 1973 Yom Kippur war and throughout 1982 in the Bekaa Valley and the South Atlantic Crisis. More recently, in 1991, EW achieved remarkable results in the Gulf War. Today's EW can therefore be accepted as "battle proven."

The benefits of EW are manifest. A decoy force can prevent massive casualties, or waste a missile harmlessly. A "jammer" can now create an entire fleet, its decoy targets fixed in position from scan to scan, and an ESM can fingerprint each individual modulator, so that a specific platform can be followed from home to over the horizon.

Once radar had become operational in 1937 with the Chain Home system, the need for *electronic countermeasures* (ECM) became apparent. Yet each time the EW response was exposed, *electronic counter-countermeasures* (ECCM) were developed. In particular an increase in frequency provided the radar with a significant ECCM advantage. However, now that stealth technology has been proven operationally by the F-117s flying unhindered over Baghdad, there are strong indications that radars will have to return to unseemly low frequencies, perhaps even as low as the 25 MHz used by the Allied radars in the late 1930s.

i

As the art of radar has evolved, so have electromagnetic countermeasures become more sophisticated, often employing the best and fastest computers. However, missiles have become covert, using *low probability of intercept* (LPI) technologies and incorporating dual-mode infrared seekers or home-on-jamming. So the EW suite must encompass *electro-optic countermeasures* (EOCM) and seek the benefits from sensor fusion techniques.

The dramatic advances in EW technology during the past decades have become cloaked in security and technical jargon. Filippo Neri has taken the complex technical concepts of the sophisticated radar and IR threats, and for each in turn developed a simple explanation of the EW solution.

He explains the principles with lavish diagrams and schematics. How easy it is to master the complex EW concepts with the help of a world expert who recognizes the need for simplicity. Negotiating the problems of security and national interests, he describes openly and frankly the chess-like relationship between poacher and gamekeeper.

Electronic defense is one technology that can be tested in peacetime, without the need to shoot down the threat. Yet there is the conundrum that when practiced in peacetime the enemy is alerted and can make ready with counter-countermeasures. However, it should be remembered that showing that one is ready to fight a war is the surest way of preventing one.

Where does EW start and finish? Can it stand apart from the combat system and be presented to the commander as an alternative? How can the operator be assured that EW is working, when there is no huge plume, white-hot metal or acoustic signature?

The answers to these and many other questions are to be found in this book of clarity and breadth. A solution to the supersonic, high-diving, fast-weaving stealthy missile is here. A hardkill solution is not.

Peter Varnish,
Director, Above Water Warfare
Defence Research Agency
Portsdown, Portsmouth, England

Acknowledgements

I wish first of all to thank Filippo Fratalocchi, President of Elettronica SpA, and Enzo Benigni, Vice-President of Elettronica SpA, for the encouragement they gave me while I was writing this book.

I wish also to thank my colleagues at Elettronica SpA and at other firms and organizations who have collaborated in the writing, and in particular the following colleagues at Elettronica SpA:

Sandro Carnevale, and the analysis group which he heads, for their help in reviewing the whole book and for their contributions on ESM processing, IR, and missile systems, Andrea De Martino for his help and for material on radar systems, Salvatore Scarfò for contributions to and revision of the material on ECM, Professor Guido Gasparrini for revision of the material on electro-optic systems, Candidoro Giannicchi for revision of material on tracking systems, and Michele Russo for contributions on telecommunications systems.

Finally, it is a pleasure to thank Admiral Mario De Arcangelis for his help and his valued advice, Gianfranco Scafè who kindly performed the onerous task of final revision, and Mr E. T. Hill who revised the English-language version.

Filippo Neri

"It is my hope that as soon as the evidently greater effectiveness of air, naval and land platforms equipped with electronic defense systems has once again been demonstrated, the armed forces of all nations will aim more at the *quality* rather than at the *quantity* of their assets. The acceptance of this concept, besides ensuring an effective defense, will lead to a significant improvement in the cost-effectiveness of military expenditure."

Filippo Fratalocchi,
Founder and President
Electronica SpA

Preface

In my work as a designer of electronic defense equipment, I have often realized that there is no book, readily available to the designer, that explains the principal functions of the different electronic warfare systems, what are the vulnerable parts of radars, what are the limitations of weapon systems, and what it is that makes an electronic defense system effective.

Taking advantage of my experience as a designer of radar and weapon systems, I thought that a single volume describing the operating principles of both weapon systems and electronic defense systems might be useful to those wishing or needing to enter the field.

The book is addressed to those who are about to start working as designers of these systems, to those who are or will become their users, and to those who administer their procurement.

The formulas and the mathematical theory have been reduced to a minimum, and readers are frequently invited to consult the appropriate references for in-depth analyses. The book can therefore be read and understood by anyone with high-school education and interest in the systems used by the armed forces.

The book is divided into eight chapters. Chapter 1 explains briefly the usefulness of electronic defense, how it is organized, and what systems it includes. In addition it describes the operational objectives of electronic defense.

Chapter 2 analyzes the sensors of weapon systems in order to highlight their merits and, above all, their limits. The objective is to help the designer to exploit their weaknesses.

Chapter 3 gives examples of artillery and missile systems

that use the electronic sensors described in Chapter 2, again with the objective of emphasizing how their effectiveness may be reduced.

Chapter 4 describes and analyzes electronic systems dedicated to passive interception, generally known as electronic support measures (ESM). Both those using radio-frequency emissions and those using infrared are described.

Chapter 5 is devoted to the more striking part of electronic defense: electronic countermeasures (ECM), i.e., the generation of signals that, by interfering with the receivers of "victim" systems, cause a degradation in the performance of the associated weapon systems.

Since weapon systems can undergo intentional jamming aimed at reducing their capabilities, they have been equipped with counter- countermeasures (ECCM) systems. These systems and their applications are described in Chapter 6, which gives an account of their effectiveness.

But the thrust and parry of countermeasures and counter-countermeasures seems to have no end, and new technologies ensure that it is now possible to attack the very operating principles of weapon systems, which makes the development of effective counter- countermeasures very difficult. Chapter 7 describes this technological advance.

Finally, to assist designers in achieving the optimum solution of the design problems described in Chapters 4 and 5, Chapter 8 lists the criteria that should be followed during the design stage of electronic defense systems. Chapter 8 also discusses the methods of evaluation and simulation that can determine whether a system is really effective. This should be particularly helpful to those who have to decide on the choice of an electronic defense system.

I hope that this book will be found useful. It is best seen as a simple reference book for the rapid evaluation and organization of material. In the end, the quality of a system and the achievement of its operational objectives will depend, as always, on the ability, the preparation, and the dedication of those who have to perform the work.

Table of Contents

4.   ELECTRONIC INTERCEPT SYSTEMS

Chapter 1

Electronic Defense

1.1 Introduction

With the passing of time, electronic technology has come to play an increasingly important role in military operations. The electronic era, and with it the first steps in the introduction of electronics into weapons, goes back to the time when radio and the radio direction finder were first used to give the platform position. The second step was the introduction of radar for the detection, and location in angle and in range, of hostile platforms, and its subsequent use to increase the accuracy of artillery. The last step, and probably the most lethal one, has been the use of electronic devices for precision guidance of missiles (Fig. 1.1).

The effectiveness of electronically guided weapon systems, expressed in terms of kill probability, has risen to values very

Figure 1.1 A missile system exploits radar signals to hit its targets with precision.

close to unity, thus leaving undefended targets little hope of escape. Consequently, almost all effective weapons now employ electronic guidance devices. However, the sophistication of today's weapon systems is such that they are rendered worthless should their electronic circuits not operate correctly. As a result, it has become essential to develop counter electronic systems capable of reducing the effectiveness of weapon guidance devices.

The fruitfulness of these countermeasure techniques has quickly become apparent. They have been developed to the point that they can seriously degrade the performance of nearly all weapon systems (Fig. 1.2). The inevitable next step has been the development of counter-countermeasures to try to restore the original effectiveness of the weapon sensors.

The techniques and technologies that lead to the construction of devices capable of electronically countering a weapon system, and to the development of counter-countermeasures,

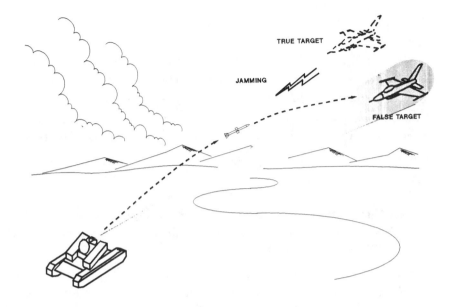

Figure 1.2 The aim of an electronic defense system is to incapacitate the enemy's weapon systems by generating electromagnetic jamming signals.

go under the name "electronic warfare." However, given the basic harmlessness of these electronic systems ("Electrons don't make holes", at least as long as no directed-energy weapons are available), the name "electronic defense" seems more appropriate.

1.2 Systems in use in the armed forces

In every country, the armed forces have at their disposal a number of weapon systems, each with a different function. The following brief survey of the missions of navy, army, and air force will help to identify the main weapon systems against which electronic defense must operate.

It should be emphasized that the aim of this survey is the identification of electronically guided weapon systems, without reference to any specific military organization. Moreover, systems pertaining specifically to nuclear warfare are outside the scope of this book; only systems used with conventional armaments will be discussed.

Generally speaking, the mission of an air force is the surveillance and defense of the sky above national territory, the mission of a navy is the surveillance of the seas surrounding national territory and protection of important sea routes, and the mission of an army is the protection of the national territory itself.

The Air Force

An air force has to provide air defense of the national territory, coordinating its own systems with those of the other armed forces. That is, the air force must:

- contribute to the survival of important centers.
- inflict losses and give attrition when attacked by an enemy.
- ensure the neutralization of important military objectives in enemy territory.
- give air support to land and sea action.
- ensure air transport.
- execute reconnaissance as necessary.

Under hostile air attack, survival of important centers and attrition of the enemy are achieved by combining the use of a surveillance (or search) radar network with the deployment of air forces in the area where an incursion has been detected.

Such a combination is called an air defense network. Air defense search radars are characterized by high sensitivity, and can detect and give early warning of, the approach of targets at long ranges. They are sometimes called *early warning radars* (EWR), and are characterized by high sophistication and reliability, as they must operate continuously in the complete range of environments.

Because of their sensitivity, they can give broad cover to the national air space. The coordination of the data they provide and the correlation with other information (data fusion) takes place in special command and control centers, where operational decisions are taken.

In time of peace, upon the detection of a suspect aircraft (i.e., one that has not spontaneously revealed its identity), the search centers warn an air unit, usually consisting of two very fast and maneuverable fighter aircraft, and (with the help of an appropriate communications system) guide it towards the position of the aircraft which has to be identified. The interceptor fighters approach the target, identify it, and either let it pass or force it to retrace its steps or to land (Fig. 1.3).

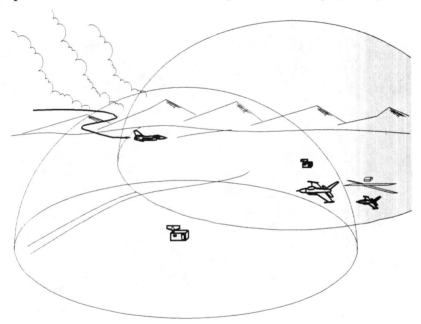

Figure 1.3 The air defense network detects and locates all aircraft penetrating into national air space.

In time of war, the procedure is different. As soon as the presence of an alien aircraft has been discovered, the fighters (Fig. 1.4) take to the air with quite different intentions. They are still guided from the ground and try to locate the target as soon as possible with their own on-board radar. An identification is made with the help of identification of friend or foe (IFF) equipment, devices for automatic recognition of friendly and hostile platforms). If the result warrants it, they lock onto the target and fire at it with their on-board weapons (usually air-to-air missiles). They then try to make a "kill assessment," that is, to determine the amount of damage suffered by the target and finally make their way back to base. If the target is not immediately hit by a long- or medium-range missile, the fighters will have to approach closer to the intruding aircraft, starting a series of dog fights either by launching short-range missiles, usually infrared-guided, or by firing their *on-board cannon*.

Figure 1.4 The task of fighter aircraft is to ensure air space superiority. The photograph shows a model of the European fighter aircraft (EFA).

Surveillance of air space can be conducted directly by air patrols or by a network of air defense radars.

The neutralization of military objectives of special importance on enemy territory is achieved by sending special strike aircraft (Fig. 1.5) and bombers. Strike aircraft make covert surprise attacks with a few units approaching the target at very low altitude. On the other hand, bomber tactics entail a powerful attack by many aircraft, aided by fighters, with radar and electro-optic sensors being employed to locate and identify

Figure 1.5 Strike aircraft are entrusted with the task of hitting important military objectives on enemy ground. The photograph shows the Tornado in its IDS version.

their ground targets.

In its air support role, the air force cooperates with ground forces to stop the advance of enemy forces. It launches air raids against advancing enemy columns and bombs their tanks and support services (the ground attack function). This function is carried out by fighter-bombers (Fig. 1.6), which are extremely maneuverable aircraft fitted with a variety of air-to-surface weapon systems.

The air force will usually have to provide also for the defense of its own bases, airfields and services.

To sum up, in order to be able to carry out the functions detailed above the air force will require the following systems:

• Surveillance and search systems, consisting of land-based search radars (Fig. 1.7) positioned on high ground (i.e., mountains, high hills) to offset the limitations of radar range at low altitude, as explained in Chapter 2.

• Airborne surveillance and search systems (Fig. 1.8). These

Figure 1.6 The high maneuverability of fighter bombers allows for quick raids against enemy ground forces. The photograph shows the Italo-Brazilian AMX fighter bomber.

systems are similar to the preceding ones but weigh less, and can therefore be carried on board aircraft having long-range capabilities without refuelling. Such systems help to solve the problem of intercepting low level targets at adequate ranges.

- Fighters, characterized by very high speed and maneuverability, fitted with:

(a) radar systems for target acquisition and tracking (airborne interceptors) (Fig. 1.9). An on-board radar of this type is set in the front of the fuselage, and protected with a tapering shell, transparent to electromagnetic waves, called a "Radome" (radar dome).

(b) long-, medium-, and short-range air-to-air missile (AAM) systems (Fig. 1.10).

(c) airborne weapon delivery systems for combat at very close range.

- Strike aircraft, for raids into hostile territory. These must be equipped with:

Figure 1.7 The early detection of targets is ensured by search radars. The photograph shows the RAT 31S search radar for land-based installations.

(a) avionic radar for target acquisition and tracking.

(b) AAM and air-to-surface missile (ASM) systems.

(c) bomb control and guidance systems.

Often this type of aircraft is fitted with a terrain-following radar for blind navigation at very low altitude.

(d) fighter bombers for ground attack.

Besides all these, mention must be made of other aircraft assigned to the following missions:

- transportation.
- patrolling.
- training.

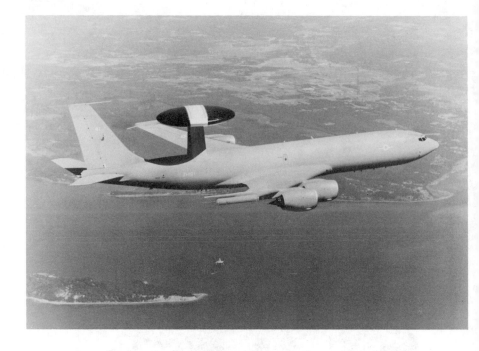

Figure 1.8 Airborne search radars permit detection of targets flying at low altitude. The photograph shows the airborne warning and control system (AWACS).

- surface-to-air missile (SAM) systems, for the defense of air-fields and other objectives of vital importance. A missile system can be long-range, for the defense of a zone or an area (area defense) (Fig. 1.11), or medium to short range for the defense of a site or a point (point defense) (Fig. 1.12). Usually, a missile defense system consists of a local search radar (sometimes called acquisition radar) which detects incoming threats, points them out target Indication (TI) to the different sensors capable of tracking them one by one, to provide the guidance to different missiles against the threatening platforms.
- firing systems or artillery antiaircraft artillery (AAA) systems equipped, as for point defense, with search (acquisition)

Figure 1.9 Military aircraft exploit sophisticated radars for accurate target detection and weapon guidance. The photograph shows the APG 65 radar.

Figure 1.10 The AAM Sidewinder.

radar and with various tracking radars capable of directing the fire of the associated cannon and machine guns to designated targets with great precision.

Figure 1.11 Long-range SAM systems are frequently employed to defend huge areas of territory. The photograph shows the SAM Patriot system.

The Navy

In brief, the main tasks of a navy are the following:

- protection of sea traffic (e.g., convoys.)
- coastal protection against potential attacks by inflicting heavy losses on the enemy in the open sea.

Convoy protection is carried out by well-armed naval vessels, which specialize in conflict with various potential attack

Figure 1.12 The defense of specially important sites is ensured by short to medium-range SAM systems. The photograph shows the Spada missile system.

systems such as submarines, aircraft, and other large ships. When some of a nation's interests lie far from its own territorial waters, it will be necessary to deploy aircraft carriers (Fig. 1.13) to ensure adequate air cover for the fleet. For the Navy, too, the system requirements will be on a par with those for air defense. In fact, what is needed is the organization of the defense of the very wide area covered by the whole fleet.

The following systems will therefore be required:

- shipborne and airborne EWR to prevent surprise attack against such a valuable target as a carrier.
- fighters and strike aircraft, equipped with medium- and long-range AAM and ASM systems.

Because of its enormous value, a carrier is usually escorted by other naval vessels such as cruisers, destroyers, and frigates. Ships of these three types constitute the usual naval armament of those nations whose strong interests are normally limited to their own territorial waters.

Cruisers (Fig. 1.14) are heavily armed, medium- to high-tonnage ships (8,000–20,000 tons displacement). They defend the formations which they escort from air, surface, and

Figure 1.13 Aircraft carriers give the necessary air cover to the fleet when it operates far from national waters.

underwater threats.

Destroyers (4,000–8,000 tons) are in practice large frigates equipped with a variety of armament.

Frigates (Fig. 1.15) are well-armed low to medium tonnage ships (1,500–4,500 tons) whose task is to provide an effective escort to other ships in convoy or formation. Often they are designed for antisubmarine warfare. Shipborne helicopters are frequently used in order to increase the effectiveness of this role.

For patrolling not too far from coasts, moderately armed small tonnage vessels are often used including corvettes (200–800 tons Fig. 1.16) and sometimes hydrofoils.

Effective patrolling is often achieved with small or medium

Figure 1.14 The Italian "all-deck" cruiser Garibaldi allowing the usage of vertical and short takeoff and landing (V-STOL) aircraft.

tonnage submarines because of their very low detectability when submerged, and when the level of noise is kept low. However, submarines and electroacoustic equipment, such as active and passive sonars, and weapon systems such as magnetically or sonar-guided torpedoes are beyond the scope of this book.

Minesweepers are responsible for the detection and neutralization of submerged mines dispersed by the enemy on major naval routes.

Operationally, in time of war the seas are patrolled by the various naval formations. Each vessel has an on-board long-range air-search system for early detection of potential air attack. For good sensitivity at medium to high altitudes, this radar operates at rather low frequencies, which, as will be seen later, does not allow good sensitivity detection at low altitudes. In order to detect surface targets, a higher-frequency radar is needed. This last requirement may sometimes be met in conjunction with the navigation radar, whose range is horizon-limited.

A naval vessel is a relatively easy target. To avoid detection,

Figure 1.15 The Mistral class frigate. The following radar-employing systems are distinguishable: 1–air-search radar, 2–navigation radar, 3–tracking radar for guidance of cannon and SAMs, 4–antiship Otomat missiles.

it has to limit its own radar and other electromagnetic emissions as much as it can.

Observed threats are assigned (in the jargon, this operation is called "designation or TI") first, if sufficiently distant, to long-range missile systems (area defense), and to SAM systems or to artillery fire (point defense) if at shorter ranges.

When a ship-to-ship engagement takes place, the first step is the launching of antiship surface-to-surface missile (SSM) systems. These are often called sea skimmers because they attempt to fly at an extremely low elevation where search radars do not see well (Fig. 1.16). Obviously, each vessel tries to be

Figure 1.16 A radar-guided, sea-skimming missile is the greatest threat to a naval vessel. The photograph shows the launching of an antiship Otomat missile and a Harpoon missile.

the first to fire. If no missiles are available, or if the target is not worthy of them, artillery systems can be used.

As can easily be seen, the main threat to a ship is the anti-ship sea-skimming missile, because of its high kill probability and its low detectability.

For their own defense, naval vessels are frequently equipped with SAM systems (Fig. 1.17) and special short-range systems, or close-in weapon systems (CIWS), that, once enabled, come into action automatically and fire at missile threats detected at the very last moment (Fig. 1.18).

To sum up, the main weapon systems for a navy are:
- Shipborne early warning surveillance and search systems.
- Airborne surveillance and search systems.
- Medium- to long-range SAM systems.

Figure 1.17 Launching of a missile of the ALBATROS system.

Figure 1.18 Short-range defense systems are the last links in a ship's defense system. The photograph shows the DARDO system and the model of the latest MYRIAD.

- Short- to medium-range SAM systems.
- Artillery or other weapon delivery systems.
- SSM systems.
- CIWS.

The Army

The task of an army is to conduct operations on the ground which will wear the enemy down by a process of attrition and repel or deter an attack.

To achieve this end, the army will make use of the usual corps:

- Infantry.
- Armored units.
- Artillery.
- Engineers.

To stop or weaken enemy forces, the army will have at its disposal ballistic or inertial-guidance SSMs and long-range artillery to strike in depth and to prevent the enemy from taking the initiative.

It will have tanks to counter enemy tanks. Those tanks will have weapon delivery systems controlled by laser rangefinders, to enable them to hit with the first shot, hopefully without having come to a halt.

The army will be provided with SAM systems (Fig. 1.19) to counter the enemy's ground attack aircraft, and with radar-guided artillery (AAA) (Fig. 1.20). A characteristic of army systems is that they must be mobile, so that they can easily follow troop movements and be redeployed frequently to avoid being detected and destroyed.

The army will also be equipped with helicopters (Fig. 1.21) able to climb swiftly and to launch wire- or infrared-guided ASMs against enemy tanks.

To help them control the battlefield, the army will use dedicated information-gathering sensors, such as radars, infrared systems, and remotely piloted vehicles (RPV) also known as unmanned air vehicles (UAV).

Figure 1.19 SAM systems are frequently used to defend ground forces against air raids. The photograph shows the Soviet SA-13 battery.

The Army will have antimortar radar systems capable of locating accurately the direction and hence the location from which projectiles are fired so as to be able to direct counter fire with precision to destroy the battery. More recently, *weapon locating radars* (WLR) have been developed to specifically locate the source of rockets.

In an army, *command, control, and communications* (C^3) systems have great importance. In fact, the army consists of a multitude of extremely mobile units whose activities are constantly in need of coordination.

To sum up, the army will mainly draw upon the following weapon systems:

- SSM systems.
- Long-, medium- and short-range artillery systems.
- Search and acquisition radar systems to detect the ground attack aircraft.
- SAM systems.
- AAA systems.

Figure 1.20 Radar-guided antiaircraft artillery has shown its great effectiveness. The photograph shows the Soviet ZSU-23.4 (RIO) system and the integrated point defense SKYGYARD 35 mm TWIN.

Figure 1.21 Thanks to its mobility, the helicopter is extremely effective against tanks. The photograph shows the A-129 antitank helicopter.

- Antimortar radars and weapon locating radars (WLR).
- Armored vehicles.
- Helicopters with wire- or infrared-guided missiles.
- Battlefield surveillance systems.

1.3 The main weapon systems

All the operations listed above are conducted in two phases, first the detection phase, and then the response phase, in which missiles or guns are used. Accordingly, the main systems employed by the armed forces against which protection is required are the following:

- Search systems.
- Missile systems.
- Artillery systems.

The functioning of all these systems is based on the use of electronic sensors. It has been amply proved that the effectiveness of a weapon system is destroyed by adequately jamming its sensors. The concept of electronic defense, the development of specialized equipment, and the mission it has to accomplish, are all consequences of this fact.

In order to understand the way in which jamming interferes with weapon systems, it is necessary to examine in more detail how the weapon systems themselves are structured and how they operate.

The performance of an air defense system depends on the capabilities of the long-range search radar associated with it. An electronic device designed to jam a detection system can interfere only with the radar sensor and its signal processing, as the ensuing data processing can take place in remote, well-protected command and control centers.

As already stated, a missile system usually consists of:

- a medium-range search radar (acquisition radar).
- a number of tracking radars each tracking one target to supply guidance data for the missile.
- a number of missile launchers.

A missile may be guided exclusively by commands from the tracking radar (command missile) or it may be launched on the basis of data supplied by the tracking radar, and then acquire signals for self-guidance to its target (homing). Homing can therefore be as follows:

- active, if the missile is fitted with a sensor (seeker) comprising a small tracking radar.
- semi-active, if the energy source is an illuminator at the missile site and the seeker is a tracking radar receiver which sees radiation reflected by the target.
- passive, if the missile has a seeker which does not require any transmitter, but detects the energy radiated by the target in the infrared, ultraviolet or microwave spectrum.

An artillery system consists of:

- a medium-range search radar (acquisition radar).
- a number of tracking radars (sometimes the tracking is achieved by optical means).
- a number of cannon and machine guns.

Here too the search radar detects and identifies the target, then designates it to a tracking radar. The tracking tadar searches, detects, acquires, and tracks the indicated target, and supplies its data to a computer which accurately computes the interceptor point and aims the weapon.

To sum up, it is apparent from the review presented up to now that all of the weapon systems that we have examined employ one of the following sensors that could be the victims of electronic jamming systems:

- Search radar.
- Tracking radar.
- Radio-frequency seeker.
- Electro-optic search systems.
- Infrared seeker.

In Chapter 2, an analysis of the way in which the sensors operate will also indicate their weaknesses and the possibilities for interfering with them. To show to what extent disturbance of a sensor is useful for electronic defense, Chapter 3 will analyze the way in which the weapon systems themselves operate.

The armed forces coordinate among themselves by extensive use of communications systems, which can be jammed. These systems will also be examined briefly in Chapter 3.

1.4 The objectives of electronic defense

In the preceding section the main means of defense and offense have been listed, based on "hard kill". This section deals with the electronic defense devices themselves, their military functions, and how by interfering with them "soft kills" of the enemy can be achieved .

The organization of electronic defense

It should be remembered that the ultimate objective of electronic defense is to minimize the effectiveness of those weapon systems which draw on electronic sensing devices for their operation. To achieve this end, the following measures are necessary:

(a) Strategic knowledge of the enemy's electronic devices. This is obtained by monitoring and studying the signals which they emit (*signal intelligence* (SIGINT)).

(b) Tactical knowledge of the enemy's devices; that is, knowledge of the distribution over an area, or around the protected point or platform, of hostile electromagnetic sources (*electronic order of battle* (EOB)). This is needed both for a defensive response for self protection or mutual protection and for an electronic offense operation *suppression of enemy air defence* (SEAD). Responses employing traditional weapons are not discussed here.

(c) Generation of *electronic counter measures* (ECM), which has as its aim the maximum reduction of the operational capabilities of enemy electronic devices, including search radars, acquisition and tracking radars, infrared systems, laser systems, and communications systems.

(d) Adoption of *electronic counter-counter measures* (ECCM). It is in fact sometimes possible to reduce or eliminate an intentionally caused disturbance or interference by incorporating filters and other special devices.

The organizational display of electronic defense, as shown in the table in Figure 1.22, is based on the above list.

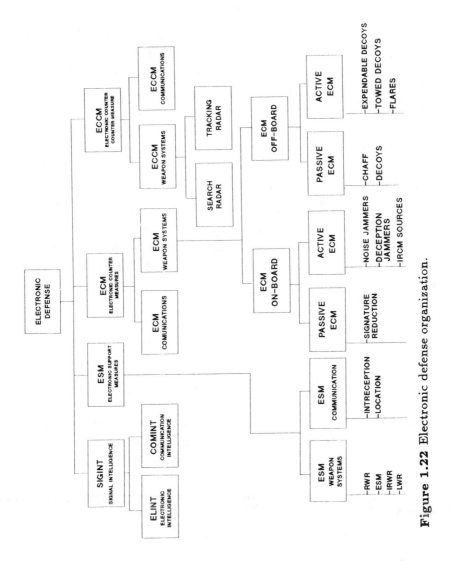

Figure 1.22 Electronic defense organization.

1.5 Electronic defense systems and their operational objective

The main electronic defense systems will be defined here according to their position in the table of organization. It should be remembered that an electronic defense system can consist of a collection of the equipment described below. For example, it is possible to have two separate electronic support measures (ESM) *and* ECM systems, or one integrated system, when both functions are performed together.

Signal intelligence (SIGINT)

The task of SIGINT systems is the acquisition of as much data as possible about the electromagnetic emissions of a potential enemy. They can be further classified into electronic intelligence (ELINT) systems, which collect radar emission data, and communication intelligence (COMINT) systems, which collect enemy communication data.

Their function is primarily a strategic one: They are essential for the identification of a potential enemy's operational procedures.

Electronic intelligence (ELINT)

This equipment must be able to define the characteristics, the time dependence, and the location of hostile electronic emissions. It should also be able to analyze the enemy's electronic signals both in time and in frequency, and to associate with them a serial number of the enemy's equipment (i.e., finger printing), sometimes even in a one-to-one relation, thus making it possible to follow the movement of the equipment.

These systems can be airborne for deep probes into the electronic scenario of a potentially hostile country. They can also be land-based, on sufficiently elevated mountain sites and on promontories or straits, for control of sea traffic (Fig. 1.23).

The collected data are usually transmitted to an analysis center which codes them suitably, memorizes them in a data

Figure 1.23 The main purpose of an ELINT system is to intercept and analyze, for strategic purposes, all the electromagnetic radiation generated in a potentially hostile country.

base, and correlates them with the information gathered by equipment of other types, or by other organizations, or at different times.

All this information, processed according to operational criteria established by the military organizations, will be used to build up special files, in which all emissions and other features of enemy equipment will be listed (libraries). From these files information is compiled to be loaded into the memories of electronic defense equipment used for detection of enemy signals.

Communication intelligence (COMINT)

These systems are similar to the preceding ones, but their task is the interception and analysis of telecommunications emissions and the identification of relevant communications networks.

Electronic support measures (ESM)

The main objective of equipment of this type of class is tactical interception. The simplest systems are those whose main function is to detect the presence of already known emitters by comparison of the intercepted signals with stored data. They are called *radar warning receivers* (RWR).

This equipment, which can instead reconstruct a very complex electromagnetic scenario, including previously unknown emitters, and can therefore contribute to an attack by identifying and detecting enemy platforms, is more sophisticated. These are the *electronic support measures* (ESM) systems.

RWR

The main features of equipment of this class are simplicity (they measure few parameters with moderate accuracy) and high reliability low weight, and low cost.

They are used to detect an imminent threat, that is, the presence in a given direction of the radar of a hostile weapon system locked on to the protected platform. They are mainly committed to aircraft defense and enable the pilot to react promptly either by an evasive maneuver, or by both a maneuver and the simultaneous launching of chaff, which consists of explosive cartridges containing millions of tiny, extremely light dipoles, capable of generating a very strong radar echo which masks the platform (see chapter 5), or by generating electronic jamming signals, or by a combination of these different techniques.

ESM

This class of equipment is characterized by medium to high complexity and sophistication. Its task is an almost real-time reconstruction of an electromagnetic scenario, which can be highly complex and previously unknown, starting from the interception of the multitude of signals crowding into its antenna. Usually, the total "traffic" consists of pulse and continuous wave signals. Pulse signals are frequently very dense (millions of pulses per second), are dispersed on bandwidths from a few hundred megahertz to a few tens of gigahertz to millimeter waves, and make use of the most varied wave forms, including pulses, modulated pulses, and so forth.

The main aim of such a system is to give a picture of the electromagnetic scenario in the environment both for self-defense, by discovering the presence of enemy platforms (ESM on a naval platform) (Fig. 1.24), and for passive surveillance of a wide area (ESM on an aerial platform or land-based network of ESM systems).

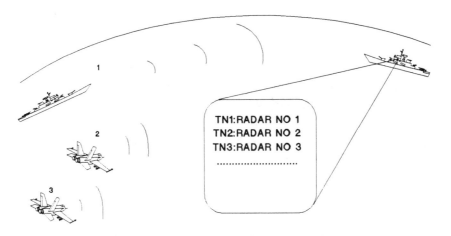

Figure 1.24 The purpose of an ESM system is to detect the presence of enemy platforms by intercepting their electromagnetic emissions.

Reconstruction of the electromagnetic environment depends both on detection of the electromagnetic signal input to the antennas and on characterization of signals in terms of carrier frequency, direction of arrival (DOA), time of arrival (TOA), pulse width (PW), amplitude, modulation on pulse (MOP), form and modulation in time, and modulation and amplitude of continuous waves (CW).

Out of this information an ESM installation must extract knowledge of the generating emitters. The process of correlating pulses, and of grouping them in possible "families" is a very complex one, called sorting or deinterleaving. Because of the variability of the signals, automatic extraction is even more difficult. Frequently wrong conclusions are reached: Emitters that do not really exist are created, and so false alarms are generated that reduce the reliability of the equipment.

In the field of military electronics, ESM automatic extraction is generally regarded as one of the most difficult problems, as the complex electromagnetic signal, which has to be extracted from a crowded and complicated background, is usually not known in advance.

ESM-COM

The aim of these systems is to intercept all enemy communications, both for location of transmitters and radio relay systems, and for detection and decoding of the messages themselves. Knowledge of enemy intentions is of the first importance to the choice of appropriate action and to the effecting of electronic countermeasures.

Infrared warning

Enemy missiles with infrared guidance do not need to radiate any RF signals because they lock onto infrared emission naturally generated from a target. This means that the presence of an infrared missle cannot be detected by any radio frequency electronic support measure equipment. In fact their detection is normally obtained by dedicated radar. But the desire to defend a platform against missile attack often conflicts with the need to keep radar turned off to avoid detection by the enemy (a "radar silence" situation). In this case, passive electro-optic sensors offer a solution. This kind of equipment is in fact capable of detecting either the aerodynamic heating or the infrared radiation produced by the booster at the time of launching.

The problem with these sensors is that the background infrared radiation usually gives a much stronger signal than the signal produced by the threat to be intercepted.

Systems which detect the infrared radiation emitted at launch are distinct from those which detect aerodynamic heating. Among the latter, the simpler surveillance or infrared vision systems, for example *forward-looking infrared* (FLIR), should be distinguished from much more complex and costly systems, capable of warning automatically, such as infrared search and track (IRST).

Laser warning receivers

The last decade has seen a proliferation of weapons either guided or controlled by a laser emitter. In tank warfare, laser rangefinders yield accurate ranges, while laser designators give precision guidance for bombs or missiles towards ground targets. The carbon dioxide laser now allows missiles to be guided towards fast-moving platforms.

Obviously, the first requirement of adequate defense against such threats is the ability to detect their presence. This is what laser warning receivers do.

Electronic countermeasures (ECM)

After this brief survey of the main types of equipment for reconnaissance of the electromagnetic environment surrounding a protected area, it is time to describe those systems whose task is the neutralization of hostile electronic systems that have been detected. Their purpose is either to conceal the protected platform or to deceive the hostile weapon system by creating spurious targets.

Chaff

A chaff system comprises a launcher that ejects cartridges. These cartridges explode within a certain distance of the protected platform and disperse a multitude of tiny dipoles into space. These dipoles remain suspended in space, producing a cloud which reflects radar signals.

Chaff generates wide corridors within which search radars are dazzled, and therefore cannot identify aircraft targets, even

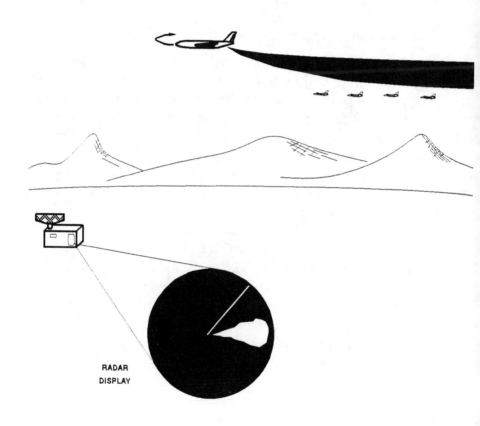

Figure 1.25 Chaff consisting of clouds of extremely light, conductive metal foil strip dipoles is used to create areas in which radar is blinded and cannot see targets.

at altitudes different from those filled with chaff (Fig. 1.25). To create these corridors, aircraft flying at great heights dispense an enormous quantity of chaff over a very wide area. Sometimes chaff is launched from a platform as a defense against an attacking weapon system. In this case, the weapon system's radar is usually deceived by the strong signal produced by the chaff and is diverted from the pursuit of the true target.

Stealth techniques

Naturally, the best way of preventing dangerous response is to avoid detection. Since the signal received by a radar is directly proportional to the *radar cross section* (RCS) presented by the platform, a drastic reduction of the strength of the radar signal produced by the protected platform is very desirable.

To this end a new technology has been developed in recent years for the study of materials and structural geometries capable of minimizing target RCS. The techniques are usually called "stealth" techniques. They are very promising; the supporters of the stealth aircraft in the United States call it "invisible."

Noise jammers

A noise jammer generates signals of the same frequency as an opponent's radar. These signals create a disturbance equivalent to a very strong thermal noise in the radar receiver. Thus the signal produced by the platform is drowned in noise and is no longer "visible" (Fig. 1.26).

Deception jammers

A deception jammer generates false radar targets. In the case of search radars, it impedes identification of the real platform. In the case of tracking radars, it ensures that the tracking and ensuing locking on of the weapon system is progressively shifted onto the false target (Fig. 1.27).

Expendable decoys

There are several types of decoy. A decoy is considered to be

NO JAMMER PRESENT

NOISE JAMMER PRESENT

Figure 1.26 The purpose of a noise jammer is to mask targets by emission of signals that create confusion on the radar display.

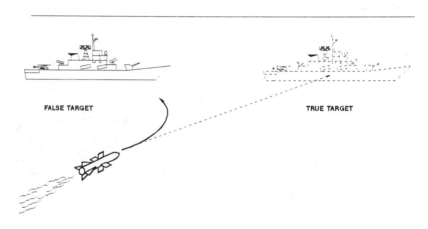

FALSE TARGET   TRUE TARGET

Figure 1.27 The purpose of a deception jammer is to protect a platform by luring enemy radars with false targets.

an object, usually ejected from the protected platform, which generates a spurious but sufficiently convincing target for enemy radar. Decoys can be passive, for example a corner reflector on a buoy, or active, that is, able to return a strongly amplified radar signal.

ECM-COM

The purpose of these systems is to generate noise signals, or interference, in order to jam the receivers of enemy telecommunications systems, thus rendering messages incomprehensible. Inability to rely on its communications systems is a major drawback for any armed force.

Infrared countermeasures (IRCM)

These are systems which prevent infrared-guided missiles from reaching the target. Currently there are two types of system, on-board and off-board.

The off-board system is a flare dispenser, which is a launcher capable of ejecting cartridges that generate an intense infrared radiation to deceive the missile heat seeker.

The on-board type is composed of modulated infrared transmitters. Since infrared seekers are quite often based on a scanning tracking system, an amplitude-modulated infrared signal can introduce huge errors into a missile trajectory.

ECM-Lasers

These systems are designed to prevent accurate rangefinding by a laser system. They either operate on the same principles as the jammers mentioned above, or dispense clouds of smoke that reduce visibility.

Electronic counter-countermeasures (ECCM)

These devices are usually added to weapon sensors to enable them to operate in an electronically hostile environment, that is, in the presence of intentional jammers, with minimal reduction of their normal capabilities.

1.6 Need for the study of weapon systems

The main purposes of electronic defense equipment have now been described. Chapters 4 and 5 will deal with actual performance, with the technical solutions, and with their distinctive characteristics. However, before contemplating action against a weapon system (the "victim" of the jamming), it is necessary to know how the weapon system operates, the principles on which it is based, its problems, and its limits. It is precisely by amplifying the problems of weapon system sensors that enemy forces can be weakened. Once the problems are known, it is easier to neutralize the system.

For example, if it is known that a radar is employed to give an angular tracking accuracy of a milliradiant, and that this is an absolute requirement for the performance of an artillery system, there is no need to prevent tracking entirely, that is, to achieve a "break lock." A disturbance introducing a ten milliradiant error will suffice to reduce the effectiveness of the weapon system satisfactorily.

Again, if it is known that a search radar guarantees valid protection only if it can detect targets at its maximum range, the use of countermeasures capable of reducing the detection range by half is enough to indicate that the objective of electronic defense has been at least partially achieved.

Chapter 2

Sensors

2.1 Introduction

The sensors of the principal weapon systems operate by exploiting either electromagnetic energy reflected by the target in the radio-frequency band, or electromagnetic energy emitted by the target in the infrared band. For the better understanding of sensors, some of the theoretical laws on which their performance depends will be recalled.

Only the concepts and formulas needed for a correct understanding of later discussions of radar systems, intercept systems, and noise and deception jamming systems will be reviewed here. For a more detailed analysis, the reader should consult the bibliography. Table 2.1 shows the electromagnetic spectrum, from radio waves to millimeter waves (whose wavelength λ is on the order of millimeters), which is of interest for radar systems.

2.2 Radar sensors

2.2.1 Review of electromagnetic signal transmission

Table 2.1
The electromagnetic spectrum of interest for radar systems.

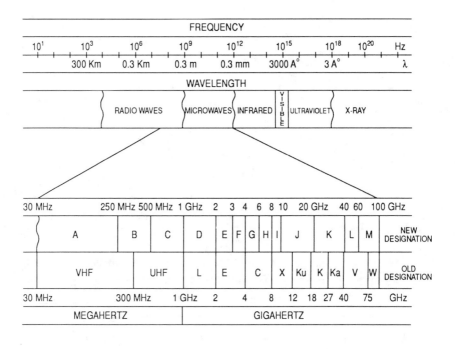

A radio-frequency signal may be generated and amplified to the power P by means of a suitable transmitter (Fig. 2.1).

Suppose that there is at A an isotropic radiator: an antenna capable of radiating uniformly in all directions a signal of power

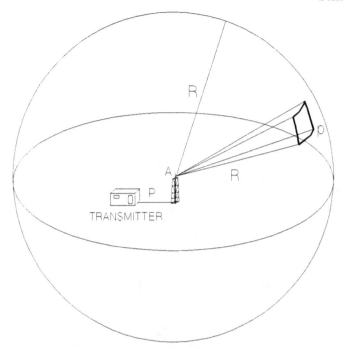

Figure 2.1 An isotropic radiator radiates electromagnetic energy equally in all directions.

P. At a distance R from A, the power transmitted will be distributed over a sphere whose surface area is $4\pi R^2$.

Suppose the distance R is large enough to be in the Fraunhofer region (far field) of the antenna, such that

$$R > \frac{2D^2}{\lambda}$$

where D is the maximum dimension of the antenna. Then the power density p will be (it is assumed that the radiant efficiency is 1, that is, all the power reaching the radiator is radiated into space)

$$p = \frac{P}{4\pi R^2}$$

Defining the radiant intensity $I(\vartheta, \varphi)$ to be the power radiated per unit solid angle (watts per steradian) in the (ϑ, φ)

direction, and recalling that the solid angle is 4π steradians, one may write, for an isotropic radiator,

$$I(\vartheta, \varphi) = \frac{P}{4\pi} = I_m$$

A non-isotropic radiator, or non-isotropic antenna, will radiate more in some directions than in others (Fig. 2.2), so that the radiant intensity $I(\vartheta, \varphi)$ will not be constant, but will vary with ϑ and φ.

The *antenna gain* G, a measure of the maximum radiative capability of the antenna, is defined by

$$G = \frac{I_{\max}(\vartheta, \varphi)}{I_m} = \frac{I_{\max}(\vartheta, \varphi)}{P/4\pi}$$

The ϑ, φ direction in which the radiant intensity is maximum ($I_{\max}$) is called the electrical axis, or boresight, of the antenna.

Assuming for simplicity that the antenna radiates all the power into the equivalent solid angle represented by the product $\vartheta_B \varphi_B$ (Fig. 2.3) where $\pm\vartheta_B/2$ and $\pm\varphi_B/2$ are the angles from the boresight at which the power is half the maximum; the radiated beam in this region is often called the $-3\,\mathrm{dB}$ beam, and the maximum radiant intensity, for ϑ_B and φ_B sufficiently small, is given by

$$I_{\max}(\vartheta, \varphi) = \frac{P}{\vartheta_B \varphi_B}$$

Substituting this expression for $I_{\max}$ into the equation for G, one obtains

$$G = \frac{P}{\vartheta_B \varphi_B} \frac{4\pi}{P} = \frac{4\pi}{\vartheta_B \varphi_B}$$

where ϑ_B and φ_B are in radians. If ϑ_B and φ_B are in degrees, then

$$G = \frac{41253}{\vartheta_B \varphi_B}$$

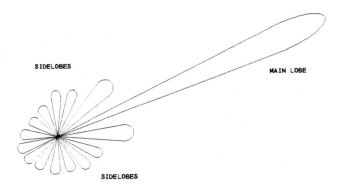

Figure 2.2 A directive radiator radiates electromagnetic energy prefer-
entially in one direction.

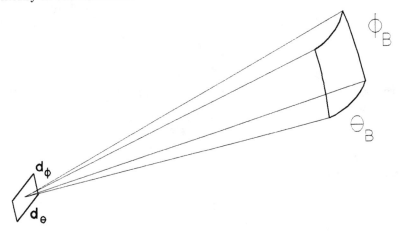

Figure 2.3 Equivalent solid angle. Frequently it is convenient to assume
that the antenna radiates only in the neighborhood of the direction of
maximum radiation.

In reality, taking account of the efficiency of the antenna, the
formulas of practical use are [1]

$$G \simeq \frac{30000}{\vartheta_B \varphi_B} \qquad \text{for higher frequencies (6 to 18\,GHz)}$$

and

$$G \simeq \frac{25000}{\vartheta_B \varphi_B} \qquad \text{for lower frequencies (1 to 6\,GHz)}$$

The pattern of the antenna beam, which is to say, the electric field radiated as a function of the angle measured from the boresight, is, for a uniformly illuminated rectangular antenna whose dimensions are $d_\vartheta \times d_\varphi$ of the form $\sin x/x$. Normalizing to maximum gain, one obtains (Fig.2.4)

$$E(\vartheta,\varphi) = \frac{\sin\left[\pi(d_\vartheta/\lambda)\sin\vartheta\right]}{\pi(d_\vartheta/\lambda)\sin\vartheta} \frac{\sin\left[\pi(d_\vartheta/\lambda)\sin\varphi\right]}{\pi(d_\vartheta/\lambda)\sin\varphi}$$

The pattern of radiated power, which is proportional to $[E(\vartheta,\varphi)]^2$ will be of the form

$$\frac{\sin^2 x}{x^2}$$

From the preceding equations, one may infer that, for each dimension, the width of the $-3\,\mathrm{dB}$ beam will be, in radians,

$$\vartheta_B = 0.88\frac{\lambda}{d}$$

or, in degrees,

$$\vartheta_B = 51\frac{\lambda}{d}$$

Recalling the definition of antenna gain, one may write

$$G = \frac{4\pi}{\vartheta_B\varphi_B} = \frac{4\pi\eta}{(\lambda/d_\vartheta)(\lambda/d_\varphi)} = \frac{4\pi\eta\, d_\vartheta\, d_\varphi}{\lambda^2}$$

where η is the efficiency of the antenna.

Thus, defining the effective area of the antenna as $A_{\mathrm{eff}} = \eta d_\vartheta d_\varphi$ one may write

$$G = \frac{4\pi A_{\mathrm{eff}}}{\lambda^2}$$

Since uniform illumination of the aperture implies a very high level for the first sidelobes (only 13 dB down from the

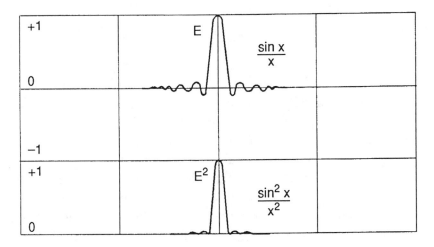

Figure 2.4 An isotropic radiator radiates electromagnetic energy equally in all directions.

main lobe), usually illuminations stronger in the central region of the aperture and weaker toward the edges are used. In this way, together with a reduction of the sidelobes, there results a loss of gain and a widening of the $-3\,\text{dB}$ beam, which is normally compensated by an increase in the dimensions of the aperture.

With the above formulas it will always be possible, if the dimensions and operating frequency of the antenna are known, to calculate the values of the antenna gain and the width of the $-3\,\text{dB}$ beam quickly and with sufficient accuracy. This is very useful in systems practice.

It should also be recalled that an antenna is normally a reciprocal device, that is, its behavior on transmission is identical to its behavior on reception. This means that the transmission gain is equal to the reception gain.

An antenna is connected to the transmitter or the receiver

by means of a transmission line capable of carrying the electromagnetic signal with only slight attenuation. The principal transmission lines are cables and waveguides. Cables make for simpler installations, but introduce more attenuation and support less power than waveguides (Fig. 2.5). Usually coaxial cables are employed for frequencies up to a few tens of gigahertz. (Here the flat, strap-like conductors for transmission of lower-frequency signals, up to a few tens of megahertz, are not considered.) The cables may be rigid, semi-rigid, or flexible. The attenuation will increase with increasing frequency; for example, it is approximately $2\,dB/m$ at $18\,GHz$.

To minimize losses in high frequency transmission, waveguides must be used. Waveguides are metal tubes, usually of rectangular cross section, which allow propagation of a signal of given frequency. A waveguide is characterized by its cut-off frequency, that is, the frequency below which the attenuation, which depends on the waveguide dimensions, increases sharply.

When signals must be carried in a very wide band, it is possible to resort to double ridge waveguides. However, they introduce a larger attenuation than standard waveguides, and are more complex.

The simplest antennas are dipoles, which are open lines of length $\lambda/2$, and whip antennas whose lengths are approximate submultiples of the wavelength (the telescopic antenna of ordinary portable FM radios is of this type). When the line is a waveguide, it is easy to use a horn antenna. The main parts constituting a complex antenna are (Fig. 2.5):

- an illuminator, that is, an energy source, generally called *feed*, consisting of a small, simple antenna (dipole, horn, etc.) illuminating the surface area of the main reflector
- a main reflector, to generate the required beam shape, generally a slice of a parabolid surface in when focus is allocated the feed.

In general, an antenna may be either fixed, as in a radio relay system, or movable, as in a radar. In the latter case, the antenna is set on a pedestal equipped with servomechanisms

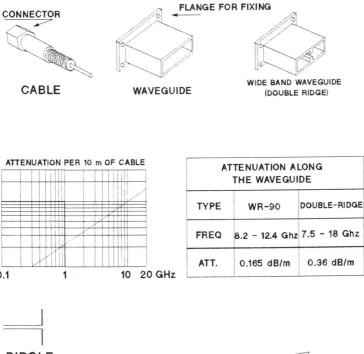

Figure 2.5 Transmission lines and main components of an antenna.

that point it in the desired direction.

2.2.2 The radar equation

A *radio detection and ranging* (RADAR) is a device capable of detecting the presence of an object, the target, in space, and of measuring its bearings in angle and in range by the use of electromagnetic waves. This is generally achieved by the generation of a pulsed signal of a certain frequency, which is radiated into space by a directive antenna capable of scanning

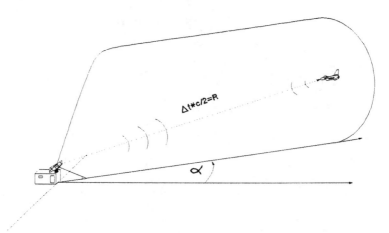

Figure 2.6 Operating principle of radar. The target range is obtained by measuring the time interval between emission of the signal by the radar and reception by it of the signal reflected from the target.

a given sector (Fig. 2.6).

When the antenna points at the target, electromagnetic energy striking the target is reflected and scattered. There is a similar situation in optics, when a beam of light strikes an object in dark surroundings. Because of the scattering, the object is visible from other directions besides the beam direction.

The electromagnetic energy reradiated in the direction of the antenna may be captured by it and passed on to an adequately sensitive receiver.

A radar measures the time Δt required for the electromagnetic pulse to cover the distance R to the target and the distance R back to the receiver; the round-trip distance is thus $2R$.

Since in empty space electromagnetic energy travels at the velocity of light $c = 3 \times 10^8$ m/s, the distance covered, which is given by the product of velocity and elapsed time, may be written

$$2R = c\Delta t$$

and the target range may be written in terms of the time delay

$$R = \frac{c\Delta t}{2}$$

Expressing the time in microseconds, one obtains

$$R_{(m)} = 150\Delta t_{(\mu s)}$$

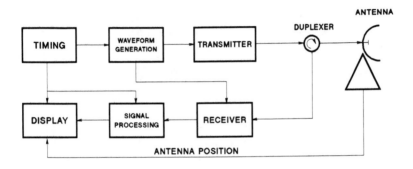

Figure 2.7 Simplified block diagram of a radar.

A radar usually consists of (Fig. 2.7):

- a timing circuit determining the times (triggers) at which pulses should be transmitted, the times at which measurements should be taken, and so forth.
- a circuit generating the frequency and the waveform to be transmitted.
- a device, called a "transmitter", providing the pulses with adequate power; simple radars employ a power tube called a "magnetron" which can also generate the waveform to be transmitted.
- a device known as a "duplexer" that channels the transmitter power output to the antenna, and the signals received by the antenna to the receiver.
- an *antenna* for the transmission of pulses and the reception of returns from the target. Antenna movement is usually controlled by servomechanisms.

- a *receiver*, usually of superheterodyne type, tuned to the transmitted frequency, which detects the received signals after *intermediate-frequency* (IF) amplification.
- a circuit for signal processing; according to the type of radar, this will be either very simple, as in radar for civil navigation, or very complex, including filters for cancellation of unwanted signals, for automatic data extraction, and so forth.
- a device, called a "display", for the presentation of data.

If the antenna is highly directive in the horizontal plane, for example an antenna with a fan-shaped beam, then energy will be reflected by the target (echoed) only when the antenna points at the target, and the target azimuth will be given by the horizontal angle of the antenna at the moment when the echo is detected (Fig. 2.8). This type of antenna is generally used for search radars such as those for air traffic surveillance, that is, *air traffic control radars* (ATCR).

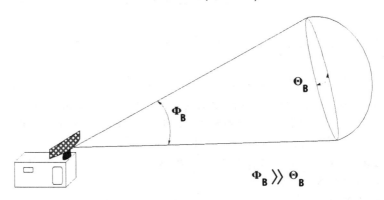

Figure 2.8 Antennas with fan-shaped beams are used in search radars. The beam is very narrow in azimuth to indicate the precise direction of the target in the horizontal plane, and very wide in elevation to provide the coverage required.

In the simplest radars, the information is presented to the operator on a screen called a *plan position indicator* (PPI) (Fig. 2.9), displaying a circular map-like presentation, with the radar at its center. It is a *cathode-ray tube* (CRT) display in which the electron beam is deflected in a direction corresponding to the angular position of the antenna. Deflection of the beam is proportional to target range; that is, the time taken by the cathode ray to shift from the center to the point representing the target is the same as the time taken by the radar pulse for its round trip to and from the target.

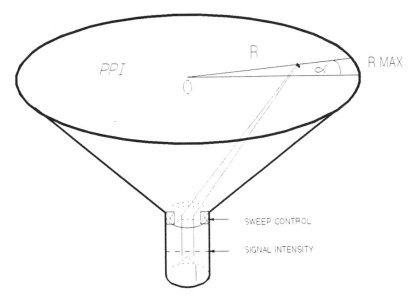

Figure 2.9 The *plan position indicator* (PPI) positions the target echoes as if on a map.

The cathode-ray intensity is directly proportional to the receiver output signal strength. In the absence of echoes it will be minimal, being due exclusively to the amplified receiver thermal noise. As soon as a target is encountered, however, the target return will produce a strong signal generating a bright blip on the fluorescent screen of the cathode-ray tube. The internal surface of the screen is lined with phosphor salts, whose illumination has an intensity and a persistence which depends on the intensity of the cathode ray. The radial distance from the center represents the range, while the angular position of the antenna with respect to a reference direction, for example, geographic North, coincides with the angular position of the electron-beam sweep.

The antenna can be highly directive in both the horizontal and the vertical plane. In such a case, it has a pencil beam, and the radar can provide the three coordinates of the target: range, azimuth, and elevation. Such an antenna is used in tracking radars, such as those employed for gun guidance. Besides the PPI, this type of radar may have an *amplitude/range* (A/R) display, in which the amplitude of the signal as a function of range is shown on a very low persistence cathode-ray tube (Fig. 2.10).

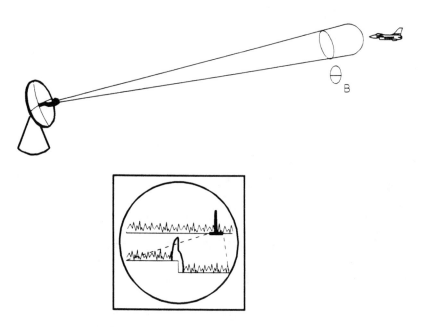

Figure 2.10 Pencil-beam antenna. A radar using this type of antenna must often deal with one single target at a time, and therefore use an amplitude/range (A/r) display.

The main feature of a radar is its detection range; that is, the maximum distance at which a target of given dimensions can be detected. To determine the detection range, a whole set of operating parameters of the radar must be known.

Assume that a pulse of power P_T is transmitted and radiated into space by means of an antenna with gain G_T. When the antenna points at a target (that is, when the target is in the $-3\,\mathrm{dB}$ beam) at range R, the transmitted pulse, of power P_T, will generate a target power density

$$p = \frac{P_T G_T}{4\pi R^2}$$

assuming a perfectly transparent atmosphere in free space. In general, the target will absorb, reflect, and scatter the incident electromagnetic pulse in a manner depending on its constituent materials, its shape and size, the radar carrier frequency (or wavelength), and the angle at which the pulse strikes the target surface.

The ratio of the power P_r reradiated in a given direction to the power density p impinging on the target is called the RCS of the target. The RCS is measured in square meters, and is frequently denoted by σ:

$$\sigma = \frac{P_r}{p}$$

The mechanism of reradiation is very complex. Each elemental area of the target of size equal to a few times the wavelength behaves as an elementary radiator almost independently of all others. The power in a certain direction is the vector sum of many elementary signals, and will strongly depend on target vibrations and movement. That is why the RCS can be calculated easily only for targets of simple geometry (sphere, cone, cylinder, and so forth [2, 3]). For more complex targets, only an average value may be defined, which varies from case to case. From the last equation, it follows that the target will reradiate toward the radar antenna a power

$$P_r = p\,\sigma = \frac{P_T G_T \sigma}{4\pi R^2}$$

This power will travel back over the distance R, producing at the radar antenna a power density

$$p_r = \frac{P_T G_T}{4\pi R^2} \quad \sigma \quad \frac{1}{4\pi R^2}$$

The radar antenna, with equivalent capture area A, will direct toward the receiver a signal power S given by

$$S = \frac{P_T G_T}{4\pi R^2} \quad \sigma \quad \frac{1}{4\pi R^2} A$$

Recalling that

$$A = \frac{G_R \lambda^2}{4\pi}$$

one obtains

$$S = \frac{P_T G_T}{4\pi R^2} \sigma \frac{1}{4\pi R^2} \frac{G_R \lambda^2}{4\pi}$$

whence

$$S = \frac{P_T G_T G_R \sigma \lambda^2}{(4\pi)^3 R^4}$$

The input signal to the radar receiver is so weak that it requires very strong amplification in order to be of use. However, the very weak thermal noise at the input to the amplifier will be amplified as well. To avoid the need of the amplifier gain calculation which is of no interest here, an "ideal" (no noise) amplifier is usually considered, and in addition to the signal an equivalent noise N is included in the input;

$$N = KTBF$$

where K is Boltzmann's constant $(1.38 \times 10^{-23}\ \text{W}/(\text{Hz}^{-1}\text{K})$, T is the standard temperature $(290\ \text{K})$, B is the receiver equivalent bandwidth, and F is the receiver noise figure.

Often it is convenient to calculate N in dBm (that is, in decibels above $1\ \text{mW}$). In such a case, if B is expressed in dBm/MHz, KT is $-114\ \text{dBm/MHz}$.

Therefore, in the radar receiver output there will be not only the signal S, but also the noise N. This is one of the major problems confronting radar.

In fact, whenever it has to be decided whether a very distant target, which would generate a very weak signal easily mistaken for noise, is present or absent, there is a risk of reaching the conclusion that a small noise is a target (a false alarm), or that a weak signal is just noise (a miss). In practice, the presence of a target will easily be detected only if its echo S is strong compared to the noise N. In what follows, the *signal-to-noise* ratio will be denoted by either SNR or S/N.

As is well known, the statistical pattern of the receiver output signals at a given range bin is as shown in Figure 2.11, where the noise alone distribution (Rayleigh distribution) and the noise plus signal distribution for signals of increasing intensity are shown.

According to the level at which a threshold has been established (either a luminous intensity set by an operator, or a voltage magnitude set by an electronic circuit), there will be a probability P_d of detecting the true signal, and a probability P_{fa} of taking noise for a signal.

On account of the nature of the RCS, the power S at the receiver input is a fluctuating signal of statistical nature, while noise too is a statistical phenomenon, so that the maximum range of a radar is a statistical fact associated with a detection probability P_d (an actual target has been detected) and a false-alarm probability P_{fa} (the threshold has been crossed by noise alone).

Since the noise power is proportional to the receiver bandwidth, it might be thought that the best way to reduce the fluctuations introduced by noise would be to narrow the receiver bandwidth as much as possible. However, this may be done only up to the point at which no significant regions of the signal spectrum are eliminated (Fig. 2.12).

According to the theory of filters (North, [4]), the maximum ratio of signal power S to noise power N is obtained when the

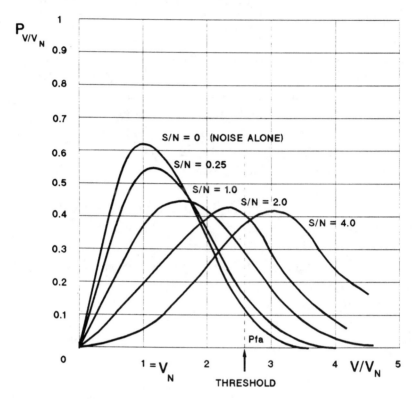

Figure 2.11 A weak signal may be mistaken for receiver thermal noise. The diagram shows the probability that a signal exceeds a given voltage level compared to the average level produced by noise.

filter is ideally matched to receive the signals during observation time. This requires, in particular, the condition that the filter bandwidth B be the reciprocal of the observation time of the signal, that is,

$$B = \frac{1}{T_{\text{obs}}}$$

so that

$$(S/N)_{\max} = \frac{S}{KTBF} = \frac{ST_{\text{obs}}}{KTF} = \frac{E}{KTF} = \frac{E}{N_0}$$

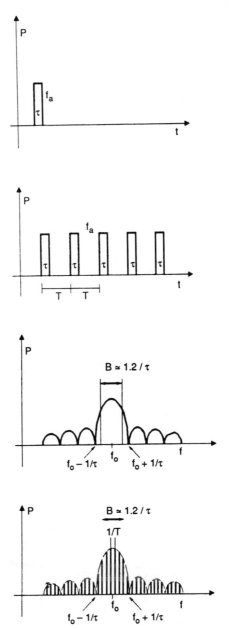

Figure 2.12 Spectrum of pulse radar signals. A continuous spectrum corresponds to a single pulse; a line spectrum, where the power is concentrated around precise frequencies, to a train of pulses, repeated with period T.

where E is the energy of the radar signal reradiated by the target toward the receiver and N_0 is the noise power density of the receiver per unit bandwidth.

If only one radar pulse of power P and length τ is transmitted, the radiated energy is given by the product $P\tau$, and the energy reradiated by the target is given by

$$E = \frac{P_T\,\tau\,G_T\,G_R\,\sigma\lambda^2}{(4\pi)^3\,R^4}$$

Assuming that a matched filter is used, one obtains

$$S/N = \frac{P_T\tau G_T G_R \sigma\lambda^2}{(4\pi)^3\,KT\,FR^4}$$

Solving with respect to R^4, one obtains

$$R^4 = \frac{P_T\tau G_T G_R \sigma\lambda^2}{(4\pi)^3\,KT\,F\,S/N}$$

To achieve maximum range, the signal will have to be detected, always by means of a matched filter, with the smallest SNR capable of giving the required detection and false-alarm probability $(S/N)_{\mathrm{Pdfa}}$[5].

Figure 2.13 shows the SNR pattern required to achieve a given detection probability P_d, and a given false-alarm probability P_{fa}. As is shown, the pattern depends also on the target characteristics: whether it be a non-fluctuating target, scan-to-scan fluctuating target, pulse-to-pulse fluctuating target, or whatever.

However, the radar would not be able to see very far if it used only a single pulse. In order to have at disposal many pulses, the antenna is made to rotate at a rate ωa, in such a way that it stays on the target for a time T_{ot} (time on target) given by

$$T_{ot} = \frac{\vartheta_B}{\omega_a}$$

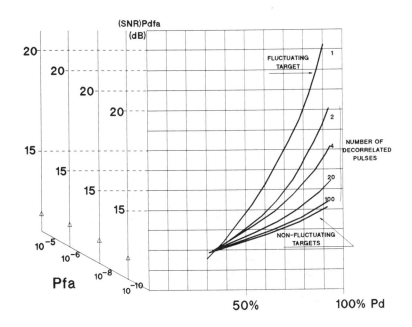

Figure 2.13 SNR needed for detecting a signal with specified detection and false-alarm probabilities, SNR_{pdfa}, as a function of the type of signal received.

In this way, the receiver will have at its disposal a number of pulses

$$N_i = F_R T_{ot}$$

where F_R is the *pulse repetition frequency* (PRF).

Rather than an in-depth analysis of the SNR over more pulses, and of its optimization, what is of interest here is a radar range equation that may be useful not so much to radar system designers as to those concerned with *electronic defense* (ED) systems. If N_i pulses impinge on the target, the energy will be $N_i P_T \tau$. This energy may be integrated in the radar receiver by means of special devices, although not perfectly, but with the introduction of losses L_i. Values of these integration losses are shown in Figure 2.14 for different cases. It follows that

$$R_{\max}^4 = \frac{N_i P_T \tau G_T G_R \sigma \lambda^2}{(4\pi)^3 KTF(S/N)_{Pdfa} L_i}$$

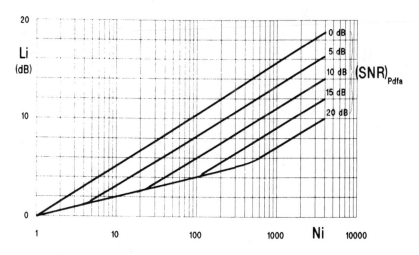

Figure 2.14 A gain is obtained by integration of time-on-target signals, but this must be reduced by the integration losses L_i.

The following additional losses have also to be considered [6]:

- transmission losses L_{Tx} in the lines connecting the transmitter to the antenna, and in elements along these lines.
- receiver losses L_{Rx} in the lines and elements between the antenna and the receiver.
- beam-shape losses L_b, which account for the fact that the target is not illuminated with a constant antenna gain during the time on target; usually it is assumed that the beam has a Gaussian-like shape, which is convenient for calculations.
- matched filter losses L_m, which account for the fact that on reception the matched filter is not an ideal filter.
- losses L_x due to the type of signal processing.

Taking these losses into account, in free space the range may be written

$$R^4_{\max} = \frac{N_i P_T \tau G_T G_R \sigma \lambda^2}{(4\pi)^3 KTF(S/N)_{pdfa} L_I L_m L_x L_{Tx} L_{Rx} L_b}$$

This is a general formula for pulse radars, although they use coded pulses to increase their range resolution. In this case, the transmitted pulse τ split into n coded elements of elementary duration τ_{el}, and the range equation may conveniently be written

$$R^4_{\max} = \frac{N_i P_T n \tau_{el} G_T G_R \sigma \lambda^2}{(4\pi)^3 KTF(S/N)_{pdfa} L_i L_m L_x L_{Tx} L_{Rx} L_b}$$

Recalling that $1/\tau_{el}$ is roughly equal to the radar passband B, the preceding equations may be written

$$R^4_{\max} = \frac{N_i P_T n G_T G_R \sigma \lambda^2}{(4\pi)^3 KTF(S/N)_{pdfa} L_i L_m L_x L_{Tx} L_{Rx} L_b}$$

Obviously, if the pulses are not coded, $n = 1$, and the radar bandwidth is approximately $1/\tau$.

For coherent *continuous wave* (CW) or pulse-doppler radars, rather than the quantity

$$N_i P_T \tau$$

it is necessary to consider the average power P_{av} and the time on target, or observation time, T_{ot}. In this case, the integration losses are negligible, and

$$R^4_{\max} = \frac{P T_{ot} G_T G_R \sigma \lambda^2}{(4\pi)^3 KTF(S/N)_{pdfa} L_m L_x L_{Tx} L_{Rx} L_b}$$

where, in general, the term L_x includes eclipsing losses, due to the loss of echoes arriving during transmission, when the receiver is switched off, and losses due to the positioning of the doppler filters.

To sum up, it should be noted that:

- range does not depend on the peak power or on the waveform used by the radar, but on the energy transmitted to the target and reradiated by it. As will be seen later, radar resolution in range and velocity depends on the waveform.

- range strongly depends on the kind of processing carried out, since the magnitude of L_x may be quite high (1 to 6 dB).

Two complex elements of the radar range equation require a more detailed discussion of (1) the radar cross section σ and (2) the equivalent noise temperature T.

2.2.2.1 The radar cross section

As stated above, the radar cross section is the ratio of the power reradiated toward the radar by the target to the power density impinging on the target.

The reradiated power is obtained by vector summation of the signals generated by the many elementary scatterers composing the target, and will depend on the wavelength of the electromagnetic signal, the position and mobility of the single scatterers, and their geometry. According to whether these vectors sum in phase or out of phase, the resulting vector may be either very large or very small.

Therefore, a moving target will reradiate toward the radar now much, now little, signal power; its visibility will thus be a statistical phenomenon, too. In fact, even a relatively large target may, when illuminated, reflect so small a signal that the radar cannot detect it. Conversely, in the next scan the reflected signal may be very strong, and therefore easily detected.

This fluctuation in the power reradiated by the target is called *scintillation*. It is a low-frequency phenomenon; in fact, roughly 90% of the fluctuation power is in a bandwidth of less than 5 Hz [7]. No scintillation manifests itself for targets of simple geometry, such as the sphere, whose RCS is known and constant, and, for wavelengths much greater than the diameter, is equal to the area of its maximum cross section (Fig. 2.15).

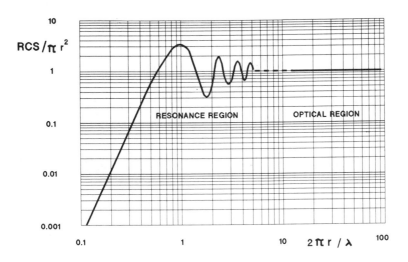

Figure 2.15 Variation with wavelength of the *radar cross section* (RCS) of a sphere of radius r.

For better understanding of the mechanism of scintillation, consider the signals produced by two elementary scatterers, $v_1(t) = \sin \omega t$ and $v_2(t) = k \sin \omega t$. It is assumed that the first signal is of unit amplitude and the second of amplitude k (Fig. 2.16).

Let the angle between the normal to the system consisting of the two reflectors and the line joining the observation point A to the center of the system be denoted by α. Then the two signals will be received at A with a phase shift φ given by

$$\varphi = \frac{2\pi}{\lambda} L \sin \alpha$$

where the path difference between the two signals has been approximated by

$$\Delta R = L \sin \alpha$$

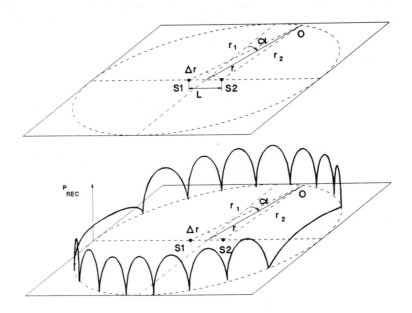

Figure 2.16 Power received at range R, produced by two interfering elementary radiators. This phenomenon underlies the mechanism of formation of the RCS of complex targets.

The total signal (in voltage) received at A will depend on the phase difference between the two elementary signals. Neglecting the common multiplier $\sin(\omega t)$ which denotes the operating frequency, one may write

$$V = |v_1 + v_2| = \sqrt{1 + 2k \cos \varphi + k^2}$$

Recall that signals should always be combined in voltage, considering their relative phase, and then converted to power by squaring. Thus if $k = 1$ and $\varphi = 0$, the power received at A is given by

$$P = (v_1 + v_2)^2 = 4P_1$$

where P_1 is the power received from a single scatterer.

If $k = 1$ and $\varphi = 180$ then the sum of the two signals vanishes, and the resulting power is zero.

Therefore for wavelengths of a few centimeters such as those usually employed in radar, a small relative motion of the two reflectors will suffice to produce a shift from the maximum to the minimum signal.

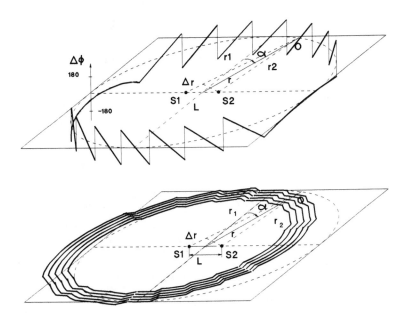

Figure 2.17 Wavefront distortion when the signals generated by two radiators are out of phase (glint).

An examination of the total radiation given by the two elementary radiators shows a very important feature. The radiated wavefront undergoes a distortion at all the points in space at which the two signals arrive shifted by nearly 180. The more equal in amplitude the two signals are, the more evident is the distortion. This phenomenon, acting on the elementary signals, causes fluctuations in the apparent angular position of the target (glint) (Fig. 2.17). Moreover, in low-altitude tracking, when both the direct echo signal and the signal reflected by the earth's surface are present, depending on the actual difference between the two path length, the two signals will sometimes be received by the radar out of phase and the distortion of the wavefront just mentioned will occur. Since every radar tracks by pointing its antenna orthogonally to the wavefront, a major pointing error will result at those points in which the two combined signals are out of phase, causing an oscillation of the tracking antenna in the vertical phase known as nodding.

From the theory of signals produced by two point sources, it follows that the pointing error, expressed as the ratio of the transverse shift to the apparent distance between the two sources, $L \cos \alpha$, is given by

$$\delta = \frac{1}{2} \frac{1 - k^2}{1 + k^2 + 2k \cos \varphi}$$

If it is assumed that the target is a set of elementary radiators, it is possible to simulate the pattern of its RCS. Figure 2.18 shows two examples of computer simulation, one of a naval target, the other of an airborne target.

The average RCS of a ship may be expressed approximately by the following equation quoted by Skolnik [8]

$$\sigma \simeq 52\sqrt{f_{\text{MHz}}} \sqrt{D_{Kt}^3}$$

where f is the frequency in megahertz and D is the ship's displacement in kilotons.

In fact, this equation must be used with care, since military vessels take special precautions to minimize their RCS.

Typical RCS values for air target are:

- for a fighter, 0.5 to 5 m^2, with scan-to-scan fluctuation.
- for a missile, 0.01 to 0.1 m^2, with no fluctuation, given the relatively simple missile geometry.

The following generally used models and definitions are taken from Swerling's studies of the RCS of targets and their consequences for radar range:

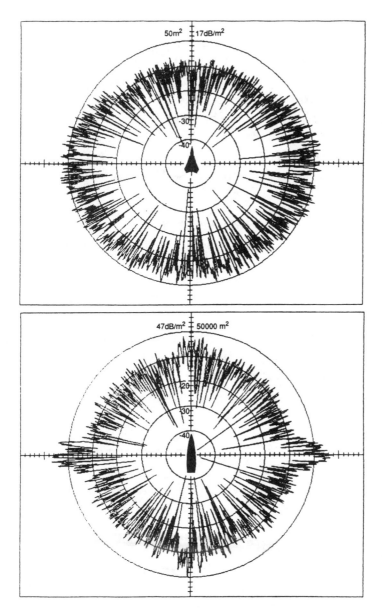

Figure 2.18 Computer simulated RCS of an aircraft and a ship. According to the way in which the single elementary scatterers are combined in phase and amplitude, a stronger or weaker RCS is obtained.

- swerling 0: Steady, non-fluctuating target.
- swerling I: Target consisting of many reflectors of comparable amplitude, with slow scan-to-scan fluctuations.
- swerling II: Target similar to the preceding one, but with fast pulse-to-pulse fluctuations.
- swerling III: Target comprising a single dominant reflector and many independent smaller reflectors, with slow scan-to-scan fluctuations.
- swerling IV: Target, similar to the preceding one, but with fast pulse-to-pulse fluctuations.

2.2.2.2 Equivalent noise temperature

Since the receiver operates at a given temperature $T(-K)$, a noise power given by

$$N = KTBF$$

will be introduced in the receiver, where B is the equivalent noise bandwidth (Fig. 2.19), F the noise figure of the receiver, and K Boltzmann's constant.

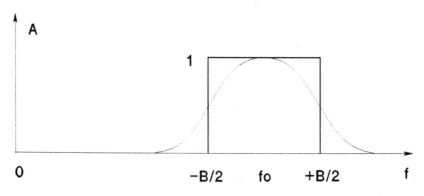

Figure 2.19 Equivalent noise band.

If the noise is expressed in dB m and B in megahertz, at the standard temperature of 290 degrees K, the product KT will be

$$KT = -114\,\text{dB}\,\text{m}/\text{MHz}$$

In calculations of radar range, an equivalent temperature, higher than the real one, is frequently introduced. Besides the noise figure, it considers also galactic and atmospheric noise, ohmic losses at the antenna, and transmission losses between the antenna and the receiver input.

The temperature introduced into the radar range equation is called the system temperature T_s. It is the sum of three contributions [9]

$$T_s = T_a + T_r + L_r T_e$$

where T_a is the temperature of the antenna, T_r the temperature of the receiving path, and T_e the temperature of the receiver.

If the temperature is given in degrees Kelvin and the losses numerically (and not in decibels!), the temperature of the antenna T_a may be expressed as

$$T_a = \frac{0.88\, T_{ai} - 254}{L_a} + T_0$$

Here T_{ai} may be deduced from Figure 2.20, L_a are the antenna losses of approximately 1 to 2%, and $T_0 = 290\,\mathrm{K}$ is the standard temperature.

The temperature of the receiving path T_r may be expressed

$$T_r = T_c(L_r - 1)$$

where T_c is the temperature of the components between the antenna and the receiver and L_r are the line losses. Finally, the receiver temperature T_e may be written

$$T_e = T_0(F_n - 1)$$

where F_n is the noise figure of the receiver (numerically and not in decibels).

Noise is always present, at all frequencies. Therefore, the wider the receiver bandwidth, the higher the noise power entering the receiver. On the other hand, a bandwidth wide

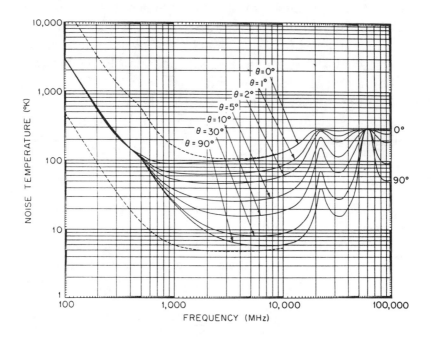

Figure 2.20 Noise temperature for an idel antenna (From L.V. Blake, *Pulse-Radar Range Calculation Work Sheet,* NRL Reports 6930 and 7010, 1969).

enough to let a significant part of the signal spectrum pass through is required.

As stated above, the receiver bandwidth is perfectly matched to the pulse when the ratio of the received signal power to the noise power is a maximum.

For a rectangular pulse of length τ, the bandwidth is roughly

$$B = \frac{1.2}{\tau}$$

For normal applications and evaluations of ED equipment, the value $KT = -144\,\text{dB}$. W/Hz may be assigned to the specific noise power, while the noise figure F and the losses of the line connecting the antenna with the receiver (L_r) are considered separately. When this is done, in the overwhelming majority of cases (frequencies over 500 MHz), the range error is negligible.

Since the receiver will consist of a number of amplifiers or circuits in cascade, which in their turn introduce noise, the total noise figure F to be considered in the radar range equation will be given by

$$F = F_1 + \frac{F_2 - 1}{G_1} + \frac{F_3 - 1}{G_1 G_2} + \cdots + \frac{F_i - 1}{G_i G_2 \ldots G_{i-1}}$$

where F_i and G_i are the noise figure and the gain of the ith circuit in the chain, respectively.

Therefore, if the first amplifier has a large enough gain, the total noise figure F will practically coincide with the noise figure for the first stage. In fact, a radar receiver usually has a pre-amplifier characterized by a low noise figure and by a gain such that the effects of the other amplifiers are negligible.

2.2.2.3 Example of a radar range calculation in free space

Usually, in order to calculate the radar range, the SNR at the receiver input, in decibels above 1 W (dB W), or in decibels above 1 mW (dB m), is plotted as a function of range, logarithmically measured on the abscissa axis, without considering the signal processing which follows (Fig. 2.21).

To do this, an easily expressed distance (for example 1, 10, or 100 km) is chosen, all parameters are converted into decibels, the following SNR equation is taken into account

$$S/N = \frac{P_{Tn} G_T G_R \sigma \lambda^2}{(4\pi)^3 K T B F R^4 L_{Tx} L_{Rx} L_b}$$

and all the positive and negative values are tabulated in two columns; finally, the values are added in each column. The algebraic sum of the two calculated values gives the SNR with respect to the chosen range. From this point, since the SNR depends on the inverse fourth power of the range, to obtain the required graph it suffices to draw a line of slope -40 dB/decade.

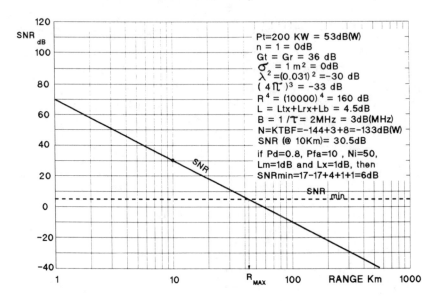

Figure 2.21 Calculation of the radar range in free space of a fluctuating target, with $P_d = 0.8$ and $P_{fa} = 10^{-6}$. Repeating the calculation for a given target at various heights, one obtains the coverage diagram.

At this point, according to the type of signal processing (number of integrated pulses, response of the MTI filter, and so forth) it is possible to calculate, for the target of interest, with specified detection and false-alarm probabilities, the minimum SNR required to determine the maximum range,

$$(S/N)_{\min} = (S/N)_{\text{Pdfa}} - 10\log N_i + L$$

where $(S/N)_{Pdfa}$ is obtained from Figure 2.13 as a function of the required P_d and P_{fa},

$$N_i = F_R T_{ot} = F_R \frac{\vartheta_B}{\omega}$$

$$L = L_m + L_x + L_i$$

Here L_i is obtained from Fig. 2.14 as a function of N_i and SNR; L_m is usually equal to $1\,\mathrm{dB}$ (or to $0\,\mathrm{dB}$ if the pulse is rectangular and the bandwidth B is $1.2/Y$); and L_x will depend on the type of radar processing and usually be between 1 and 6 dB. The distance corresponding to minimum SNR gives the radar range with the target, the P_d, and the P_{fa} in question.

This procedure is particularly useful in the design phase. In fact, parameters such as transmitter power, antenna gain, and so forth, can generally be but little modified, while, on the other hand, in order to achieve the required performance, the designer can intervene strongly concerning signal processing.

2.2.3 Radar equation in the operational environment

The above discussion refers to radar range calculated in free space, and therefore in an ideal case. In practice, however, the actual operational environment should be considered. Essentially, the environment is responsible for:

(1) additional signal attenuation arising from the transmission medium.

(2) generation of unwanted signals which clutter the radar screen (PPI), because of rough terrain, the presence of point obstacles such as fences and houses, sea waves, and scattering by clouds and rain.

(3) intensification and attenuation of the signal (lobing) due to the presence of rays reflected by the sea or the ground, when the roughness of the surface is small compared to the wavelength of the radar, and presents a high reflection coefficient (multipath).

(4) extremely strong signal attenuation because of the horizon, or the presence of mountains, hills, and so forth.

(5) anomalous propagation of the signal, in special atmospheric conditions (ducting), which attenuates the signal less than usual.

2.2.3.1 Atmospheric attenuation

The atmosphere is a gaseous medium. It will therefore attenuate the RF signal because of resonant absorption by molecules of the gases that are present, mainly oxygen and water vapor.

Figure 2.22 shows the two-way path atmospheric attenuation coefficient in decibels per kilometer (for a given range, the signal will make a two-way round trip).

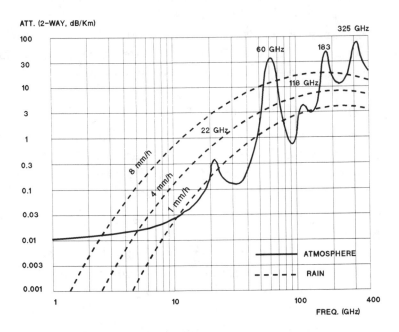

Figure 2.22 Two-way atmopsheric attenuation coefficient and additional attenuation coefficient in rain. The respective one-way coefficients are equal to one half of those show.n

The same figure shows the additional attenuation caused by rain; in this case, only the path length in which there is rain should be considered, not the full radar range. Attenuations which are due to the fact that signals pass through the atmosphere represent atmospheric losses L_{atm}, and must be included in the radar range calculation.

2.2.3.2 Clutter

When the strength of the signals produced by clutter is greater than the strength of the signals produced by targets of interest, the latter are masked and cannot be detected. An effective way to evaluate the impact of clutter is to measure its equivalent radar cross section, in order to compare it with that of true targets. In what follows, radar cross sections relative to ground, sea, and rain clutter will be calculated. The RCS of chaff will be discussed in section 5.5.1.2.

Ground Clutter
    When the ground is illuminated by the radar signal, it will, because of its unevenness, scatter the signal in all directions, including the direction back to the radar (Fig. 2.23).

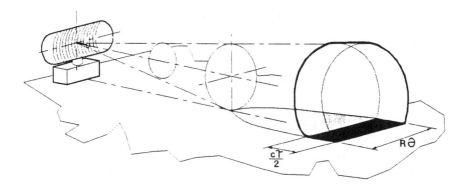

Figure 2.23 The ground corresponding to a radar cell gives an echo which clutters a radar PPI and may mask the presence of a target.

At a range R, a target T will generate a signal proportional to its RCS. In order to be detected clearly, such a signal will have to compete with the signal produced by clutter in the same radar cell with radial dimension $c\tau/2$ and transverse dimension $R\tan\vartheta \simeq R\vartheta$ (R in meters and ϑ in radians).
    If the area of the cell is multiplied by the *ground reflectivity* σ_{0g}, which depends on the type of ground, the angle at which it is seen from the radar (grazing angle), the frequency, the polarization, and so forth, the RCS σ_g of ground clutter is obtained.

For grazing angles far from 90 degrees, where the ground tends to behave more like a mirror than a scatterer [10],

$$\sigma_{0g} = \gamma \sin \psi$$

where ψ is the grazing angle and γ takes into account the scattering capability of the surface at that frequency and polarization.

Usually, for hilly ground covered with trees, one may assume for the reflectivity coefficient γ the value

$$\gamma = \frac{0.00032}{\lambda}$$

from which it may be deduced that ground clutter is less at low frequencies.

The RCS of ground clutter, competing with the target RCS, is given by

$$\sigma_g = \frac{c\tau}{2} R \frac{\vartheta_B}{\sqrt{2}} (\sin \psi) \gamma$$

The factor $\sqrt{2}$ in the denominator refers to the fact that the antenna gain is not constant in the $-3\,\text{dB}$ beam, but has an approximately Gaussian shape. As for the grazing angle, assuming a spherical earth, one has

$$\sin \psi = \frac{Z_1}{R} - \frac{R}{2a_c}$$

where Z_1 is the height of the antenna from the ground, and a_c is the equivalent radius of curvature of the earth, which, for the radio waves of interest, is 8.5×10^6 m.

For Z_1 sufficiently small, and R not excessively large,

$$\sin \psi = \frac{Z_1}{R}$$

Substituting one obtains

$$\sigma_g = \frac{c\tau}{2} \frac{\vartheta_B}{\sqrt{2}} Z_1 \gamma$$

Sea Clutter

In line with what has been said above about ground clutter, the equivalent surface of sea clutter (Fig. 2.24) may be expressed

$$\sigma_s = \vartheta_B \frac{R}{\sqrt{2}} \frac{c\tau}{\sqrt{2}} \sigma_{os}$$

where σ_{os}, the reflectivity coefficient of the sea, depends on the state of the sea, the grazing angle, and the wavelength, according to the equation [10]

$$\sigma_{os} = \frac{10^{0.6} K_B \sin \psi}{2.51 \times 10^6 \lambda}$$

in which K_B is the constant on the Beaufort scale showing the state of the sea, ψ is the grazing angle, and λ is the wavelength. Substituting for $\sin \psi$ the value found above, one obtains

$$\sigma_s = \vartheta_B \frac{R}{\sqrt{2}} \frac{c\tau}{2} 10^{0.6} \frac{K_B \left[(Z_1/R) - (R/2a_e) \right]}{2.51 \times 10^6 \lambda}$$

It appears that the RCS of sea clutter is practically constant up to the horizon and then decreases rapidly. For grazing angles near 90 degrees, the surface of the sea tends to behave like a perfectly reflecting surface.

Rain Clutter

Clutter due to rainfall, unlike sea or ground clutter, is volumetric. In fact rainfall, the source of the cluttering signals, is distributed in the radar volume cell, as shown in Figure 2.25.

The radar cell volume may be expressed, by arguments similar to those used for ground clutter, by

$$v = \frac{c\tau}{2} R \vartheta_B R \varphi_B$$

SEA STATE SS	WIND SCALE K_B = SS +1	WAVE HEIGHT (rms) σ_h
1	2	0.003 m
3	4	0.2 m
5	6	0.7 m
7	8	1.7 m

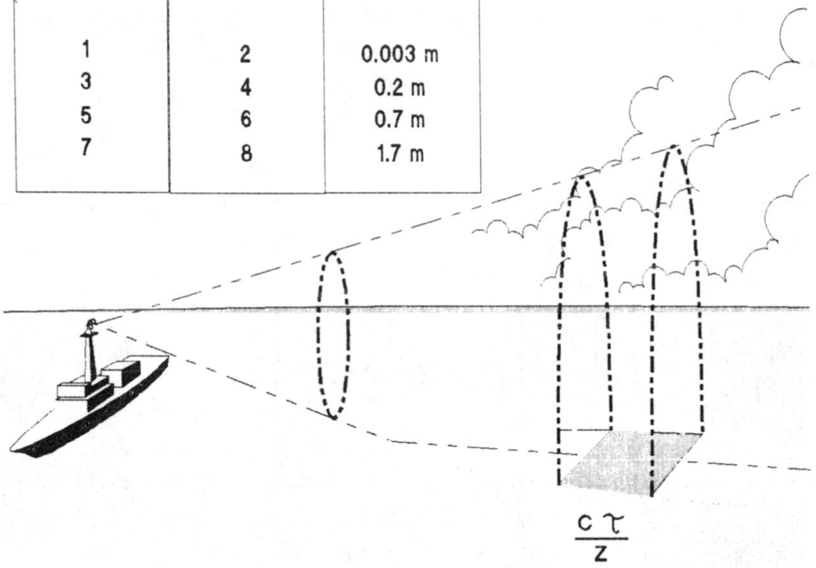

Figure 2.24 Sea clutter arises from the scattering of the radar signal by the sea waves corresponding to one radar cell. In the table are shown the values usually assumed for the heights of waves according to the state of the sea.

Taking into account the factor $1/\sqrt{2}$ in both azimuth and elevation, and denoting by η (in square meters per cubic meter) the reflectivity coefficient of the rain, one may say that

$$\sigma_r = \frac{c\tau}{2} R^2 \frac{\vartheta_B \varphi_B}{2} \eta$$

The volumetric reflectivity coefficient η may be expressed [11]

$$\eta = 6 \times 10^{-14} r^{1.6} \lambda^{-4}$$

where r (rain) expresses the amount of rainfall in millimeters per hour. The accepted values are $r = 1$ or 2 mm/h for light

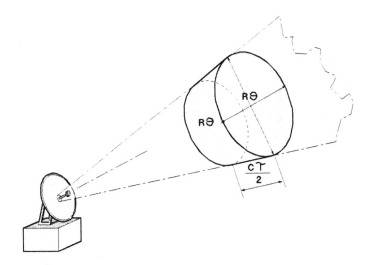

Figure 2.25 Rain clutter is produced when raindrops in a radar volume cell scatter a radar signal.

rainfall, $r = 4$ mm/h for heavy rainfall, and $r = 6$ mm/h for heavy downpours.

2.2.3.3 Lobing

In the operational environment, the radar range is strongly influenced by the ground or sea surface. These surfaces, if not excessively rough with respect to the wavelength, behave like mirrors; consequently, the radar will see not only the true target but also its reflected image (Fig. 2.26).

Because of the difference in path length, ΔR, between the direct ray and the reflected one, there will be a phase shift $\Delta \varphi$ which will depend on the geometry, and a phase shift $varphi_s$ due to the reflection. The total phase shift between the direct and the reflected rays is given by

$$\varphi = \varphi_s + \Delta \varphi \simeq \varphi_s + \frac{2\pi}{\lambda} \Delta R$$

With varying geometry, according to whether the direct and reflected rays are combined in phase or out of phase, the signal will be either intensified or attenuated, with a resulting

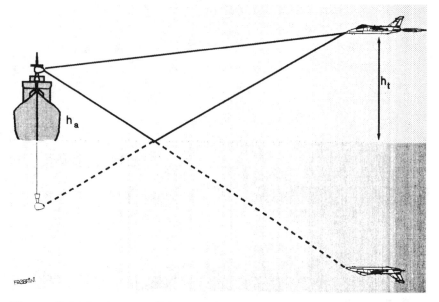

Figure 2.26 In some conditions, such as that shown in the figure, the echo signal reaches the radar by two paths, direct and reflected (multipath).

increase or decrease in the range assessment. The extent of the phenomenon will depend essentially on the value of the reflection coefficient, which, in its turn, will depend on the nature of the ground or the state of the sea.

From the theory of electromagnetic wave propagation, the complex reflection coefficient $\vec{\rho}$ may be expressed [12]

$$\vec{\rho} = \rho_s \rho_r \rho_d \exp(j\varphi_s)$$

where $\rho_s \times e^{j\varphi_s}$ is the specular reflection coefficient; ρ_r is the scattering coefficient due to the roughness of the surface; and ρ_d is a factor expressing the attenuation of the reflected ray because of the divergence due to the curvature of the earth.

The expression for the specular reflection coefficient depends on the polarization. For horizontal polarization it is

$$\rho_s \exp(j\varphi_s) = \frac{\sin\psi - \sqrt{\epsilon_c}}{\sin\psi + \sqrt{\epsilon_c}}$$

and for vertical polarization

$$\rho_s \exp(j\varphi_s) = \frac{\sqrt{\epsilon_c}\sin\psi - 1}{\sqrt{\epsilon_c}\sin\psi + 1}$$

where ψ is the grazing angle, and

$$\epsilon_c = \epsilon_r - j\,60\lambda\sigma_c$$

is the complex dielectric constant whose real part is the relative dielectric constant ϵ_r, and whose imaginary part is a function of the wavelength λ and of the electrical conductivity ϵ_c of the surface.

Figure 2.27 shows the dependence on grazing angle of the specular reflection coefficient and of the phase shift due to the surface of the sea, at different wavelengths, in vertical and horizontal polarization.

For the surface of the sea, the factor $\rho_r (< 1)$, which takes into account the roughness of the surface, may be characterized by an rms value σ_h for the height of the waves, and can be expressed

$$\rho_r = \exp[-2(2\pi\sigma_h\psi/\lambda)^2]$$

The divergence factor $\rho_d (< 1)$, which takes into account the curvature of the earth's surface, need be considered only for large distances. The signal radiated by the radar will reach the target both by the direct path and by reflection.

The electric fields of the two signals will be combined taking into account the reflection coefficient and their relative phase

$$E = E_d\cos(\omega t) + \bar{\rho}E_d\cos(\omega t + \varphi) = Re[E_d e^{j\omega t}(1 + \rho e^{j\varphi})]$$

If it is assumed that $\varphi = 1$, the electric field will vanish whenever

$$\varphi = (2K + 1)\pi$$

where $K = 1, 2, 3$, etc. and will be doubled whenever

$$\varphi = 2K\pi$$

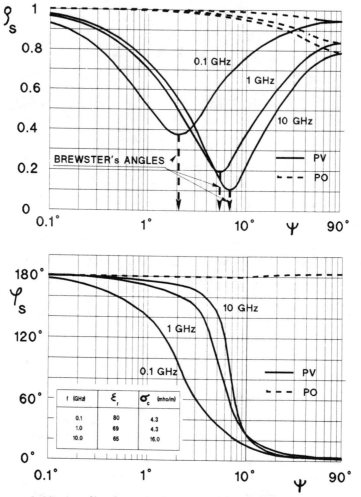

Figure 2.27 Amplitude and phase of the reflection coefficient, in horizontal (HP) and vertical (VP) polarization, calculated for calm sea, for various grazing angles, and for different frequencies.

The ratio of the electric field incident on the target in the actual conditions of propagation to the electric field in free space, where only the direct ray exists, defines the propagation factor F_p, which takes into account the actual conditions of propagation

$$F_p = \left| \frac{E_d e^{j\omega t}(1 - \rho e^{j\varphi})}{E_d e^{j\omega t}} \right| = |(1 - \rho e^{j\varphi})|$$

For $\rho = 1$, the power of the signal relative to the maxima will be four times that in free space.

It follows that the power density impinging on the target in the actual conditions will be

$$p = p_{FS} F_p^2$$

where p_{FS} is the power density in the ideal free space case. The signal reradiated by the target will reach the radar both directly and by reflection; to correct with respect to the ideal case, here too it is necessary to multiply by the function F. To sum up, the radar signal in the real-world situation is expressed by

$$S = S_{FS} F_p^4$$

where S_{FS} is the radar signal in free space.

Finally the radar range equation becomes

$$R_{max}^4 = \left[\frac{N_i P_T N G_T G_R \sigma \lambda^2}{(4\pi)^3 K T F B (S/N)_{Pdfa} L_t L_m L_x L_{Tx} L_{Rx} L_{atm}} \right] F_p^4$$

Signal, and therefore power, peaks will occur when $\varphi = 2K\pi$. They may be up to 16 times the free space value. Minima will be present for $\varphi = (2K + 1)\pi$.

For vertical polarization and small grazing angles, the phase shift caused by reflection, φ_s is equal to π. In this case, the condition for maxima may be expressed as

$$\varphi = \pi + \frac{2\pi}{\lambda} \Delta R = 2k\pi$$

Observing that from Figure 2.26 R may be expressed as

$$\Delta R \simeq 2h_a \sin \vartheta_t \simeq 2h_a \frac{h_t}{R}$$

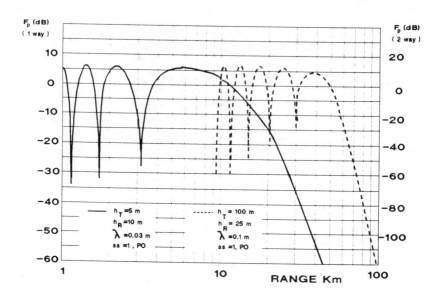

Figure 2.28 Propagation factor F_p^2 (one way) and F_p^4 (two way).

one obtains the condition for maxima

$$2k\pi = \pi \left(1 + 2\frac{2h_a h_t}{\lambda R} \right)$$

that is

$$\frac{4h_a h_t}{\lambda R} = 2k - 1$$

In the same way, it is found that the condition for minima is

$$(2k + 1)\pi = \pi \left(1 + 2\frac{2h_a h_t}{\lambda R} \right)$$

that is

$$k = \frac{2h_a h_t}{\lambda R}$$

Figure 2.28 shows the values of F_p for a number of ranges, target, and antenna heights at different frequencies. Figure 2.29 shows a diagram of radar coverage modified by the lobing effect.

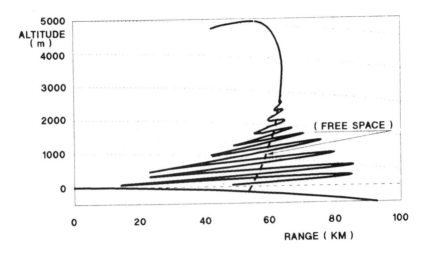

Figure 2.29 Lobing in the radar coverage diagram in the presence of multipath.

2.2.3.4 The radar horizon

Because of the sphericity of the earth, radar emissions cannot reach points on the earth's surface which are very distant from the source. In free space, radar rays (that is, rays normal to the wavefront), like optical rays, are straight lines. However, it may be shown that, as they pass through the earth's atmosphere, with its varying index of refraction, usually decreasing with height, radar rays curve downward. This means that, heights being equal, the radar horizon is beyond the optical horizon.

In the case of a standard atmosphere, it has been demonstrated that rays can be treated as straight lines by considering the earth's surface as a spherical surface with an equivalent radius four-thirds the radius of the real earth [13]. Consequently, in the radar band,

$$R = \frac{4}{3} R_T = \frac{4}{3} \times 6.382 \times 10^6 = 8.488 \times 10^6$$

From Figure 2.30 it is easy to compute the distance R_H to the radar horizon:

$$R_H = \sqrt{,(R_e + H_R)^2 - R_e^2} \simeq \sqrt{2 H_R R_e}$$

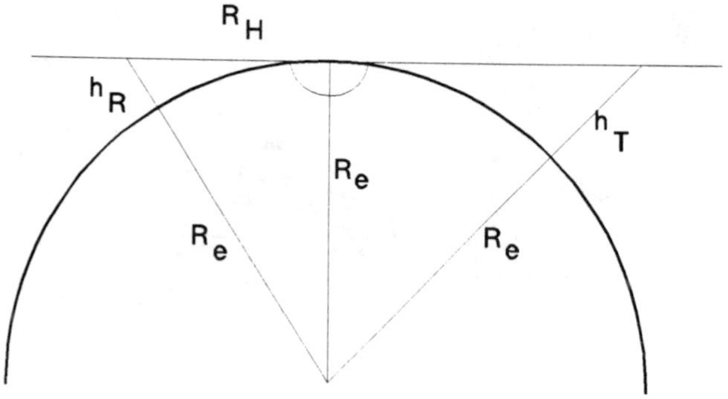

Figure 2.30 Radar horizon.

Radar rays, like optical rays, may be intercepted by non-transparent obstacles (the shadow effect). However, in the region shadowed by the obstacle, an electromagnetic field may still be detected, because of diffraction by the edges of the obstacle.

The theoretical treatment of this phenomenon derives from Huyghen's principle, shown in Figure 2.31, and is based on the consideration that the obstacle suppresses a fraction of the wavefront reaching it, while its edges diffract the rest.

The theoretical results [14] may be synthesized as in Figure 2.32, where the attenuation is calculated as a function of the dimensionless parameter v composed from the parameters shown in Figure 2.31 according to the formula

$$v + \pm h \sqrt{\frac{2}{\lambda}\left(\frac{1}{d_1} + \frac{1}{d_2}\right)} = \sqrt{2\frac{h}{\lambda}(\alpha_1 + \alpha_2)}$$

The positive sign is used for v when point H is outside the obstacle, that is, when the obstacle does not intercept the line of sight. A similar theory may be applied to the curvature of

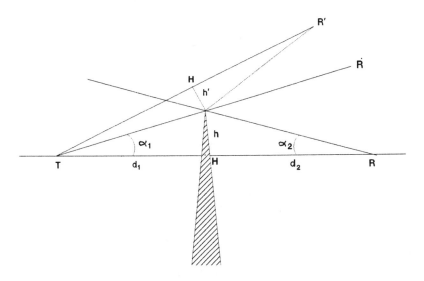

Figure 2.31 The effect of obstacles on propagation.

the earth. Usually, the attenuation effect over the horizon is considered by suitably modifying the factor F_P.

2.2.3.5 Ducting effect

Unlike the lobing effect, which never fails to occur in the presence of a reflecting surface, the so-called ducting effect occurs only in particular atmospheric conditions. It is due to the inversion in altitude of the thermal gradient, and to the ensuing anomaly in the index of refraction.

A kind of waveguide is formed between the surface of the sea and the layer of air above it, which channels the signal, preventing its normal rapid attenuation with the square of the distance [15] (Fig. 2.33).

2.2.3.6 Radar range in the operational environment

In order to calculate the radar range in real-world situations, all the phenomena described above must be taken into account. Once the pattern of the SNR in free space has been calculated by the method described above (Fig. 2.34), it has to be modified by the introduction of the additional attenuation due to the atmosphere and to potential rainfall. The *clutter-to-noise*

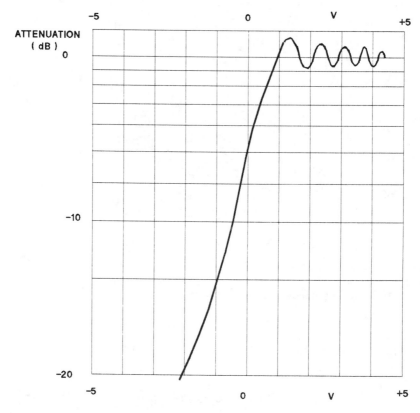

Figure 2.32 Attenuation by an obstacle.

ratio (CNR) of the signal, possibly attenuated by the use of *moving target indication* (MTI) or frequency agility, must be plotted. The SNR has to be modified, by means of the propagation function F_p, to take into account possible reflecting surfaces and the radar horizon.

The correct visibility is achieved when the SNR is higher than both the minimum SNR and the clutter residues. If the calculation is iterated for various target heights, it yields a radar coverage diagram for the actual operational environment.

2.2.4 Radar techniques

Sophisticated techniques have now been developed to overcome the problems caused by the operational environment and the

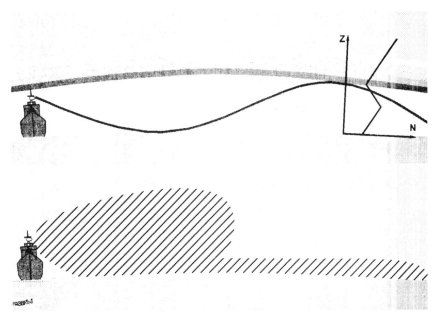

Figure 2.33 Ducting effect. The radar signal is trapped between the surface of the sea and an overhanging air layer, and is therefore only slightly attenuated with increase in range.

behavior of target RCS, such as clutter, sea reflections, and target fluctuations. The main techniques are the following:

- *Moving target indication* (MTI), to minimize echoes due to clutter.
- *Constant false-alarm rate* (CFAR); that is, receivers capable of matching received signals to keep the false-alarm rate constant.
- Frequency agility; that is, change of transmitter carrier frequency on a pulse-to-pulse or batch-to-batch (of pulses) basis.
- Pulse compression; that is, coded pulse transmission to increase range resolution.

2.2.4.1 Moving target indicator

MTI devices are frequency filters based on the doppler effect. They pass target returns with a certain radial velocity, and

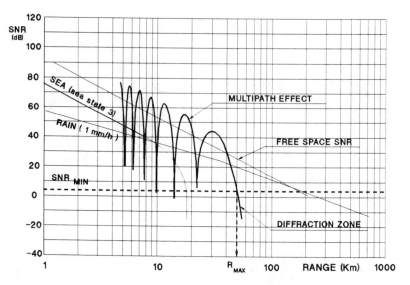

Figure 2.34 Calculation of radar range in an operational environment.

attenuate returns from fixed or slow moving targets and from background clutter.

When a radar pulse of frequency f impinges on an aircraft which is moving with a certain radial velocity V_R, the reradiated echo will be received at the radar with a frequency

$$f_r = f + f_d$$

where

$$f_d = 2\frac{V_R}{\lambda}$$

is the doppler frequency and

$$\lambda = \frac{c}{f}$$

is the wavelength of the radar carrier.

By detecting radar pulses coherently, that is, bringing them into the baseband by means of a local oscillator *coherent* with the generator of the transmitted waveform, it is possible to

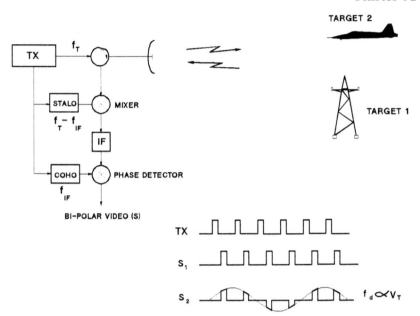

Figure 2.35 Bipolar video output of a coherent radar. The amplitude is constant for a stationary target, but oscillates at doppler frequency for a moving target.

obtain a bipolar video display such as that shown in Figure 2.35, where pulses are transmitted with a PRF

$$F_R = \frac{1}{T}$$

The doppler frequency is evidently much lower than the radar frequency. When these pulses are sent through the circuit of Figure 2.36, whose frequency response is shown, echoes from a nonmoving target, with zero doppler frequency, are attenuated, while echoes from a moving target pass the filter.

As can be seen in the figure, when

$$f_d = \frac{1}{T} = F_R$$

the filter output is again zero. The corresponding velocity is called the "blind speed". For higher velocities, echoes will

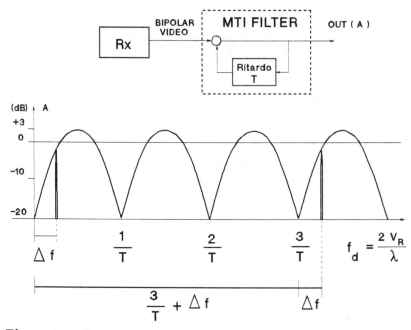

Figure 2.36 Response curve of a single-delay-line MTI. A radar using a low PRF (large T) measures velocities ambiguously if they yield doppler frequencies higher than the PRF itself.

again pass, since the filter is of iterative type and has a succession of passbands like the spans of a bridge. When the target velocities generate doppler frequencies in the first passband, the measured doppler frequency f_{dmeas} indicates unambiguously the radial velocity of the target. However, since a pulse radar measures the doppler frequency by sampling at the pulse repetition frequency F_R, one cannot tell whether the measured doppler frequency is the one which corresponds to the radial velocity, or just the difference between that and a multiple of F_R

$$f_{d\,meas} + nF_R$$

To eliminate this velocity ambiguity it would be necessary to use a PRF high enough to be above the doppler frequency generated by targets flying at the fastest expected speed.

However, the use of a very high PRF would cause the dual

phenomenon called "range ambiguity". In fact, all targets at a range shorter than

$$R_{\max} = \frac{c}{2}\frac{1}{F_R} = \frac{c}{2}T$$

are in a one-to-one correspondence with the range as measured by the radar. But targets at a range

$$R_1 = \frac{c}{2}(T + \Delta T)$$

or

$$R_n = \frac{c}{2}(nT + \Delta T)$$

would appear to the radar to be at a range

$$R = \frac{c}{2}\Delta t$$

To avoid range ambiguity, the PRF should be low enough to ensure that targets of interest are all within $R_{\max}$.

In practice, the radar must deal either with velocity ambiguity or with range ambiguity. In order to avoid velocity ambiguity, $1/T$ should be sufficiently high, but if $1/T = F_R$ is very high, range ambiguity will ensue, unless the radar range is enormously reduced. For example, at a frequency of 3 GHz, for a target moving at Mach 1 (velocity $= 300$ m/s), one obtains

$$f_d = \frac{2V_R}{\lambda} = \frac{600}{0.1} = 6\,\text{kHz}$$

A radar designed for a 150 km range, corresponding to an echo return time of one millisecond, must have a PRF less than 1000 Hz in order to avoid range ambiguity. But in such a case, in order to avoid velocity ambiguity, the maximum acceptable target speed must be equivalent to 1000 Hz; that is, only one-sixth of the required velocity.

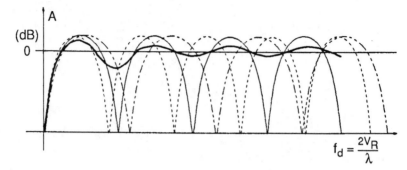

Figure 2.37 The average output of the MTI does not exhibit blind speeds if a variety of PRFS are used.

To eliminate the blind speed problem, one may adopt the PRF staggering technique which entails a variation, according to certain rules, of the radar PRF. In this way, it is possible to "fill the holes" in the MTI filter for doppler frequencies corresponding to multiples of the PRF (Fig. 2.37).

The preceding example uses a filter with a single delay line. Such a filter is a simple MTI delay-line canceler, which needs two radar returns, that is, two *pulse repetition intervals* (PRI), in order to reach the steady-state condition. The filter is very simple, but there is a disadvantage: Since ground clutter may consist of echoes produced by trees blowing in the wind, the clutter is not fixed and will have a spectrum of a certain width (Fig. 2.38).

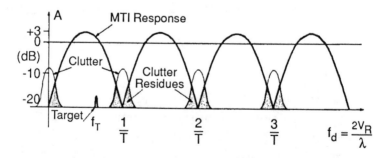

Figure 2.38 Since clutter has a relatively broad spectrum, a good MTI must display a sufficiently wide null around the clutter carrier frequency.

This means that the shaded region in the figure will not be completely cancelled and might yield a signal stronger than the signal from the target of interest. MTI filters with more delay lines need to be used, in which the various elements with different delays are weighted in such a way that the filter response is matched to the clutter to be cancelled.

Figure 2.39 shows the situation for an MTI double canceler (two single-delay-line cancelers), which requires three samples for a useful output. The improved cancellation behavior is apparent.

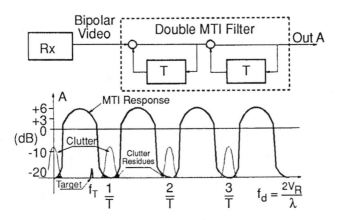

Figure 2.39 A double-delay canceler permits better clutter cancellation, with relatively broad spectrum, compared to a single-delay MTI.

The more complex the MTI, the better the clutter cancellation. However, as the number of samples needed for the steady-state condition of the MTI increases, it may become too high, and for example, deny the radar the possibility of exploiting other techniques, such as pulse batch frequency agility.

The purpose of an MTI device is to detect a target that produces a signal weaker than the clutter signal. The *subclutter visibility* (SCV) achieved by an MTI is defined as the clutter-to-target ratio present at the input of the MTI that permits target detection at the output with the same detection and false-alarm probabilities as for the noise. Here it is assumed

N = Number of delay lines
λ = Wavelength (m)
f_r = Pulse repetition frequency (s $^{-1}$)

σ_v = Standard deviation of clutter spectrum (m/s)	
Wooded Hills	0.02 - 0.4
Sea Chaff	0.4 - 1.5
Rain	1.5 - 4

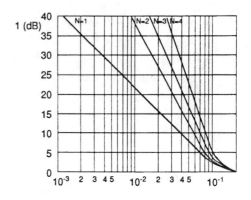

Figure 2.40 Improvement factor and MTI types.

that the clutter residues are decorrelated, in the same way as in the case of noise. The effectiveness of an MTI filter is usually measured by its improvement factor I, defined as the enhancement of the signal-to-clutter ratio (S/C) achieved in passing the MTI filter:

$$I = \frac{S_0/C_0}{S_i/C_i} = \frac{C_i}{S_i}\frac{S_0}{C_0}$$

As remarked earlier, S_o/C_o must be equal to the minimum SNR, so that

$$I = \frac{C_i}{S_i}(SNR)_{\min} = (SCV)(SNR)_{\min}$$

The improvement factor possible for a radar depends both on elements internal to the radar, such as the stability in time and phase of its receive-transmit circuits, and on external elements, such as the stability of the clutter to be canceled, which is in its turn, a function of the MTI type (single, double, triple, etc.).

In Figure 2.40, various types of MTI and various frequency responses are shown. The table gives the possible improvement factor as a function of the type of filter [16].

If the number of samples necessary for implementing the MTI is high, this may conflict with the use of frequency agility. In order to cancel the clutter completely, the radar will have to maintain all its parameters, and in particular its frequency, rigorously constant for the whole period in which the MTI is operating. If the radar frequency changes, the clutter amplitude changes. This means that the output of the MTI cannot be accepted as correct until once again all the samples needed to implement it have been processed. Figure 2.41 shows the situation for a radar with a double-delay canceler, changing frequency every 10 pulses, where the information loss is 20%.

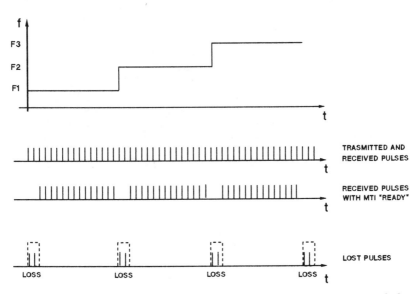

Figure 2.41 MTI by pulse group: frequency agility may be used, but some pulses, needed to have the MTI, in the steady-state condition are lost.

On the basis of Figure 2.35, it may be shown that if the doppler signal is rectified before being processed by the MTI filter, there are some initial phases and doppler frequencies for which the filter response to two successive pulses is zero, as if it were fixed clutter. Thus, useful signals with these doppler frequencies become less visible, because of the phase. This phenomenon is known as *blind phase*, and introduces an average loss of 3 dB.

This loss is avoided in more sophisticated radars by the use of a coherent detector, with both in-phase (I) and quadrature (Q) outputs (Fig. 2.42). At the filter output these signals are recombined to give the modulus

$$s = \sqrt{I^2 + Q^2}$$

In this way the lost 3 dB is recovered, but the MTI circuit must be duplicated, since now the Q channel has to be processed as well.

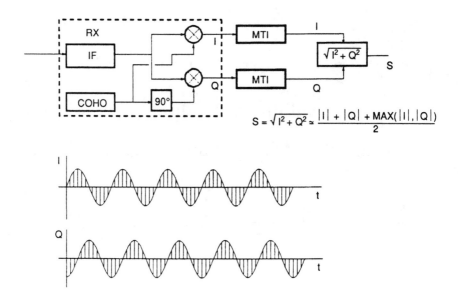

Figure 2.42 Elimination of blind phases.

2.2.4.2 CFAR receivers

In simple radars, for example civil navigation radars, an operator observes the signal intensity on the cathode-ray tube of the PPI, considers the persistence, the duration in angle and in range, and so forth, of the blip and decides whether it is, or is not, a target. Obviously, the operator needs some time to come to a decision.

Such a situation is unacceptable in a military radar warning center, which has to control a region of radius between 100 and 200 km, and to detect airborne threats, especially when threats may be flying at low altitude, at very high speed, and in the presence of friendly aircraft.

In these conditions, to decide which are the true threats, to select the most dangerous ones, and to designate them to the various weapon systems so that they can react (acquire the target, set tracking loops, set guns or launchers, and fire) on time, is a task too onerous to be done manually by an operator. To perform this mission properly the radar has to be equipped with more sophisticated devices.

First of all, at the receiver output, the received and detected signals are compared with a first threshold. This may be a fixed or, more usually, an adaptive threshold. Its function is to ensure that signal processing is performed exclusively on signals having high probability of representing a true target.

In the most sophisticated radars, the first detector is followed by a complex circuit called an "automatic detector". With this device, the operator is no longer confronted on the PPI by a "raw" video that needs interpreting, but by a synthetic video, which is the result of various integrations and correlations. The operator can now be confident that all the visible signals represent true targets. Thus the operator, or a group of operators each controlling a sector of the radar screen, appraises the situation, if it is not excessively complex, with the aid of powerful computers. Threat assessment, weapon assignment, and so forth, may be speedily performed by these computers on the basis of the radar detector outputs.

The computers in these warning centers must perform a large number of calculations. In particular, the computer section most closely linked to the radar will have to perform its calculations starting from each detection at the output of the radar automatic detector. Since the computer is able to perform a qualitatively limited, although quantitatively very high, number of operations per second (computational capability is expressed in *millions of instructions per second* (MIPS), it is necessary to set a limit to the computer input by reducing the number of false alarms as much as possible.

CFAR receivers are often used for this purpose. They maintain a constant false-alarm rate at the radar output, perhaps at the cost of some desensitization. A CFAR device frequently used in radar is the autogate. This receiver uses an array of delay lines to compare the signal present at the central output with a threshold obtained by averaging the signals on either side of the radar range cell of interest and multiplying the mean value by a factor α (Fig. 2.43). In this way, the threshold is adaptive, which means that if the noise in the receiver increases, or if there are clutter residues, the threshold increases as well, thus keeping the false-alarm probability constant [17].

Without a device of this kind, the presence of a jammer in a certain region would be enough to saturate the automatic detector, or the computer that follows it, thus totally degrading the operation of the radar.

It is important that designers of electronic warfare equipment should clearly understand this. To operate properly against modern threats, flying at high speed and at low level, an air defense radar center must be equipped with automatic detectors and powerful computers. A countermeasure forcing a radar to do without those functions, such as MTI, CFAR threshold, and automatic detection that enable it to perform its mission correctly, is an effective countermeasure. To be effective, a countermeasure should always address a radar's mission and prevent its performance, or at least its successful performance.

2.2.4.3 Frequency agility

The *frequency agility* technique exploits a change in radar

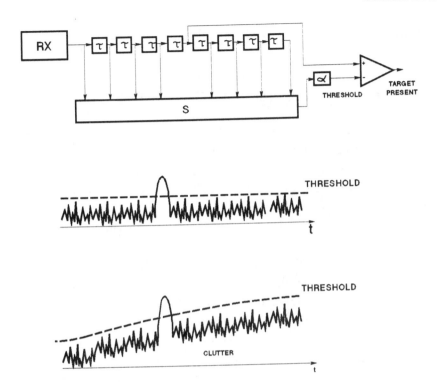

Figure 2.43 An Autogate. This device allows generation of an adaptive threshold above noise or clutter residues, and thus confers CFAR characteristics on the radar receiver.

carrier frequency on a pulse-to-pulse or batch-to-batch basis, within a band that may be more than 10% of the central frequency. A special case is frequency diversity, that is, the use of preset frequencies inside the band.

Frequency agility, although it entails severe technical difficulties, offers enormous improvements in performance compared to fixed frequency operation. These are:

a) an increase in range, other parameters being equal, of up to 35%.
b) clutter reduction in radars without MTI.
c) glint reduction in tracking radars.
d) reduction or nullification of lobing.
e) reduction of nodding in tracking radar.

f) reduction of the effectiveness of jamming.

Range Increase with Frequency Agility

As was remarked above, radar range depends on the minimum SNR required to achieve a given detection probability P_d and a given false-alarm probability P_{fa}. One such SNR value, after integration, is shown in Figure 2.13; as can be seen, it depends on the type of target to be detected. This could be either a nonfluctuating target, such as a sphere or a missile, that is, a target of simple and not excessively inhomogeneous geometry, or a scintillating target, that is, one with a scan-to-scan fluctuation (Swerling I). In the latter case, during a scan of the target, the relative positions of the target's elementary scatterers may be such as to generate signals that combine in phase in the radar receiver, thus producing a very strong echo. However, during the next scan, as a result of a change of relative position, the signals produced by the elementary scatterers may combine out of phase to give a very weak echo signal. This type of target is called a "scan-to-scan fluctuating target."

When high detection probabilities are required, mean RCS and all other parameters of the radar being equal, the probability of detecting a nonfluctuating target is much higher than the probability of detecting a scan-to-scan fluctuating one. Moreover, if during the radar time on target the target assumes all the RCS values, as is the case with fast-fluctuating targets, the radar may integrate the various reflected signals to yield an integrated signal equivalent in value to that of a nonfluctuating target (Fig. 2.44).

Since pulse-to-pulse frequency agility causes a change of wavelength, its effect is such that, during the time on target, a scan-to-scan fluctuating target becomes a pulse-to-pulse fluctuating target (Swerling II or IV). However, this is true only if the difference in frequency is sufficiently high. The greater the size of the target and the more inhomogeneous it is, the greater the decorrelating effect of frequency agility. The

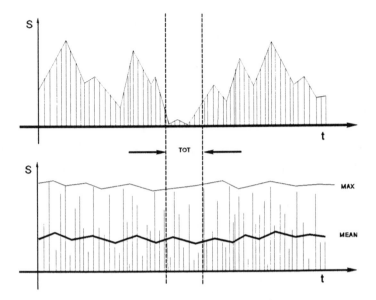

Figure 2.44 Effect of frequency agility on detection probability. During the time on target, visibility of the target mean RCS is ensured, even with fluctuating targets.

correlation index may be written [18]

$$\rho(\Delta f) = \frac{\sin^2[2\pi\Delta f(l\cos\alpha)/c]}{2\pi\Delta f(L\cos\alpha)^2}$$

where α is the angle between the line of sight and the normal to the dimension L, and c is the speed of light (and of radar signals).

Therefore, for a given mode of frequency variation in time, it is possible to check how many returns from a slowly fluctuating target, during the time on target, may be considered to be decorrelated. If all of them are, the target is known as a "pulse-to-pulse fluctuating target".

From Figure 2.12 it may be deduced that for an 80% detection probability, transformation of a fluctuating target into a fast-fluctuating target results in a reduction of the required minimum SNR by 5.5 dB. All other parameters being equal, such a reduction yields a 35% increase in range.

Clutter Reduction

For the reasons just given, it happens, in the case of (extended, not single-point) ground clutter, and of sea and rain clutter, that target return signals with an almost constant amplitude during the time on target are converted into pulse-to-pulse fluctuating signals, just as in the case of noise (Fig. 2.45). It may therefore occur that for a slightly fluctuating target, such as a missile, against an extended clutter background, the radar integrator gains more on the target (N_i) than on the clutter ($\sqrt{N_i}$). In such cases, reduction of the clutter by frequency agility may be as high as 10 or 11 dB.

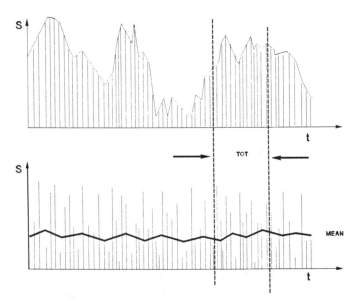

Figure 2.45 Pulse–to–clutter decorrelation by means of frequency agility. After integration over the time on target, target visibility is increased, compared to clutter.

Frequency agility is very useful in modern navigation radars not provided with MTI, to decorrelate clutter returns, thus reducing the intensity of clutter compared to targets of interest.

Glint Reduction

A target consists of a set of elementary scatterers. Therefore the apparent center of radar reflection, or effective origin of the radar echo, does not coincide with the physical center, but fluctuates around it, causing an angular error (glint) which may be written

$$\epsilon_g = \frac{1}{3}\frac{L}{R}$$

as will be shown later. Usually, this is a low-frequency fluctuation, its spectrum being in a band which in practice extends below 5 Hz.

Figure 2.46 shows the time dependence of the glint error. Such a pattern depends on the frequency; changing the transmission frequency may change the pattern completely. When the pulse-to-pulse frequency is changed, such an error takes all possible positive and negative values, and when these errors are averaged, taking into account the time constant of the tracking loop, their effect vanishes.

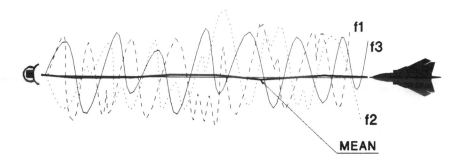

Figure 2.46 Glint reduction by means of frequency agility. The apparent radar center, averaged over all the frequencies used, tends to coincide with the center of the target.

Lobing Reduction

As has already been pointed out, the angles at which minimum and maximum values of radar coverage occur, because of the presence of a reflecting surface such as the sea (lobing), depend on the value of the wavelength λ. Therefore if the carrier frequency is changed, the conditions for these extreme values are decorrelated, thus bringing the actual coverage diagram closer to the free-space one.

The angular positions of the nulls and peaks in the coverage diagram depend, in the presence of the sea, on the frequency employed. Averaging over frequency the signals received during the time on target, one obtains the frequency agility coverage diagram shown in Figure 2.47, which, as can be seen, ensures a more uniform visibility.

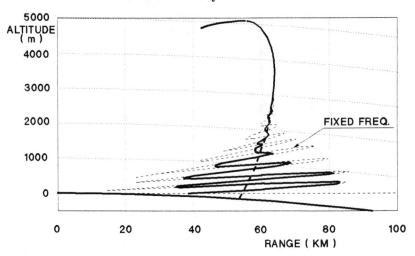

Figure 2.47 Reduction of lobing by means of frequency agility. Unfortunately frequency agility has little effect at lower altitudes.

Nodding Reduction

As noted earlier, when two sources interfering at the same frequency go out of phase, and the difference in amplitude is small, there is a distortion of the wavefront such that the apparent origin of the combined signal does not coincide with that of the two sources, but may appear farther away.

This phenomenon occurs when a tracking radar tracks an aircraft in flight over a reflecting surface such as the sea. Because of the motion of the target, the direct and reflected signals are sometimes out of phase, giving the radar erroneous information concerning the elevation angle of the target. This phenomenon is called nodding. It will be seen below that this is one of the major problems confronting tracking radars (section 2.2.6.4).

Since the phase condition depends on wavelength, a variation of the carrier frequency will have a decorrelating effect on the phenomenon, attenuating its effects, as will be shown in more detail later [19].

Reducing the Effect of Countermeasures

Frequency agility is useful as a counter-countermeasure against noise jamming. The jammer will be forced to spread its power over the frequency agile band, while the radar bandwidth, defined by the bandwidth of the intermediate frequency chain, remains unaltered. The effect of frequency agility on noise jamming may be so high as to counter the effectiveness of the jamming completely, as will be seen when ECCM systems are discussed in chapter 6 (section 6.2.1.7).

2.2.4.4 Pulse compression

If high peak power transmission cannot be achieved, or is not wanted, but both long-range and high-range resolution are desired, it is possible to resort to the *pulse compression* technique.

This technique entails transmitting pulses of low peak power, but long duration τ, which contains a code consisting of n elements of duration $\tau_{el} = \tau/n$, allowing the receiver to recognize the code elements, realign them in time, add them coherently, and generate an output pulse characterized by high intensity (the output SNR is n times the input SNR) and duration equal to the duration of the code element (Fig. 2.48).

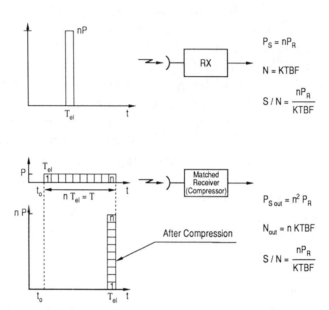

Figure 2.48 Equivalence between a coded low peak power pulse and a high peak power pulse of equal energy.

Radars which exploit this technique are called "pulse compression radars", or "coded radars". They employ *modulation on pulse* (MOP) of the transmitted pulses according to a certain code. The spectrum of the coded pulse occupies a band whose width is approximately equal to the reciprocal period of the coded element:

$$B = \frac{1}{\tau_{el}}$$

Modulation may be *phase modulation on pulse* (PMOP) or *frequency modulation on pulse* (FMOP).

It is necessary to resort to pulse compression when, for example, the radar must have *low probability of intercept* (LPI) characteristics in order to avoid detection by hostile ED equipment. This technique is also needed in sophisticated radar that must have optimum clutter cancellation performance, and at the same time be frequency agile on a burst-to-burst basis over broad bands (higher than 5%).

To achieve good MTI capabilities, the radar should use - *coherent chains*, where all the oscillators, both those used for signal generation and those used as local oscillators for conversion to baseband, are driven by a single, very stable quartz oscillator.

As final power amplifiers for generating the required power, klystrons can be used. Klystrons are able to amplify the signal greatly while maintaining good phase stability. However, these tubes usually give good performance only over limited frequency bandwidths (less than 5%), and are not suitable when larger band widths are required (more than about 10%).

Consequently, the use of *traveling-wave tubes* (TWT), which allow high mean powers, are highly stable in phase, and can cover bands exceeding 10%, is more appropriate. The peak power of a TWT is directly related to the anode-cathode voltage. In order to avoid too high a voltage and make such transmitters feasible, long-duration pulses should be used so that the peak power can be limited, and the average power better exploited.

Since the radar range resolution cell is of dimension $c\tau/2$, the radar will not have a high resolution capability when τ is long. Moreover, as stated in the preceding section, the amount of clutter to be canceled increases with pulse duration.

The pulse compression technique solves both the problem of range resolution and the problem of clutter in the radar cell. In fact, because of the decorrelation of the various elementary returns, after the matched receiver the resolution and the amount of clutter will be the same as those derived for the elementary pulse duration.

The matched receiver, or compressor, allows for coherent summation, in voltage, of the elements of the code, yielding at the output a useful signal of peak power n^2 times the input peak power. In the recombination process noise is summed as well, but being incoherent, it will contribute to the output a power only n times higher.

The SNR at the output of the compressor will therefore be n times the SNR at the input, and the pulse will be of duration τ_{el}. This is the same as saying that, for a given range and range resolution cell, a pulse compression radar with peak power P and n coded elements is the same as a non-coded radar with peak power nP and pulse duration equal to the duration of the code element

$$\tau_{el} = \frac{\tau}{n}$$

Figure 2.49, showing a phase code at the input and at the output of the compressor, explains this.

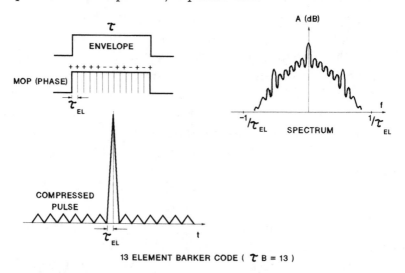

Figure 2.49 Example of a phase code with its spectrum.

The figure also shows a negative feature of this technique, that is, the presence of sidelobes, or unwanted signals before and after the echo of the compressed signal. Many studies deal with the choice of the best codes, and others deal with the weighting functions to be used in the recombination of the elements. Their purpose is the reduction of sidelobes [20].

Position codes should also be mentioned. These are codes realized by the transmission of groups of pulses spaced a few microseconds apart in a certain sequence and repeated at the

PRF. Again, the pulses may contain a phase or a frequency code. Figure 2.50 shows an example of a frequency code. Figure 2.51 shows a possible block diagram for a coherent chain radar exploiting a phase code. Phase codes giving optimumly uniform sidelobes are called Barker codes; their maximum length is 13.

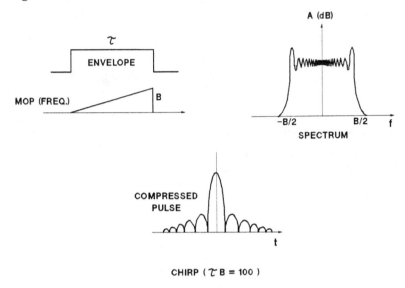

Figure 2.50 Example of a frequency code (chirp) with its spectrum.

A frequency-coded pulse consists of n elements of duration τ_{el}, each with a different carrier frequency. In the receiver the pulses are realigned in time by means of a delay line of duration equal to the duration of the pulse minus a code element, with intermediate outputs separated by τ_{el}, and coherently summed to yield at the output, a compressed pulse of amplitude n times higher than the input amplitude, and duration n times shorter (Fig. 2.52).

In practice, a linear frequency modulation (*chirp*) waveform is quite often used, and the compressor is a dispersive line (i.e., a line which delays each signal by an amount proportional to its frequency) (Fig. 2.53).

It can be demonstrated that the number of code elements is

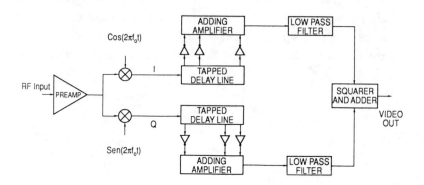

Figure 2.51 Block diagram of a pulse compression radar receiver.

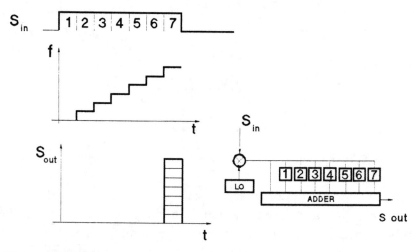

Figure 2.52 Generation of a frequency code.

given by the ratio of the matched receiver bandwidth to the bandwidth corresponding to the non-coded long pulse $B_s = 1/\tau$, i.e.,

$$n = \frac{B}{B_s} = \tau B$$

The product τB, or more generally, *time-bandwidth product (TB)* defines the gain of the code; being of general validity, it

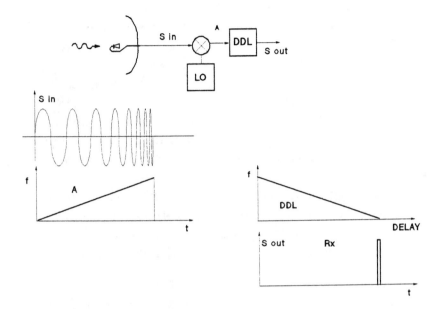

Figure 2.53 Linear frequency modulation code (chirp). Frequently, dispersive delay lines, which introduce a delay proportional to the frequency of the signal, are used as matched receivers (compressors).

is used also for phase codes.

There are coded radars whose TB is equal to a low integer (7, 13, etc.), and very high resolution radars whose TB is higher than 10,000. In the latter case, the band occupied by the signal is usually so wide that the radars are known as "spread-spectrum" radars. The form of the code may be so complex that the waveform of the transmitted pulse looks like the waveform of thermal noise; it is referred to as a "noise-like waveform".

The advantages of the pulse compression technique are the following:

- low peak power, and therefore good LPI characteristics.
- resolution equal to the code element resolution .
- clutter cell equal to the code element clutter cell .
- suppression of signals not corresponding to a code.

The disadvantages are as follows:

- the presence of sidelobes and, consequently, limited dynamic range. It is in fact impossible to detect a second target whose echo signal is n times lower, at ranges inside the sidelobes of the first one, unless particular weighting devices are used.
- a minimum radar range of $c\tau/2$, which is therefore usually very high. This means that for short range visibility the radar must from time to time transmit short, noncoded pulses, thus increasing the complexity of the receiver.
- the complexity of the processing circuits.

In general, a coded radar is very sophisticated and is used only when the complexity of the operational environment and the required performance do not permit operation with a less expensive and more traditional, simple pulse technique.

2.2.5 Search radar

Search radar is often distinguished from surveillance radar. The former is a radar which detects and identifies a target and then designates it to a missile or artillery battery. The latter is used to control a very wide region of air space. *Air traffic control radar* (ATCR) is of the second type. In what follows, this distinction will not be rigorously maintained. The mission of each piece of equipment will be made clear in context.

Search radars must satisfy two requirements: very long range (hundreds of kilometers) and very high coverage of spatial volume. High resolution in range and in angle (at least in azimuth) is also required so that targets close to each other can be clearly detected and distinguished.

Because of the long range required, such a radar uses fairly low frequency bands: usually L and S, sometimes UHF, in special cases VHF, or even HF. As has been stated above, at these frequencies atmospheric attenuation is not excessive and clutter reflectivity is low.

When entrusted with early warning responsibilities, they are positioned on rather high sites to minimize lobing and clutter problems and to give sufficient coverage against very low-flying targets.

For radars installed at fixed sites, extremely high range and resolutions are achievable since weight and volume limitations are not an issue. In particular, the antenna can be of considerable size, thus providing high gain and a very narrow beam.

For radars that have to be mobile, or mounted on-board ship, a compromise between performance, weight, and volume must usually be reached. When search radars are mounted on airborne platforms, the problems posed by weight and volume are of paramount importance.

In addition to strategic centers, tactical centers also require search radar installations. Here the mission of the search radar is to detect incoming enemy targets, to evaluate the level of the threat, and finally to designate the targets to the different missile or artillery batteries, coordinating their reactions and avoiding either confusion or overlap.

Since they must perform round the clock surveillance of the assigned air space, search radars must be particularly reliable. That is, their *mean time between failures* (MTBF) should be high; to provide for failure, they should be equipped with duplicated/triplicated back-up circuits; and, finally, their *time to repair* (TTR) should be extremely short.

To maximize the range, a number of circuits have been designed which aim at minimizing thermal noise in the receiver by means of amplifiers with extremely low noise figures, decrease of receiver temperature, and so forth.

Besides dealing with the problem of range, search radars must also cope effectively with the problems of clutter. To this end, search radars are equipped with powerful clutter cancelers (MTI). Where clutter is so strong that it is outside standard dynamic range, clutter maps or clutter contours are utilized. That is, regions are determined where the radar must give up detection entirely or else resort to special devices, for example, adaptive attenuation, which brings the clutter back within the dynamic range.

Finally, the number of targets may be very high; of those, many are friendly and many are hostile. Targets often fluctuate from scan to scan, and therefore appear and disappear. Many clutter residues may be displayed on the PPI, and at times incoming threats fly at low altitude and very high speed, so that reaction time must be very short to minimize the risk that they will be detected too late. To counter this, target detection and threat evaluation cannot be left to an operator looking at a PPI display. These functions must be accomplished by sophisticated electronic circuits: radar automatic detectors deal with target detections, while subsequent processing operations are entrusted to suitable computers.

Once the target has been detected, the plots are sent to the computer, which correlates them, opens the tracking channels, and provides velocity, direction, and other data concerning each target. The computer correlates data provided by the search radar (also called the primary radar) with data provided by an *identification friend or foe* (IFF) device, if fitted (also called a *secondary surveillance radar (SSR)*, and assesses the level of threat. This operation is also called *threat evaluation and weapon assignment* (TEWA). All this is to enable the decision makers to act appropriately.

Search radars are usually characterized by very long range, anticlutter thrust, and high discrimination in azimuth. To achieve these objectives, these radars usually operate in low bands (L or S) where the intensity of the clutter to be canceled is lower and atmospheric attenuation is also lower. They are equipped with antennas of considerable size, particularly in the horizontal plane.

In elevation, the beam is usually shaped according to the cosecant-squared law. In practice, airborne targets cannot fly at extremely high altitudes; in general they do not exceed a certain maximum height H_{max}, which is on the order of 20,000 to $30,000\,m$, while from Figure 2.54 it is evident that the higher the altitude, the shorter will be the maximum range at which an aircraft can be detected. Therefore at high elevations the antenna gain may be reduced, thus reducing the amplitude of unwanted returns from those directions.

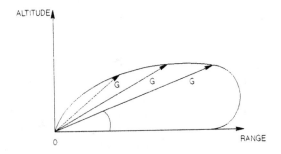

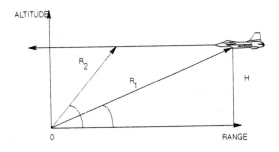

Figure 2.54 Cosecant-squared antennas achieve optimal antenna gain in elevation.

Since in the radar equation the dependence of the maximum range on the antenna gain is of the form

$$R_1^4 = k_R G^2(\eta_1)$$

one obtains

$$\frac{G(\eta)}{G(\eta_1)} = \frac{R^2}{R_1^2}$$

From the figure, one may write

$$H_{\max} = R_1 \sin \eta_1 = R \sin \eta$$

or

$$R = R_1 \frac{\sin \eta_1}{\sin \eta}$$

and substituting, one obtains

$$\frac{G(\eta)}{G(\eta_1)} = \frac{\sin^2(\eta_1)}{\sin^2(\eta)}$$

and finally, by setting

$$G(\eta_1)\sin^2(\eta_1) = k$$

one obtains

$$G(\eta) = k\cosec^2\eta$$

Antennas with cosecant-squared beams are achieved by suitably shaping either the feed or the main reflector of the antenna.

In terrestrial applications, a horizontal polarization is generally chosen because clutter returns from trees and fences are lower in this polarization. In naval applications, on the other hand, vertical polarization is sometimes chosen in order to reduce lobing as far as possible; in this polarization the reflection coefficient of the sea decreases rapidly with grazing angle.

2.2.5.1 Types of search radar

There are several types of search or surveillance radar suitable for installation either at fixed sites on land, or on-board mobile platforms on the ground, at sea or in space.

Search radars may be subdivided into the following categories:

- 2-D radars, capable of providing two target coordinates: range and azimuth.
- 3-D radars, capable of providing all three target coordinates: range, azimuth, and elevation.
- bistatic radars, complex radars in which the transmitter is not colocated with the receiver, but is installed at a distance.
- *synthetic aperture radars (SAR)*, very sophisticated, special airborne radars, providing detailed "pictures" of the area observed, even when visibility is poor due to bad weather, the performance being compatible with the relative atmospheric attenuations.

2.2.5.2 Automatic detectors

As has briefly been noted earlier (Section 2.2.4.2), automatic detectors perform the double function of maximising the detection of useful signals, and reducing to the minimum subsequent processing of unwanted signals, such as false targets produced by thermal noise, clutter residues, or intentional jamming.

In terms of decision theory, this function requires the maximum probability of detection P_{dmax}, and the minimum probability of false alarm P_{famin}, for signals that have crossed the primary threshold and so represent potential targets. The probability P_d may be increased by increasing the ratio of the energy E of the signal to the energy produced in equal time by receiver noise N_o.

For a given radar-to-target range, this ratio is a maximum when the radar receives all the power back scattered by the target during the period of illumination (time on target, T_{ot}). During this period, the target is illuminated by $N_i = T_{ot}FR$ pulses transmitted by the radar.

Each received pulse has an elementary probability P_d of crossing the first threshold, predetermined on the basis of a given false-alarm probability. P_d is a function of the ratio of signal power to noise over a single pulse.

Considering that the signal echoes are separated by the PRI, while false threshold crossings occur randomly and asynchronously, it is possible to exploit this property of echo synchronism by means of PRI delays to sum the N useful returns, thus increasing the detection probability.

Many types of automatic detectors, based on the properties mentioned, but differing in the mechanics of their design, have been realized in practice.

The-most used detectors are:
- accumulator detectors.
- moving window detectors.

Accumulator Detectors

Figure 2.55 shows a diagram of this type of detector. It

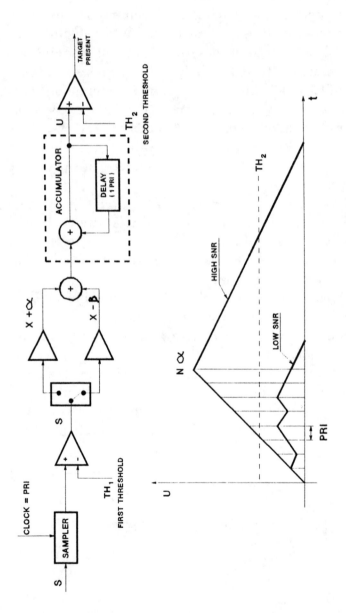

Figure 2.55 An accumulator detector. An automatic radar detector interprets the echo signals and decides autonomously whether a target is present or absent.

consists of an accumulator (that is, an adder with memory) containing a delay equal to the PRI, and an amplification logic that assigns a weight α to each signal crossing the first threshold Th_1 (detection), and a weight β to each failed detection. The instantaneous sum at the accumulator output is compared with a second threshold Th_2, in order to determine the presence or absence of a target.

To increase the probability of detection for low values of the SNR, characterized by many missed detections, it is customary to put $\alpha > \beta$; usually, $\alpha = 3$, $\beta = 1$. The figure shows typical responses of detectors of this type.

Optimization of the parameters α, β in the design of these detectors must take into account:

- the length of the time response of the accumulator in a dense environment to avoid lack of angular discrimination for close targets.
- the possibility that a distant single target may be split in two because of a long sequence of missed detections.

A discussion of the theory of these detectors is outside the scope of this book. For further information, interested readers should consult the bibliography [21].

Moving Window Detectors

Figure 2.56 shows a diagram of a moving window detector. Here, the storage period is fixed and equal to N_i (PRI), where N_i is the number of pulses during the time on target. This period varies continuously through the whole of the azimuth scanned by the rotating antenna, hence the name "moving window".

The performance of this type of detector may easily be deduced from the following considerations.

For each range R, the target, here assumed to be a point target, generates a sequence of pulses characterized by a detection

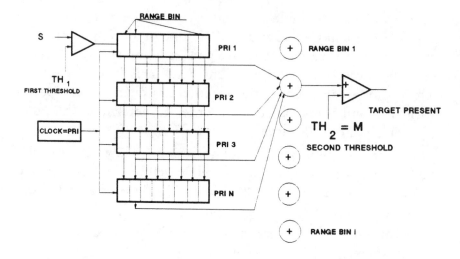

Figure 2.56 Moving window detector.

probability (see Fig. 2.11)

$$p_d = \int_{Th_1}^{\infty} P(v, SNR)dv$$

On the other hand, the false-alarm probability at the output of the first threshold is

$$p_{fa} = \int_{Th_1}^{\infty} P_{\text{Rayleigh}}(v)dv = \exp(-v^2/Th_1^2)$$

The detection event at the output of the detector is represented by $v \leq M$ (second threshold), and is characterized by a detection probability P_D estimated over all of the sequences of N pulses containing at least M detections.

One obtains the expression

$$P_D = \sum_{k=M}^{N} \binom{N}{k} p_d^k (1 - p_d)^{N-k}$$

Analogously, for the false-alarm probability at the output of the detector, one obtains

$$P_{FA} = \sum_{k=M}^{N} \binom{N}{k} p_{fa}^{k}(1 - p_{fa})^{N-k}$$

Thus for $P_d > 0.6$, the detection probability at the output of the detector increases, while, for $P_{fa} < 0.1$, the false-alarm probability at the output of the detector is greatly reduced.

For example, for $N = 10$, $M = 6$, $P_d = 0.6$ and $P_{fa} = 0.1$,

$$P_D = \sum_{k=6}^{10} \binom{10}{k} (0.6)^{k}(0.4)^{10-k} = 0.633$$

and

$$P_{FA} = \sum_{k=6}^{10} \binom{10}{k} (0.1)^{k}(0.9)^{10-k} = 1.47 \times 10^{-4}$$

Reference [22] displays the graphs and nomograms needed for detector design, according to the different types of targets to be detected (Swerling I, II, III, IV), in such a way as to obtain the required P_d and P_{fa}.

There is a question of precision in angle of the detected target: since the true position of the target in terms of pulse sequence corresponds to $N/2$ pulses, and the second threshold of the moving window detector is also about $N/2$ pulses, it is customary to assign to the target an azimuth corresponding to the leading edge of the detector output. If different values are used for the second threshold, the azimuth value should be extracted following computation of the mean azimuth between those values that correspond to the leading and lagging edges of the detector, corrected by a constant, or bias, function of Th_2.

As for the resolution in angle of two targets at the same range R, the targets must be at least $2N_i$ pulses distant from each other, that is, separated by at least $2\theta_B$. In general the presence of two targets close together may be recognised from the period of time for which the secondary threshold is exceeded.

2.2.5.3 2-D Radar

As noted earlier, 2-D radars can provide only the range and azimuth coordinates of the target. They may be either of medium complexity, as are ATCRs, or relatively simple, as are the navigation radars of merchant ships and fishing boats. In the latter case, however, in view of the more modest requirements and the shortage of space, frequencies in X band (9.3 to 9.4 GHz) are used, or, when heavy rain is expected, in S band. To control movement in harbors, very short pulses, of 50 to 100 ns, corresponding to a resolution of 7.5 to 15 m, are used. In this way, piers and other platforms in the harbor will be clearly discriminated on the PPI.

2-D radars may also be of high complexity, as are military search radars. Here, 2-D radars are used to detect potential enemy targets at the greatest possible range to permit the preparation of an adequate defense. They may be used in an air defense network, or for the defense of a more or less limited area. In this case, they must provide for TEWA and must therefore designate to the weapon system the enemy target to be destroyed by the transmission of the two coordinates, range and azimuth (2-D designation).

However, in order to detect and acquire the target, the weapon system radar will have to conduct a search in the vertical plane with its pencil-beam antenna, thus losing precious time.

Examples of 2-D radars are ATCR, air defense radar, naval warning radar, and airborne intercept radar. The Hawk-Eye radar, operating at very low frequencies (UHF), and the AWACS radar, characterized by a revolving antenna with ultra-low sidelobes, belong to this last category.

2.2.5.4 3-D Radars

3-D radars measure the three coordinates of the target:

range, azimuth, and elevation. They usually are very complex and sophisticated radars, but present several advantages. For example, they reduce reaction time, since they can designate targets to the associated weapon systems in all the three coordinates. Depending on the way in which beam scanning is performed, they may be divided into two main categories:

- radars with antennas scanning by mechanical methods in azimuth, and by various other methods in elevation.
- radars with antennas scanning electronically in azimuth and elevation, in all directions within the coverage area (planar phased-array radars) [23, 24].

The first category, based on mechanical antenna rotation in azimuth, is much more common than the second. In fact, it is much simpler and less expensive than the phased array type. Radars in this category are distinguished by the technique used for beam scanning in elevation. Among these techniques are:

- fixed beams contiguous in elevation (stacked beams), continuously illuminated or illuminated by time division within the time on target. Stacked beams are used only for reception, while for transmission, one single lobe covering the whole elevation sector is generated.
- beam scanned, in steps or continuously, in elevation by linear modulation of the transmission frequency. This modulation may be applied within the pulse (within-pulse scanning) or by time division within the antenna time on target in azimuth; this is *frequency scanning* (FRESCAN).

The principle of frequency scanning may be illustrated by an antenna consisting of an array of slotted waveguides arranged in rows at distance d from one another, fed from the side or the center by means of a serpentine feed of loop length ΔL (Fig. 2.57). The slots, suitably spaced apart, are the elementary radiators in this case.

The relative phase shift between corresponding slots in two

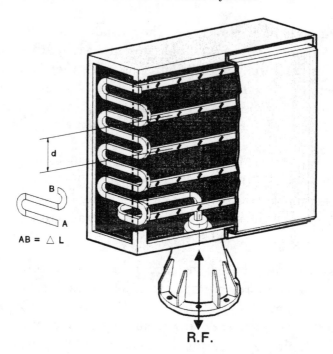

Figure 2.57 A 3-D radar supplies the three coordinates of a target. The figure shows the antenna of a FRESCAN 3-D radar. The pencil beam searches in elevation by variations in transmission frequency.

adjacent rows is

$$\varphi = 2\pi f_1 \Delta t = 2\pi f_1 \frac{\Delta L}{c}$$

When the transmission frequency is increased by Δf from f_1 to f_2, the phase shift differential between waveguides is given by

$$\Delta \varphi = 2\pi \Delta f \frac{\Delta L}{c}$$

The antenna boresight is shifted correspondingly by an amount ΔL such that

$$\frac{2\pi}{\lambda_2} d \sin \Delta \vartheta = 2\pi \Delta f \frac{\Delta L}{c}$$

that is, for small angles where $\sin \Delta \vartheta$ may be approximated by $\Delta \vartheta$,

$$\Delta \vartheta = \frac{\Delta f \Delta L}{f_2 d} = k \Delta f$$

Therefore, if the transmission frequency is steadily increased, the antenna beam will scan towards increasing angles in elevation.

The time-division illumination of the various beams available in elevation permits one also to shape the coverage diagram optimising the detection ranges, as for example in the cosecant-squared shaping of 2-D radars.

In 3-D radars, rather than shaping the antenna pattern in elevation, it is generally preferable to use a differential modulation of the illumination time. One exploits the property that the range depends on the energy scattered by the target, which, for radars operating at constant power, means that the range R is related to the target illumination time T.

In the first approximation, one obtains

$$R^4(\eta_1) = k_R T(\eta_1)$$

whence

$$\frac{T(\eta)}{T(\eta_1)} = \frac{R^4(\eta)}{R^4(\eta_1)} = \frac{\sin^4(\eta_1)}{\sin^4(\eta)}$$

or

$$T(\eta) = k \cosec^4(\eta)$$

Figure 2.58 shows a typical pattern of division of illumination time between the various elevation cells, within a time on target in azimuth.

- Electronic scanning in elevation by the *phased-array* technique. In this case the antenna consists of rows of transmit-receive radiating elements in which the relative phase between adjacent rows may be varied during the time on target to obtain the required scanning in the vertical plane.

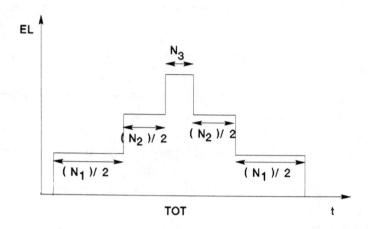

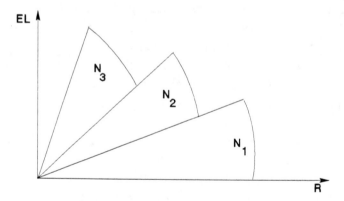

Figure 2.58 In electronic scanning radars, cosecant-squared coverage may be obtained by suitable adjustment of the time on target.

The phase shift is arranged so as to generate electromagnetic fields whose wavefronts are in phase only in the chosen direction (Fig. 2.60).

- 3-D characteristics may be achieved also by a radar with two antennas installed back-to-back. One of these produces a vertical fan beam, while the other has a fan beam inclined at about 30 degrees. This is a *V-beam radar*. Target echoes are received by the two antennas with a delay differential proportional to the target altitude (Fig. 2.59). Echoes of a

Figure 2.59 3–D radar of V–beam type, installed on board a Soviet naval unit.

target at zero altitude are received by the second antenna with a delay corresponding to a 180-degree scan. Echoes of a target at higher altitude are received with an additional delay arising from the inclination of the beam. The target altitude may be deduced from this delay.

The second category of 3-D radar, the complete planar phased array, is the last stage in the evolution of search radars. In this case the radar generally uses one or more fixed antennas, each covering 90 to 120 in azimuth, comprising a large array of transmit-receive elements capable of producing the phase shifts required to generate and steer the desired beam.

The power may be generated in a concentrated way in the transmitter and then distributed to the various elements, or may be generated in a locally distributed way in each transmit-receive module. In the latter case, the arrays are called "active planar phased arrays," and the modules may be solid state.

Beam pointing is carried out by suitably coordinating the relative phase shifts of the individual modules of the planar -

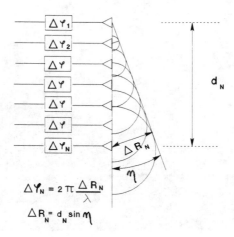

$$\triangle \Psi_N = 2\pi \frac{\triangle R_N}{\lambda}$$

$$\triangle R_N = d_N \sin \eta$$

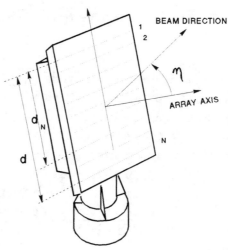

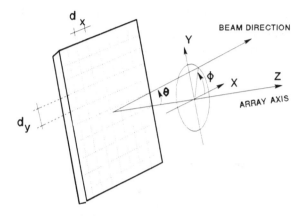

Figure 2.60 A 3-D radar may scan by means of an antenna consisting of an array of transmit-receive elements. The relative phase shift is varied either between rows of elements (linear phased array), or from element to element (planar phased array).

array so that the electromagnetic fields which they produce are in phase only in the required direction, thus achieving maximum gain in that direction, and negligible gain in all others.

With the geometry of Fig. 2.60, it may be shown [24] that the differential phase shift between adjacent elements is of the form

$$\Delta\Psi_{truepc} = \frac{2\pi}{\lambda}d_x \sin\vartheta \cos\varphi$$

for horizontal modules, and of the form

$$\Delta\Psi_y = \frac{2\pi}{\lambda}d_y \sin\vartheta \sin\varphi$$

for vertical modules.

Because of its high cost, this type of radar is not yet widely used. Installations are found typically in the inventories of the superpowers, both on land (Cobra Dane, Pave Paws) and on board ship (Aegis Spy-1 on ships of the Ticonderoga class, and an analogous system on the Kirov).

2.2.5.5 Bistatic and multistatic radars

Any radar, for example a continuous wave radar, in which electrical isolation requires that the transmitting antenna be physically separate from the receiving antenna, might be regarded as of bistatic type. Here, however, the only systems to be considered are those in which the transmitter and receiver are separated by appreciable distances [25].

This technique, hitherto hardly used, is presently being examined with renewed interest with a view to wider employment because of the spread of:

- more and more effective radar jamming systems.
- antiradiation missiles (ARM).
- stealth aircraft.

When transmitter and receiver are far apart, a jamming signal must be radiated over a very broad angle in order to have a high probability of interfering with reception. However, in this way the jamming power actually radiated toward the receiver is substantially reduced, which in turn reduces the effectiveness of the jamming.

The advantages of a bistatic radar network against ARMs may easily be inferred. The separation of transmitter and receiver leads to the establishment of radar networks consisting of M transmitters serving N receivers ($N > M$). These are multistatic radars. In these networks, the transmitters, widely spaced, may radiate sequentially, disorienting ARM receivers.

Separation between transmitter and receiver may be beneficial also for stealth aircraft detection. In fact, because these aircraft are capable of reducing their RCS mainly in the direction of arrival of the electromagnetic wave, if a stealth aircraft is illuminated from one direction, and the scattered radiation observed from another direction, the low visibility characteristic is substantially degraded.

2.2.5.6 Synthetic aperture radars

Synthetic aperture radar (SAR) is a special class of airborne radar. SAR has a normal antenna to transmit signals and receive echoes, but exploits the motion of the aircraft to simulate a large antenna consisting of n radiating elements, as if it were a linear phased array; thus it is capable of extremely high angular resolution.

Assume that a pulse is transmitted from position A, and that echo signals from various range bins characterised by amplitude and phase are placed in memory. Repeat this operation at points B, C,..., N (Fig. 2.61) coinciding with the position occupied by the aircraft at the various instants of transmission.

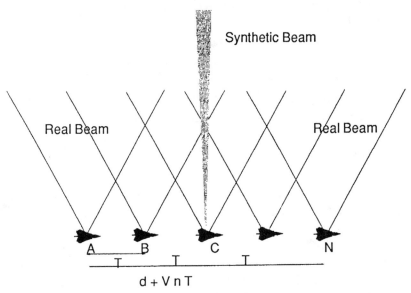

Figure 2.61 A *synthetic aperture radar* (SAR) has a small anetnna but behaves as if it had an extremely large antenna by exploiting relative motion of platform and target.

If all the signals in memory are finally recombined in amplitude and phase, taking into account the variations in doppler frequency arising from the geometry of the situation, they behave as if they were all arriving simultaneously at a uniformly illuminated antenna extending from A to N.

By adding suitable weighting to the signals before they are recombined, it is possible to achieve the required low sidelobes and beamwidth characteristics as well.

If the distance d from A to N is on the order of 50 m, and the wavelength is 5 cm, the resulting synthetic antenna beam is given by

$$\vartheta \simeq 51 \frac{\lambda}{2d} = 51 \frac{0.05}{100} = 0.0255° K$$

The factor 2 in the denominator, which is absent in the expression for a real antenna, arises from the fact that since pulses are transmitted sequentially by the elements of the array, the phase shift of the signal between adjacent elements is twice the phase shift in real arrays [26].

Thanks to the high resolution achievable, this type of radar may be used to make "photographs" of sites of importance, such as harbors and airfields, from great heights, even at night and in the presence of clouds. In this way it is possible, for example, to ascertain whether a certain ship is or is not in a harbor.

SAR performance may be achieved even with stationary radar. In this case, however, the target must be moving. This technique is known as *inverse SAR*, or ISAR (Fig. 2.61).

2.2.5.7 Search radar and ecm

Jammers are quite often expected to perform at a much higher level than is actually necessary for certain aspects, for example as regards the *effective radiated power* (ERP), while the countering of quite ordinary radar techniques, such as frequency agility and coded pulses, is neglected.

Effective jammers are those capable of:

(1) significantly reducing maximum radar range; in fact, a 50% reduction in range may be considered as a victory for the jammer because not enough time is left to the radar for a normal TEWA performance.

(2) tuning very rapidly (within the pulse!) to the frequency used by the radar to counter the group-to-group frequency agility. In fact, a good warning radar will have to use MTI,

which certainly needs more pulses, and therefore more PRI, to reach to steady-state condition.

(3) producing coherent noise and deception jamming for self or mutual protection.

(4) altering the code within the pulse; in this way they may succeed in concealing their presence and the presence of other friendly platforms, because the radar cannot recognize the code.

(5) altering the CFAR thresholds of the radar detector. This will cause either a significant desensitization of the receiver, or an excessive number of false alarms. In the first case, there will be a drastic reduction in range. In the second, computers dealing with the TEWA will be rapidly saturated, thus forcing the radar to abandon its automatic detection function, which will be assigned to the human operator, significantly degrading performance.

2.2.6 Tracking radars

The main mission of a tracking radar is to provide accurate information about the target range, azimuth, and elevation. This information is used by weapon systems for precision aiming of guns, or for missile guidance.

A tracking radar begins operation after a search radar has detected the presence of a target, estimated its threat, and designated it by its approximate coordinates in azimuth and range (2-D designation) to a fire control system, which is guided by the tracking radar itself. As soon as target designation has been received, the tracking radar points its very narrow pencil beam in the given approximate direction and starts its own search to acquire the target (Fig. 2.62). A tracking radar generally utilises a pencil beam because the angular tracking accuracy, as will be discussed later in more detail, is inversely proportional to the beamwidth.

Sophisticated search radars designate the target in range, azimuth, and elevation (3-D designation), which considerably shortens the time needed by the tracking radar to acquire the target.

As soon as the target has been acquired, the tracking radar switches to its tracking mode. After an initial transient, it starts providing the weapon system with the precise coordinates of the target needed to open fire stet.

It is worth recalling that the precision required of a tracking radar is on the order of 0.1 to 1 milliradians rms (root mean square) in angle, corresponding to errors of 0.1 to 1 m at a 1 km range, and of less than 1 m rms in range. Higher values would reduce drastically the effectiveness of the artillery system.

Factors of two types set a limit to the attainable precision: internal factors, depending on the type of radar system, and external factors, depending on the characteristics of target and environment.

At present, several types of tracking radar are available. They differ in precision, sophistication, performance, and cost. In increasing order of these factors they are:

- *Conical scan radar.* The direction of the antenna beam does not coincide with the boresight (mechanical axis), but revolves around it seeking the target direction.
- *Lobe switching (sequential lobing) radar.* The antenna lobe can assume four different positions around the boresight.
- *Conical scan on receive only (COSRO) radar.* For transmission, the antenna beam is fixed. For reception, the signal is modulated, as in conical scan. Thus the enemy is deprived of the information provided by conical scan.
- *Lobing on receive only (LORO) radar.* Operates like sequential lobing radar, but only on reception.
- *Monopulse radar.* The radar transmits simultaneously with four feeds, achieving an equivalent beam equal to the sum of the four elementary beams (Σ beam). On reception, the radar can generate three signals: one obtained by adding up the four elementary beams (Σ); one obtained by the difference between the "left" and the "right" beams (Δ_{az}); and one obtained by difference between the "up" and "down"

beams (Δ_{el}). With the Δ signal, the radar is capable of tracking in range. While with the two Δ signals, it is capable of generating the pointing "errors" in azimuth and elevation necessary for angle tracking.

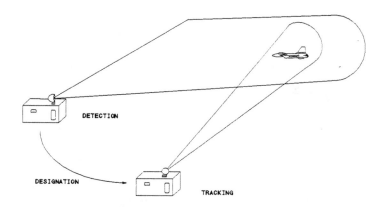

Figure 2.62 A tracking radar acquires the target designated by the search radar and tracks it with precision.

2.2.6.1 Conical scan radar

This radar derives its name from the fact that it generates its angle tracking signals by revolving its pencil beam so that the direction of maximum gain traces out a cone in space (Fig. 2.63).

A target along the axis of the cone will be illuminated with constant but not maximum G, while targets located elsewhere, for example in A, will be illuminated alternately with higher or lower G at the scan rate, which is called the conical scanning frequency f_s. The axis of the cone, the boresight, is the pointing direction, or axis, of the antenna. The angle ϑ_q between the direction of maximum gain and the boresight is called the "squint angle."

When the radar transmits a train of pulses at a certain PRF towards two targets, one of which is aligned with the boresight,

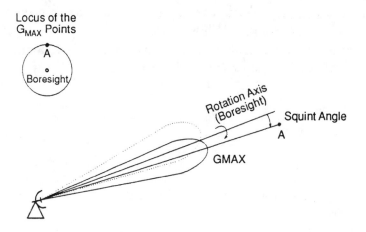

Figure 2.63 The principle of conical-scanning for tracking in angle is at the same time simple and very effective.

while the other is off the axis, the received echoes are of the form shown in Figure 2.64.

The extent of the modulation will indicate how far off the boresight the target is, while the phase of the modulation with respect to some reference value will yield its direction.

It is easy to show [27] that the signal giving the angular displacement of the target from the axis, often called the angle-error signal may be written

$$\epsilon = k_s \vartheta_t \sin(2\pi f_s t + \varphi)$$

where k_s is a constant, called the angular gradient, which depends on the antenna configuration (squint angle, width of the ϑ_B lobe, and so forth), ϑ_t is the angle between the target direction and the boresight, f_s is the scanning frequency, and φ is a phase that, by comparison with the reference phase, determines the error components in azimuth and elevation.

Normally an antenna is easily oriented by movements around a vertical and a horizontal axis, so that using servos, it is possible to correct the current orientation and track the target automatically.

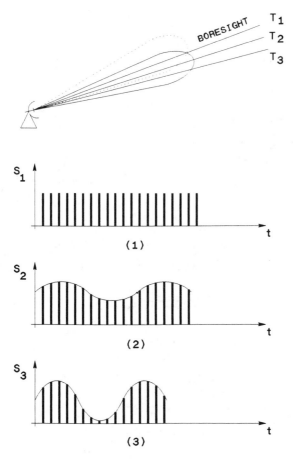

Figure 2.64 Signals received by a conical-scan radar from a target aligned with the antenna mechanical axis (boresight) (1), a target off the tracking axis (2), and a target far off the tracking axis.

The degree of modulation for a given angular error depends on the antenna beamwidth ϑ_B, the beam shape, and the squint angle.

The smaller ϑ_q, the less sensitive the system: For large angular shifts there is little modulation. An increase in ϑ_q increases the angular sensitivity of the system, or the angular gradient k_s, but the losses L_k (crossover losses) are also increased (Fig. 2.65) [28].

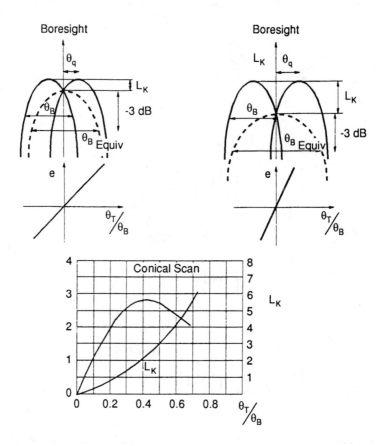

Figure 2.65 Angular gradient and cross-over losses. With the increase of squint, there is an increase in angular sensitivity but also an increase of losses, partly compensated by widening of the total equivalent beam.

The equivalent gain of the antenna system, or mean value of the gain with respect to direction, is shown by the dotted lines in Figure 2.65. L_k denotes the loss on boresight of the equivalent beam with respect to the rotating beam. The figure shows also k_s and L_k as functions of ϑ_q/ϑ_B. Usually ϑ_q is limited to values of the order of $0.3\,\vartheta_B$ to $0.4\,\vartheta_B$, with a two-way loss of roughly 3 to 6 dB.

All the echoes received by a conical scan tracking radar will be more or less modulated. However, since there is only one target of interest, the angular information needed for tracking will have to be extracted from that target alone, ignoring all other information. The target must therefore be isolated in order to deal only with it. This is achieved by positioning a gate on the range axis of the receiver, so as to allow output signals only within the gate. The initial position of the gate may be decided either by the radar operator or by the designation sent from the search radar that first detected the target. Later, the tracking radar itself will automatically position the gate around the target. After each transmitted pulse, the tracking radar will receive the signals as shown in Figure 2.66.

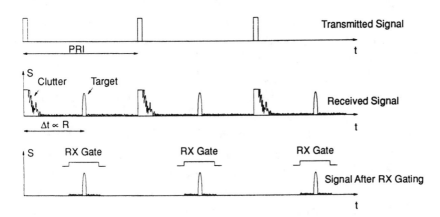

Figure 2.66 A gated receiver permits observation of the relevant range only, ignoring all the rest.

The detected (or video) signal at the receiver output is sent to the range tracking system. There, a time discriminator (*split gate*) circuit, which estimates the distribution of the signal between two gates called the *early* and *late* gates, continuously identifies the target range by precision measurements of the time elapsed from the emission of the radar pulse to the switching between early gate and late gate.

Range Tracking

Range tracking is usually performed after a linear receiver provided with *automatic gain control* (AGC). Alternatively a hard-limited receiver, or a logarithmic receiver, might be used. It is of importance that a tracking loop be designed and implemented with a phase margin such that it can track correctly, notwithstanding the residual fluctuations of the target amplitude.

As mentioned above, the split gate, which is capable of measuring the signal energy present at two adjacent gates, operates as a sensor in the range tracking loop. The split gate may be realized either in an analog or in a digital configuration. If analog (Fig. 2.67), it may consist of two circuits capable of charging a capacitor with a current directly proportional to the amplitude of the input signal. This charging is enabled only in a time interval called "early" for the signal sensor preceding the gate center, and "late" for the sensor following it (see figure). If the two gates straddle the signal, the two capacitors are charged to the same voltage and the gate positioning error is zero.

If the gate position is advanced or delayed with respect to the radar center of the target, the late circuit will charge to a voltage higher or lower than the early circuit. This will generate an error signal at the discriminator output showing that the range measured by the radar, or the time elapsed from the transmission of the pulse, is lower or higher than the correct value.

In the most recent radars, the split gate is digital. It calculates the difference between early gate samples and late gate samples, but the operating principle is the same.

The range error thus generated is sent to a second or higher order loop, in order to ensure tracking of the target. An example of a loop is shown in Figure 2.67. This type of loop is capable of tracking the targets maneuvering in range with acceleration of up to 40 g, and accepts a large amplitude variation of the input signal; it can, for example, operate correctly even with a saturated receiver.

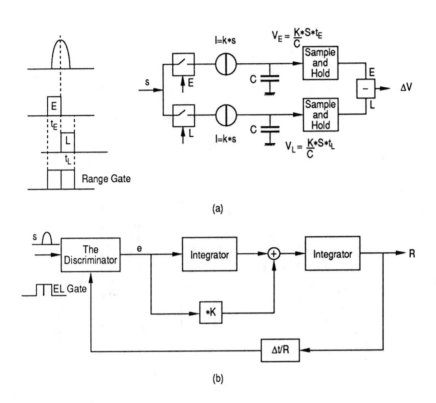

(a)

(b)

Figure 2.67 (a) Block diagram of a time discriminator (split gate) (b) A typical block diagram of a range tracking loop.

Angle Tracking

In a conical scan radar, tracking in angle is performed by a circuit called a "coherent detector" that detects pointing errors (Fig. 2.68).

The signal detected at the receiver output is sent to the range tracking circuit which gives the time t_r of the maximum signal. At this instant, a sample and hold circuit is enabled, which samples and memorizes the signal amplitude. In this

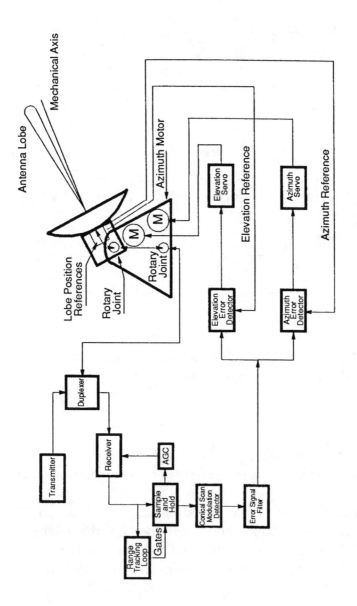

Figure 2.68 Block diagram of a conical-scan tracking radar.

way a sine wave is reconstructed from the series of echo pulses received. The amplitude of the sine wave is proportional to the angular deviation from boresight to target, and the phase specifies the error direction.

Generally, the motor that rotates the antenna feed, or some equivalent system, includes a small electrical generator capable of providing two sinusoidal signals both at the beam scan frequency but 90 degrees out of phase. These reference signals indicate the position of the antenna beam.

Seen from a distance, the locus of the G_{max} points of a conical scan radar antenna is a circle (Fig. 2.69). By plotting the motion of this point in azimuth and elevation against the time, one obtains two sinusoids 90 degrees out of phase, called the reference azimuth and reference elevation.

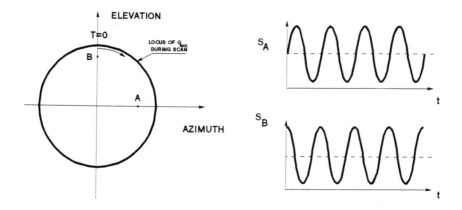

Figure 2.69 Locus of conical-scan antenna maxima, and generation of reference signals in aximuth and elevation.

A target at A, shifted only in azimuth, will generate at the radar an amplitude modulated signal, in phase with the reference azimuth and 90 degrees out of phase with respect to the reference elevation. The opposite will happen with a target at B.

A way of realizing the coherent detector, capable of detecting coherent modulations with respect to the reference azimuth and elevation, is shown in Figure 2.70. As may be seen, a signal shifted only in azimuth yields an error signal only from the detector coherent in azimuth, and not from the detector coherent in elevation.

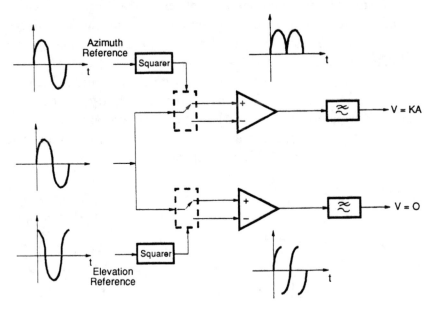

Figure 2.70 Coherent detector typically used in conical-scan radar.

It should be noted that for correct operation of the angular tracking device, the signal at the coherent receiver input must not be limited, for then modulation information is lost. This differs from the range tracking case. Neither must it be too low, otherwise the device would emit too weak a signal. The linear receiver is usually provided with an AGC circuit to ensure this.

When the SNR of the signal becomes low, the angular gradient tends to disappear, because the thermal noise generated after the antenna, and therefore lacking modulation, becomes preponderant (Fig. 2.71).

The angle-error signals are sent respectively to the two azimuth and elevation servo systems, which move the antenna so as to minimize the errors. The angular loops that provide the

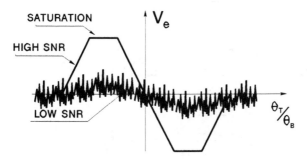

Figure 2.71 Angular error signal at the putput of a coherent detector with variation of the target position (open loop). For low SNR where thermal noise is strong, gradient suppression occurs.

angular tracking may be more or less complex. The resulting errors are described in the following sections.

Acquisition

Once it has received its designation, that is, coarse data about range and azimuth, the tracking radar must find the target and start pursuing it. To do this, the antenna and the range gate are moved onto the designated azimuth and range coordinates, respectively. This stage is generally called *rephasing*. Since the coordinates are usually inaccurate and since there is usually no elevation data unless the designation is made by a 3-D radar, the tracking radar must start a local search over the three coordinates to detect the target independently (Fig. 2.74).

The search in range is performed by opening a rather wide gate around the designation data (typically ± 1000 m) and by verifying from the detection circuits that the SNR is high enough for a "target present" declaration; in which case, the acquisition threshold has been crossed (Figs. 2.72 and 2.73).

It should be noted that the acquisition stage does not end with target detection, but after another short interval during which the radar automatically implements its tracking loops.

The threshold is usually of the CFAR type and is automatic, but the operator also may take a decision that a target is

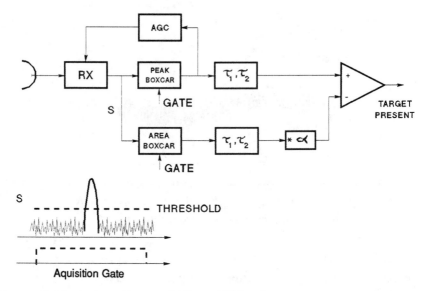

Figure 2.72 Automatic acquisition. While the antenna scans the assigned angular sector, the circuit shown automatically detects the presence of the target in the beam.

present and force the radar to declare target present so as to start the automatic tracking of the signal present in the gate.

This is done only in the most complex cases, for example, when an accurate selection among various close targets is required. More generally, automatic acquisition is used; this has enormous advantages in speed of target acquisition. A typical circuit for automatic acquisition is shown in Figure 2.72. The acquisition range gate enables the charging of two different circuits: a *peak boxcar*, able to output a dc voltage equal to the peak signal present in the gate, and an *area boxcar*, able to output a dc voltage proportional to the mean gate signal.

The peak boxcar output signal is also used to operate the AGC, thus maintaining the signal dynamics.

The integrating circuits that follow have a time constant matched to the time on target predicted during the acquisition search. This ensures achievement of the integration gain predicted for acquisition with the minimum SNR for the desired P_d and P_{fa}. When a signal exceeds this threshold, the

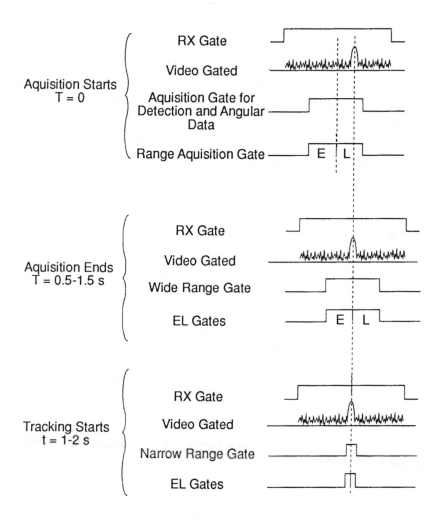

Figure 2.73 Gates in acquisition and tracking modes.

searching motion of the antenna is usually interrupted so that a second threshold with a longer time constant may confirm the actual presence of the target in the gate. If this second confirmation fails, the radar continues in acquisition mode. If confirmation is obtained, the radar can start automatic tracking.

During the acquisition stage, the range gates change as shown in Figure 2.73. The gates are generated by a complex timing circuit interacting with the range tracking-circuit. Finally the radar is able to concentrate on the target to be tracked.

At first, very wide gates are used, since the exact location of the target is not known; the designation data may well be inaccurate. As soon as a signal exceeds the acquisition threshold, the gates are moved by the range-tracking circuit to center on the target. After the time needed for this rephasing, the gates are reduced either to intermediate values, if the SNR is not very high, or to minimum values equal to roughly 1.2 times the width of the radar pulse. The block diagram of Figure 2.68 shows the relations between the fundamental circuits of a tracking radar.

The antenna scanning patterns may be of the most varied types, depending on the system, but all serve to explore the region where the probability of detecting the target is highest. Figure 2.74 illustrates some typical search patterns: the TV, or raster scan; the bar, or nodding scan; the box scan; and the spiral scan. When initial detection is by optical devices, the designation data are azimuth and elevation. In such a case, only a search in range is required in order to acquire the target.

Auxiliary Circuits

During its search, the radar receiver operates at maximum gain, if necessary using *sensitivity time control* (STC) and *fast time constant* (FTC) devices. STC is used to reduce receiver gain at shorter ranges in order to avoid severe saturation caused by clutter and by the signal itself. FTC is a

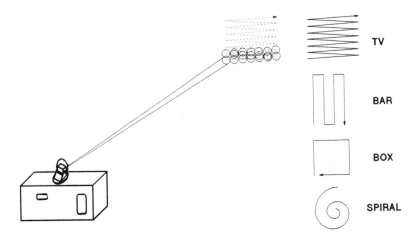

Figure 2.74 Typical acquisition movements.

differentiating circuit whose purpose is to avoid presentation of extended clutter on the PPI by presenting only its fronts. To cancel clutter, sophisticated radars resort to MTI circuits.

From the acquisition stage, after designation, onwards, the gated receiver, if of linear type, is controlled by the AGC to avoid saturation in the coherent receiver (Fig. 2.75). In logarithmic receivers, the AGC circuit is absent.

As stated earlier, conical-scan radars, which extract angular information from amplitude modulation, normally use linear receivers controlled by AGC to limit the dynamics of echo signals which otherwise would fluctuate too much.

Linear receivers with AGC are used especially in tracking radars where specific performance is required, such as, for example, strict linearity of angular gradients. In general, it is not easy to deal with instantaneous dynamics higher than 40 dB. For this reason an AGC is called in to compensate for:

(1) variations of the target echo signal power as a function of range: this power varies as $1/R^4$, and if tracking were required from 40,000 to 200 m the echo power would vary by 92 dB!

(2) variations arising from the diversity of targets, including missiles $(0.1\,\mathrm{m}^2)$, aircraft $(10\,\mathrm{m}^2)$, and ships $(10,000\,\mathrm{m}^2)$.

(3) target fluctuations arising from scintillation, which may

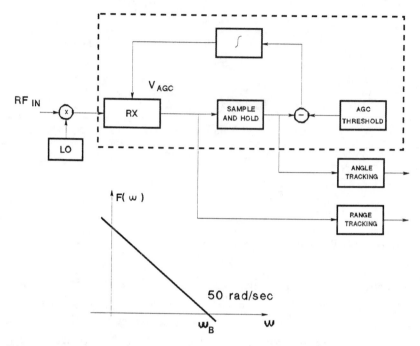

Figure 2.75 Example of an AGC circuit and the related open-loop transfer function.

be as high as 30 to 40 dB.

(4) fluctuations when the signal is mixed with clutter.

A very fast AGC is needed to keep the receiver always within its dynamic range. However, considering that in conical-scan radars angular information is obtained from amplitude modulation, it is necessary to avoid AGC compensation of this modulation. Accordingly, the conical-scan frequency is made as high as possible, but if the modulation arising from the angular error is to be reconstructed faithfully, the conical-scan frequency must be much lower than the PRF, say a tenth. Conical-scan frequencies are generally in the range of 30–200 Hz.

In order that the amplitude modulation arising from scanning should not be attenuated excessively by the AGC, it is necessary to limit the cutoff frequency of the AGC loop to roughly one decade below the scanning frequency, for example, to between 3 and 20 Hz. When the AGC cannot be very fast, this can present problems to the radar during transients, such as those arising in the acquisition phase, because the angular data, suppressed by receiver saturation, will be missing (Fig. 2.76).

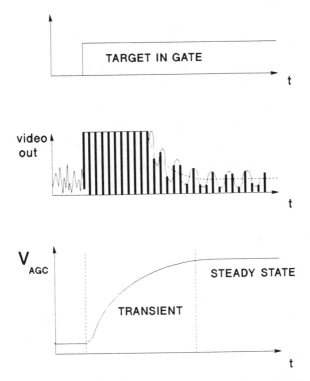

Figure 2.76 AGC in transients. As long as the ACG function is not implemented, a conical-scan radar cannot succeed in extracting angle-error signals

2.2.6.2 Monopulse tracking radar

Monopulse radar is so called because, in contrast to scan systems, it is capable of extracting the needed angular information from a single-pulse return [29]. A monopulse radar performs much better than other types of tracking radar, but this is paid for in cost and complexity of both antenna and receiver.

The antenna feed is more or less complex, but may always be thought of as comprising four radiating elements which form four elementary beams. These beams are combined by a microwave network consisting of hybrid junctions in such a way that (Fig. 2.77):

- the output Σ is the sum of the four elementary beams in amplitude and phase:
- the output Δ_{az} is the difference in amplitude and phase between the combined beams $(A+B)$ and $(C+D)$:
- the output Δ_{el} is the difference in amplitude and phase between the combined beams $(A+C)$ and $(B+D)$.

The three signals Σ, Δ_{az} and Δ_{el}, generated in the antenna, are amplified by three receivers matched in amplitude and phase, which means that the amplitude relationships and the relative phase shifts are maintained within very narrow tolerances.

Monopulse radars operate by either amplitude or phase comparison of the received signals (Fig. 2.77). Within the scope of this book these may be considered to be completely equivalent.

Figure 2.78 shows a block diagram of a monopulse radar. A transmitter sends the pulses to the antenna monopulse network entering the sum channel (Σ) through a circulator. The echo signals are amplified at intermediate frequency in three identical channels. The sum signal is used to demodulate the two Δ signals coherently, yielding the following relations

$$\Delta_{az} = \frac{\vec{\Sigma}\vec{\Delta}_{az}}{\Sigma^2}$$

$$\Delta_{el} = \frac{\vec{\Sigma}\vec{\Delta}_{el}}{\Sigma^2}$$

where $\vec{\Sigma}$ and $\vec{\Delta}$ denote the phasors of the respective signals, and the symbol x the scalar product between the phasors.

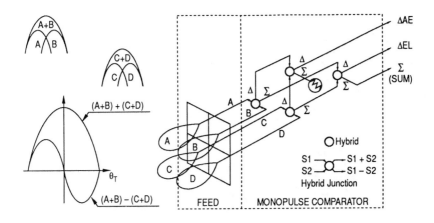

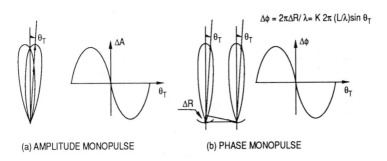

(a) AMPLITUDE MONOPULSE   (b) PHASE MONOPULSE

Figure 2.77 Monopulse radar antenna and comparator. When the beams from the four feedhorns are combined as shown, it is possible to obtain the angle-error signals within the pulses. Monopulse radars operate by either amplitude (a) or phase (b) comparison.

The division by Σ^2 is achieved by means of the AGC which operates equally on the three matched receivers. The degree of matching for a good radar is on the order of $\pm 0.5\,\mathrm{dB}$ in amplitude and $3°$ to $5°$ in phase. Thus it is possible to obtain angular precisions on the order of 0.1 to 0.2 milliradians.

The sum signal, after detection, is used for the generation of the AGC signals, for the automatic detection circuits (as in the case of the conical-scan radars) and for range tracking. Range

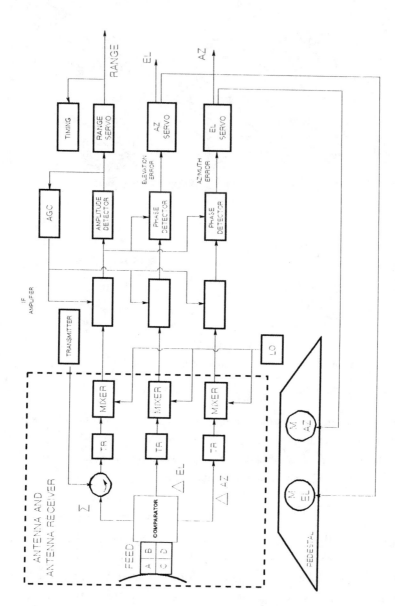

Figure 2.78 Block diagram of a monopulse radar. The higher complexity is compensated by performance decidedly superior to that attainable with a conical-scan radar.

tracking circuits determine the sampling time for the angular errors appearing at the output of the coherent detectors as pulses whose amplitude is proportional to the error, and whose sign depends on the phase shift between Σ and Δ (Fig. 2.79).

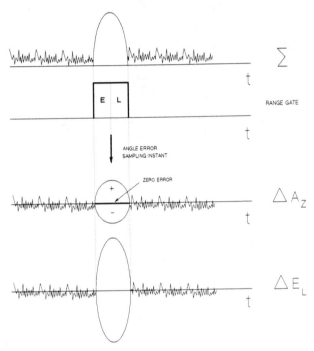

Figure 2.79 Angle-error signals in monopulse.

Sample and hold circuits memorize the pointing error of the tracked target, and transform it into a dc signal that may be used by the servo systems that steer the antenna in azimuth and elevation.

In a monopulse radar, the AGC may have a very wide band. The only limitation is set by the possible presence of MTI. In this case, the band is limited to a frequency of 30 to 50 Hz to avoid fast modulation on clutter (hardly cancelled by the MTI) generated by the AGC, which is trying to compensate the target scintillation and to be at least 10 times as high as that of the sevo loop. Monopulse radars that do not require precise linearity of the angular gradient often use three logarithmic receivers without AGC.

It should be emphasized that the AGC usually presents less of a problem in a monopulse radar than in a conical-scan radar. Apart from the fact that the AGC can be wide-band, angular information may be extracted even from a saturated receiver. Angular errors are given by

$$\epsilon = \frac{\vec{\Sigma}\vec{\Delta}}{\Sigma^2}$$

so that if the sum channel is saturated, ϵ will be quite high but still of the correct sign (Fig. 2.80).

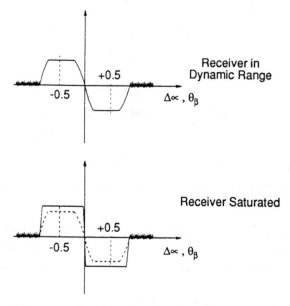

Figure 2.80 Effect of saturations in monopulse. In contrast to the conical-scan radar case, angular information is maintained despite saturation of the receiver.

If the angular loop is well designed, the effect of the saturated receiver will be a simple high-frequency vibration in the line of sight, totally harmless to tracking accuracy.

The angular gradient K_m, which is the slope of the error signal in the neighborhood of the boresight, is higher in a monopulse radar than in a conical-scan radar [30]. Fig. 2.81 shows K_m as a function of beam squint. Here, as in the case of conical-scan radar, there is an increase of system sensitivity and, simultaneously, of losses, with increase of squint. The best compromise is usually found for $\vartheta_q/\vartheta_B \simeq 0.3 - 0.4$.

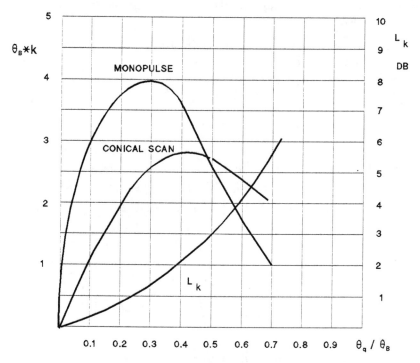

Figure 2.81 The angular gradient of a monopulse radar is higher than the gradient of a conical-scan radar.

When the radar is equipped with a fast *analog-to-digital* (A/D) converter, the entire range axis may be digitized. In order to avoid excessive losses, at least two samples are normally used for each range bin. For example, if the radar pulse duration is 200 ns, it will be necessary to digitize samples taken every 100 ns. In this case it is also possible to analyze bin by bin to determine immediately the range at which a target has been detected. When radars are used in weapon systems dedicated to short-range point defense, or *close-in weapon systems* (CIWS), quite often a warning belt is used. In fact, since these systems are the last link in the chain of defense, they must be capable of intervening automatically with an extremely short reaction time. To this end, a belt is established around the protected platform and, as soon as a target is detected within it, the system automatically switches to its target tracking mode, ready to fire.

Figures 2.82 and 2.83 show block diagrams of more sophisticated tracking radars. The former shows a coherent monopulse tracking radar with digital processing and MTI. The latter is similar, but includes pulse compression.

2.2.6.3 Range and precision in tracking radars

While for a search radar the most important feature is maximum detection range, precise processing of target coordinates is the main requirement for a tracking radar. In general, the range of a tracking radar against most targets substantially exceeds weapon range.

Range is given by

$$R_{\max} = \left[\frac{N_i P_T n G_T G_R \sigma \lambda^2}{(4\pi)^3 KTBF\,(S/N)_{Pdfa}\,L} \right]^{\frac{1}{4}}$$

where

$$L = L_i + L_x + L_{Tx} + L_{Rx} + \dots$$

is the sum of all losses, and N_i is the number of pulses integrated in the time on target, during acquisition.

In tracking radars with automatic acquisition, the maximum number of pulses that can be integrated is fixed, as it depends on the integrator, which is dimensioned to guarantee good operation at the maximum angular-scanning velocity.

When operation is not in free space, it is usually necessary to consider effects of the earth's surface and of the atmosphere. Here, the considerations of section 2.2.5 concerning

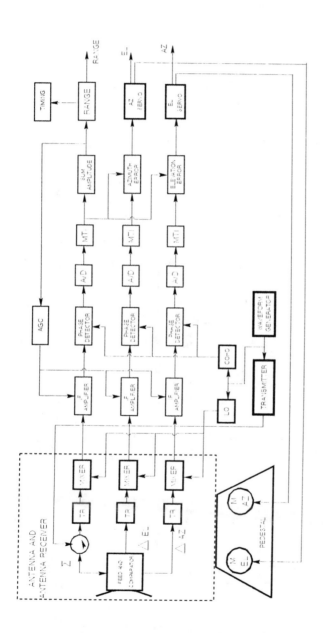

Figure 2.82 A monopulse radar allows large improvement factors for the MTI. The figure shows a possible block diagram.

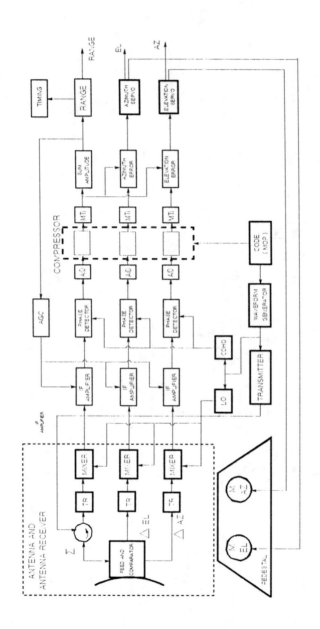

Figure 2.83 Block diagram of a monopulse radar with pulse compression and MTI. A radar of this type achieves high performance both in clear conditions and in the presence of clutter or ECM.

search radars still apply and both clutter and multipath effects must be considered. However, tracking precision in range and angle requires a special discussion.

In free space, precision depends essentially on four factors, namely, SNR, scintillation, glint, or fluctuation of apparent position, in angle and in range, and target accelerations, both radial and transverse.

Range Tracking Errors

Errors influencing range measurements made by a tracking radar [31, 32] arise essentially from the following causes:

(1) thermal noise.
(2) range glint.
(3) radial acceleration.
(4) scale linearity and alignment.

Thermal Noise

In practice, range determination by means of the early gate-late gate, or split gate method exploits the difference between the correlation functions of the early and late gates with the received signal. This is shown in Figure 2.84, where, to simplify matters, a rectangular received echo and early and late gates equal to a half pulse have been assumed.

From the theory it follows that the the zero crossing point at low SNR fluctuates because of thermal noise. The rms error in range because of thermal noise is given by

$$\sigma_n = \frac{c\tau}{2} \frac{1}{k\sqrt{2S/N.F_R/B_s}}$$

where k is a factor between 1 and 2, which depends on how the split-gate circuit has been realized; S/N is the signal-to-noise ratio derived from the range equation; F_R is the repetition rate (PRF); and B_s is the bandwidth of the range tracking loop.

More sophisticated radars exploit bands adapted to the SNR, in the sense that when the SNR is very low, the bandwidth of the loop is reduced (Fig. 2.85).

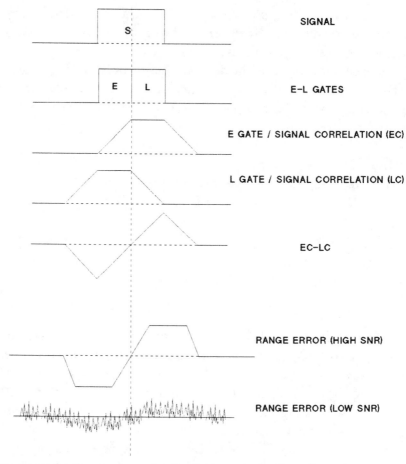

Figure 2.84 Range error signals at the output of a split gate.

Range Glint

Since the target consists of a series of elementary scatterers that are recombined in amplitude and phase, if its length L is non-negligible compared to the pulse duration, the phenomenon of range glint occurs (Fig. 2.86).

To a sufficient approximation, the range glint has an rms value

$$\sigma_s = \frac{1}{6}$$

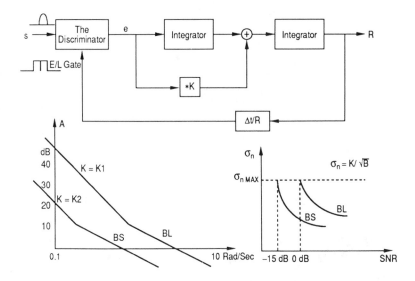

Figure 2.85 The tracking loop bandwidth can be adapted in order to reduce noise at low SNR. The figure shows an example of an adaptive range-tracking loop.

where L is the length of the target, projected onto the line of sight.

Radial Acceleration

When the target accelerates radially, there is a tracking error equal to the target acceleration divided by the acceleration constant k_a of the servo

$$\sigma_a = \frac{a}{k_a}$$

If the target acceleration is large, it is necessary to ensure that the servo system is of a type which yields acceptable residual errors. Normally k_a should be such that, for the maximum predicted target acceleration, the echo does not come out of the split gate; or rather that the error in range is not greater than half the width of either gate.

In checking for errors of this type, the relevant acceleration

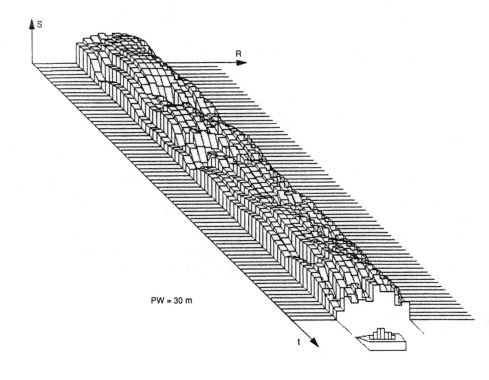

Figure 2.86 Scintillation along the range axis (R) in the time (t) of the signal coming from an extended target produces fluctuations in the measured range, called glint in range.

is the statistical one that the target may have in its attack mode, taking into account possible evasive maneuvers.

Scale and Calibration

A linear scale error is present only in old analog systems. Today it hardly exists and therefore may be neglected. Moreover, present methods for initial alignment in range are such that calibration errors may also be ignored.

Total Error in Range

Since the errors mentioned are independent statistical errors, the total error σ_R of the range-tracking loop can be calculated, with sufficient precision, by rms summation of the values found above, namely

$$\sigma_R = \sqrt{\sigma_n^2 + \sigma_g^2 + \sigma_a^2}$$

Angular Tracking Errors

There are many contributors to the total tracking error in angle [31, 32]. They include

(1) thermal noise.
(2) angular glint.
(3) scintillation.
(4) target angular acceleration.
(5) manufacturing quality.

Thermal Noise

The effects of thermal noise are apparent in the angular loop just as in the range-tracking loop. The thermal noise error is

$$\sigma_n = \frac{\vartheta_B}{k\sqrt{(2S/N)(F_R/B_s)}}$$

where ϑ_B is the 3-dB beamwidth, F_R is the PRF, B_s is the bandwidth of the servo system, S/N is the signal-to-noise ratio derived from the range equation, and k takes into account the angular gradient; it varies according to whether a conical-scan ($k_s \simeq 1.3$) or a monopulse ($k_m \simeq 1.7$) radar is used.

Angular Glint

As seen in section 2.2.4.3, it sometimes looks as if the apparent phase front were coming from a point which could even be outside the target geometry, rather than from its physical center. The rms error, in radians, is given by

$$\sigma_g = \frac{1}{3}\frac{L}{R}$$

where L is the maximum transverse dimension of the target seen by the radar, and R is its range.

Scintillation

This type of error exists only in scanning radars, conical or sequential. After the levelling action of the AGC, residual fluctuations could be interpreted as angular shifts of the target.

Assuming that the number of decorrelated samples is on the order of one thousand, the rms value of this error is given by

$$\sigma_s \simeq 0.01 \vartheta_B$$

For radars of the *track-while-scan* (TWS) type, which have to determine the angular position from the amplitude pattern of the received signal by exploiting only the few samples collected during time on target, one obtains the value of σ_s, which is generally in the range $0.1\,\vartheta_B$–$0.2\,\vartheta_B$.

Manufacturing Quality

In a monopulse radar the three receivers, $\Sigma, \Delta az, \Delta el$, must be kept well matched. This will depend on the quality of the components, which is to say, on the cost of the equipment. The errors arising from matching of a monopulse radar are less than one hundredth of the beamwidth ($\simeq 0.005 \vartheta_B$).

Another angular error present in both scanning and monopulse radars should be mentioned. It is caused by the shift of the boresight with frequency modulation. An alignment made at a certain frequency may shift by $0.01\vartheta_B$ to $0.001\vartheta_B$ at a different frequency.

Finally, it should be recalled that the servo in its turn introduces unwanted errors. Typical values for these errors, which depend on servo quality, are $0.005\vartheta_B$ to $0.001\vartheta_B$.

Target Acceleration

The angular tracking loop is characterized by a given k_v, k_a, and so forth. Figure 2.87 shows the accelerations seen by a radar when a target makes a pass; once the maximum acceptable errors have been established, it is easy to determine the characteristics of the servo system. When the target manoeuvers with an angular acceleration $\dot{\omega}$, the radar angular error will be

$$\sigma_a = \frac{\dot{\omega}}{k_a}$$

Alignment

With the means available today, this type of error is negligible; its magnitude is approximately $0.001\vartheta_B$.

Total Angular Error

The total angular error is the rms sum of the errors mentioned above:

$$\sigma_{at} = \sqrt{\sigma_n^2 + \sigma_g^2 + \sigma_s^2 + \sigma_a^2 + \cdots}$$

It is interesting to see how this error depends on range. At long range, thermal noise will dominate (low SNR), while at short range, glint will dominate (Fig. 2.88).

Angular Glint Reduction

Recalling the mechanism by which glint is formed, one sees that, if the radar changes frequency, the apparent radar center

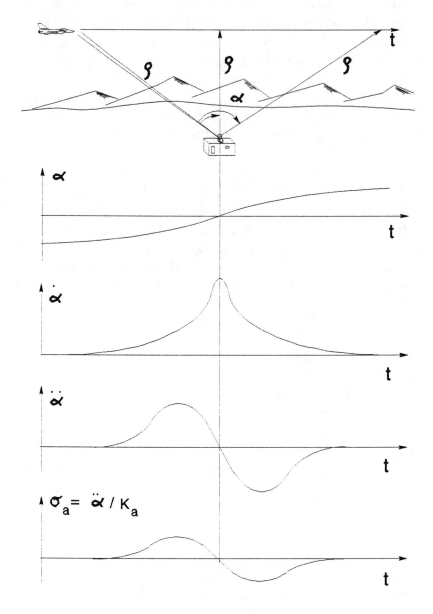

Figure 2.87 Angular accelerations in the target pass course. It should be remembered that there are also accelerations of a higher order.

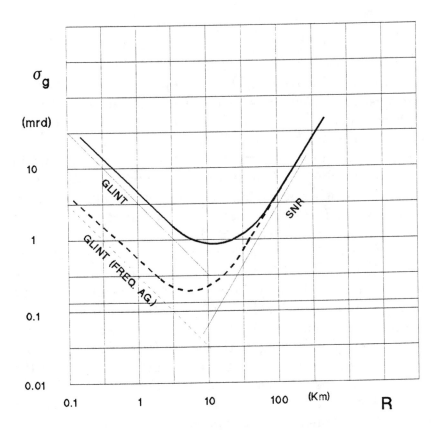

Figure 2.88 Total angular errors. At short range glint error dominates, while at longer range the error due to low SNR dominates. At medium range, errors due to servos, etc., dominate. The use of frequency agility substantially reduces the error due to glint in frequency agile mode (under dotted line).

of the target will shift (Fig. 2.88). At frequency f_1 there will be a certain pattern, in general different from the pattern at frequency f_n.

Therefore, if the radar is frequency agile, with either pulse-by-pulse or burst-by-burst frequency changes, the tracking will hold on the mean apparent radar center because of the narrowness of the servo system bandwidth B_s, thus reducing the error due to glint by a factor of between 10 and 100.

2.2.6.4 Tracking errors in the operational environment

What has been said up to now about tracking errors is valid in free space. In practice, it is the operational environment that must be considered, which as is well known causes essentially two phenomena: unwanted returns (clutter), and surface reflections (multipath) [31, 32].

Because the beam is very narrow both in azimuth and in elevation, clutter is not in general very serious for a tracking radar, except when small targets flying at low altitudes have to be tracked. If in such a case the clutter were too severe, it would have to be reduced by MTI filters.

At low altitude, when either the ground or the surface of the sea is illuminated by the beam, two reflected signals are seen, one coming directly from the target, the other via the earth's surface. The signal arriving by reflection from the surface is called a multipath signal. As remarked above, this has two consequences: lobing of the beam (Section 2.2.3.3), and, in tracking radars, nodding; that is, antenna oscillation in the vertical plane that prevents use of a tracking radar at low altitude (Section 2.2.4.3). In fact, at low altitude, the radar receives in its antenna beam two similar signals, one coming from the true target and one reflected by the surface. The two rays, the true one and the reflected one, follow different paths. When the two rays reach the receiver out of phase, there is a strong distortion of the phase front, as has already been remarked in the context of lobing.

The apparent radar center of the target will be in the direction of the source of the stronger signal, usually the true target, but shifted much higher (Fig. 2.89).

Writing the direct signal in the form

$$s_d = se^{j\omega t}$$

one may express the reflected signal by adding a phase shift:

$$s_r = \rho se^{j(\omega t + \varphi)}$$

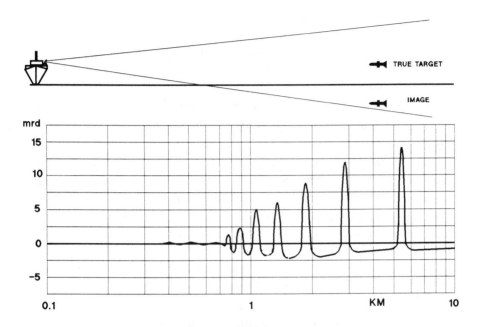

Figure 2.89 When a radar tracks a target at low altitude, depending on the strength of the reflection from ground or sea (multipath) an error called nodding occurs.

The phase shift φ will depend on the path difference ΔR between the two rays and on the phase shift φ_r occurring on reflection.

ΔR may be written in the form

$$\Delta R \simeq \frac{2h_R h_T}{R}$$

and, therefore, assuming $\varphi_r \simeq \pi$,

$$\varphi = \frac{2\pi}{\lambda}\frac{2h_R h_T}{R} + \pi$$

If

$$\varphi = (2k+1)\pi$$

for some integer k, then

$$\frac{4\pi h_R h_T}{\lambda R} = 2k\pi$$

or

$$\frac{2h_R h_T}{\lambda R} = k$$

in which case the direct and reflected signals are out of phase. The resulting signal is attenuated, and furthermore, the apparent center of the returns from the target is shifted toward the stronger signal.

Denoting by Σ_d and Δ_d the antenna weighting of the direct signals, and by Σ_r and Δ_r the weighting of the reflected signals, and applying Carnot's theorem to the two signals, direct and reflected, expressed in vector form, in the multipath situation one finds that the sum signal may be written

$$\Sigma = \sqrt{\Sigma_d^2 + \rho^2 \Sigma_r^2 + 2\rho \Sigma_d \Sigma_r \cos \varphi}$$

and similarly the signal Δ_{el} may be expressed [33]

$$\Delta_{el} = \frac{\Sigma_d \Delta_d + \rho^2 \Delta_r + \rho(\Sigma_d \Delta_r + \Sigma_r \Delta_d) \cos \varphi}{\Sigma_d^2 + \rho^2 \Sigma_r^2 + 2\rho \Sigma_d \Sigma_r \cos \varphi}$$

A radar tracking a target flying at a constant low altitude over the sea is shown in Figure 2.89. The positions of the nodding peaks, which may be expressed

$$R = \frac{2h_R h_T}{k\lambda}$$

depend on λ, so that frequency agility has a decorrelating effect, and its use yields clear benefits in angular precision at low altitude. Figure 2.90 shows the reduction in error when frequency agility is used.

2.2.6.5 Tracking radar and ECM

From the above discussion, it is possible to understand how to attain high performance in a tracking radar and, more importantly, how it may be degraded by ECM.

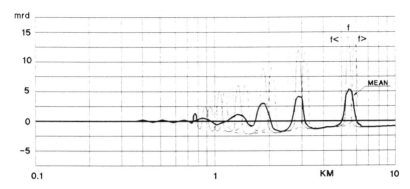

Figure 2.90 Since the positions of the error peaks arising from nodding depend on the transmission frequency, the use of frequency agility permits strong decorrelation and therefore a strong reduction of their effect.

In the first place, weapon systems will tend more and more to use monopulse radar.

To minimize the effects of clutter, MTI will be employed. When strong clutter is expected, *coherent chains* should be employed to achieve strong cancellations.

Pulse compression is desirable for high range resolution and good LPI characteristics.

Frequency agility should also be exploited, if possible compatibly with MTI.

To identify objectives for ECM against a tracking radar, one needs in the first place to know the expected missions of the radar: that is, whether it is going to be exploited for artillery or for missile guidance. Discussion of these topics is therefore postponed to chapter 3.

2.2.7 Airborne radars (interceptors)

Airborne radars are in a special class because, given the limited amount of space and weight available on board combat aircraft, they are often forced to perform several different operational roles [34, 35, 36].

First of all, they must perform the role of search radars, gathering accurate information about the possible location of enemy targets, information which may be either missing or incomplete.

Secondly, they must track several threats simultaneously, although without the highest precision, so as to be able to launch long- and medium-range missiles, and to point an illuminator for semiactive-guided missiles.

Finally, they must be able to track single aerial threats in order to direct the fire of their on-board weapons against these threats.

The functions of an airborne radar depend first of all on its specific role: whether air-to-air, air-to-ground, or air-to-sea.

In the air-to-air role, the operational modes of the radar may be the following:

- *Search* toward either higher altitudes (look up) or lower altitudes (look down). When the radar is searching for closing targets at maximum ranges (look-horizon) a velocity search mode is often used: In this case the detection is based only on the output from the doppler filters bank (FFT) whilst the range is ignored. When it is not necessary to detect targets at maximum ranges a *range-while-scan* mode can be used in order to detect targets with both the velocity and range information.

- Track-while-scan, with low precision tracking of multiple targets.
- Single-target tracking, with precision tracking of one target at a time.
- Illumination, for semi-active missile guidance.
- Fire control.

In the air-to-ground role, other specific operational modes should be mentioned:

- Ground mapping, to draw maps of the underlying territory for identification of characteristic points and zones of operation. This function can be performed in two ways, called "real beam ground mapping" and "doppler beam ground mapping." The first gives a resolution in angle that coincides with the angular resolution of the antenna beam. With the second, also known as "doppler beam sharpening, the angular resolution is increased by means of doppler processing, usually realized with *fast Fourier transform* (FFT) techniques. This is possible because the doppler velocity of the ground, as measured from the aircraft, depends on the angle of observation of the ground. This dependence, from one "slice" of terrain subtended at the radar to the next, permits the beam to be split into many parts, each characterized by its own doppler shift, thus increasing angular resolution.

- Terrain avoidance and terrain following, altimetric mapping of the terrain ahead in order to fly automatically on a low altitude path.
- Precision velocity update, an accurate measurement of the speed of the aircraft with respect to the ground.
- Ground moving target indication and tracking; that is, detection and tracking of land targets for precision launching of bombs or missiles.

A glance at Figure 2.91 will explain the difficulties confronting an airborne radar. In general, there is no problem in look-up mode. Clutter is absent or comes only from the sidelobes. The radar may use a low PRF to determine long ranges unambiguously, and possibly a single-delay MTI to detect targets of interest.

But as soon as the radar looks toward the horizon, clutter generated by the ground can be much stronger than the echo of a target of interest. For this reason, the radar must exploit the doppler effect and very powerful processing, based on spectral analysis (FFT), in order to discriminate between clutter and target.

It has already been remarked that good clutter cancellation is obtained by maintaining high stability of the emitted frequencies (coherence) and by using a high PRF, which, among other things, allows unambiguous extraction of the doppler frequency, and therefore the target velocity. A radar with these characteristics is called a "pulse doppler radar." Obviously,

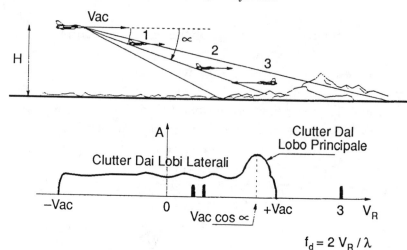

Figure 2.91 Signals and clutter seen by an airborne radar as a function of the relative velocity.

an ambiguity in range will arise, but several methods permit determination of the range.

The following are the PRFs typically exploited by airborne radars in the X band:

- 0.25 to 4 kHz for low PRF.
- 10 to 20 kHz for medium PRF.
- 100 to 300 kHz for high PRF.

Airborne radars can often exploit all three of these PRFs. In this case, the low PRF will be used in the look-up mode if altitude and atmospheric conditions do not create strong clutter returns. When the radar looks far towards the horizon in search of fast-flying closing targets, the high PRF can be very useful because the incoming targets will be in a spectral region devoid of clutter. When the radar looks down, it may at times be necessary to reach a compromise between unambiguous range, clutter folded into the different range cells (second or third track clutter, etc.), and clutter cancellation capability, so that often a medium PRF is used. In fact, as shown in the figure, in the look-down mode, targets that try to escape and are at a range shorter than the range of clutter, are better seen with a medium PRF than with a high PRF. A higher PRF would in fact cause the folding of clutter, including clutter further away than the target, in the first cell, thus increasing the need for clutter cancellation.

Figure 2.92 shows the situation in the look-down mode for three types of target, while Figure 2.93 illustrates the situation, for the same types of target, in the look-horizon mode. The first target is opening, but at a speed lower than the speed of the aircraft, and is at a range shorter than that of ground clutter as seen from the main lobe. The second target is like the first, but immersed in clutter. The third is an incoming target. For the two situations, clutter distribution in range and velocity is shown, together with the way in which the clutter is folded in the first range and velocity cells.

From an analysis of the figure, in which ground-type clutter alone has been assumed, it is possible to understand when a low, medium, or high PRF should be chosen.

2.3 Infrared sensors

2.3.1 *Review of radiant energy*

Here again, only the concepts and formulas needed for an understanding of the subsequent discussion of weapon systems which exploit radiant energy will be reviewed.

Most of the systems of interest based on the exploitation of radiant energy use the *infrared* (IR) band. That is why this particular band will receive special attention, although all the concepts relating to radiation from a body, and its detection, could be extended to visible and ultraviolet light. For a more detailed discussion, the interested reader should consult the bibliography [37, 38, 39].

With reference to Table 2.2, of particular interest for this discussion is the portion of the IR spectrum characterized by atmospheric windows; that is, regions in which the atmosphere

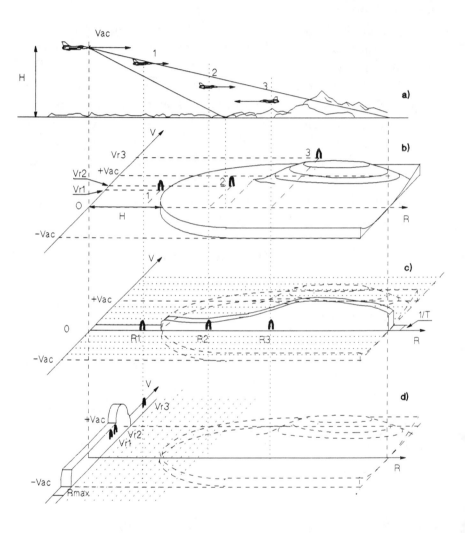

Figure 2.92 Signals and clutter in a pulse Doppler radar, in the look-down mode. (a) A typical situation; (b) Map of returns on the range/velocity plane; (c) Folding of clutter over the range axis (low PRF); (d) Concentration of returns on the velocity axis (high PRF).

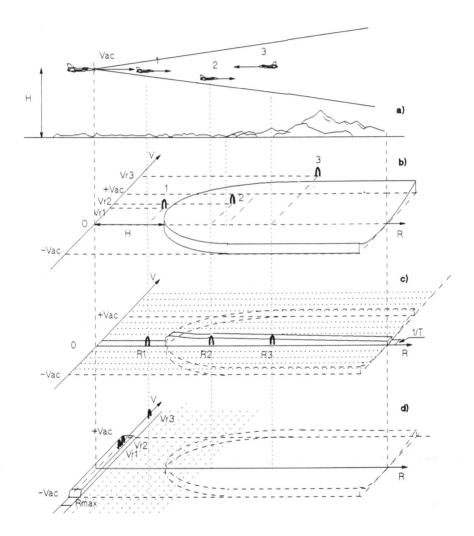

Figure 2.93 Signals and clutter in a pulse Doppler radar, in the look-horizon mode.

Table 2.2

The IR spectrum of interest. The representation in terms of
increasing wavelengths is usually to be preferred.

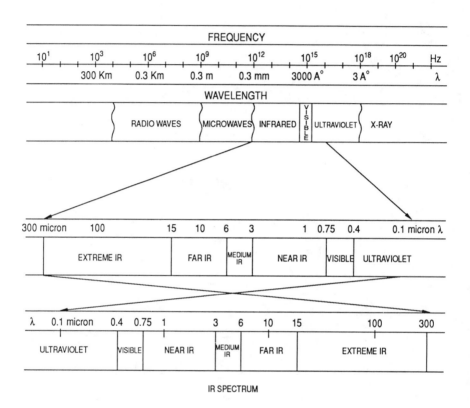

IR SPECTRUM

is transparent to radiation. The bands in which atmospheric
windows exist are as follows:

- *short-wave infrared* (SWIR), characterized by wavelengths λ
  from 0.8 to 3 m.

- *medium-wave infrared* (MWIR), characterized by wavelengths λ from 3 to 5 m.

- *long-wave infrared* (LWIR), characterized by wavelengths λ
  from 8 to 12 m.

2.3.1.1 Radiation of bodies

For a proper treatment of the topic, it is convenient to introduce the following standard definitions (Fig. 2.94). The radiant flux P is defined to be the total energy radiated by a body in all directions in unit time, where P is total energy radiated in unit time. Radiant flux is measured in watts (W). Radiant flux per unit solid angle is denoted by J and is called radiant intensity

$$J = \frac{P}{\Omega}$$

where Ω is the solid angle subtended by a sphere at its centre: $= 4\pi$ steradians. Radiant intensity is measured in watts per steradian (W/sr). Radiant flux per unit area of a source is called "radiant emittance" and is denoted by W

$$W = \frac{P}{S}$$

Radiant emittance is measured in watts per square centimeter (W/cm^2). Radiant flux per unit solid angle per unit area of radiating surface is called *radiance* and is denoted by N

$$N = \frac{P}{\Omega S}$$

Radiance is measured in watts per square centimeter per steradian (W cm^2, sr^{-1}). The radiant flux incident on a surface of area A, (i.e., the energy incident), not emitted, in unit time, is called "irradiance" and is denoted by H

$$H = \frac{P}{A}$$

Irradiance is measured in the same units as radiant emittance, (i.e., watts per square centimeter (W/cm^2)).

These quantities refer to the radiant energy integrated over all wavelengths. If the radiant energy is examined only at

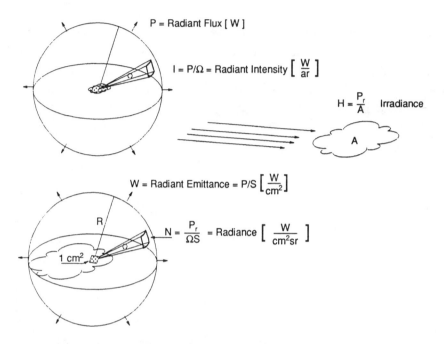

Figure 2.94 Definitions relevant to the propagation of radiant energy.

a particular wavelength, the definitions are the same, but a subscript λ is usually added to denote this case.

Before introducing the physical laws by which radiant energy is regulated, it should be mentioned that often it is convenient to refer to a "blackbody," an ideal body that absorbs all radiation falling upon it, with no reflection or retransmission. Blackbodies have the additional property of being the best possible radiators.

Planck demonstrated that a blackbody at a given temperature emits radiation of all wavelengths (Fig. 2.95). In particular, radiant emittance at the various wavelengths, integrated in the one micron (μm) band, is given by Planck's law

$$W(\lambda) = \frac{C_1}{\lambda^5 \left(\exp(C_2/\lambda t) - 1\right)}$$

where
$$C_1 = 2\pi hc^2 = 3.741 \times 10^4 \ \text{W cm}^{-2} \mu\text{m}^4$$

is the first radiation constant,

$$C_2 = \frac{ch}{k} = 1.438 \times 10^4 \ \mu\text{mK}$$

is the second radiation constant, $c = 3 \times 10^8$ m/s is the velocity of light, $h = 6.625 \times 10^{34}$ Ws^2 is Planck's constant, and $k = 1.380 \times 10^{23}$ WsK^{-1} is Boltzmann's constant.

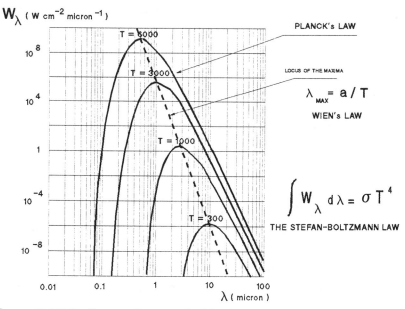

Figure 2.95 Radiant emittance of a blackbody at different temperatures as a function of wavelength (Planck's law). Shift of the maximum radiant emittance as a function of temperature (Wien's law). Total radiant emittance (Stefan-Boltzmann law).

Integrating radiant emittance with respect to wavelength over the whole spectrum, one obtains the total radiant emittance

$$\int_0^\infty W(\lambda)d\lambda = W$$

The Stefan-Boltzmann law states that

$$W = \sigma T^4$$

where

$$\sigma = 5.67 \times 10^{-12} \text{ W cm}^{-2} \text{K}^{-4}$$

is the Stefan-Boltzmann constant.

This law emphasizes the strong dependence of radiant emittance on temperature; when the absolute temperature of a body is doubled, its radiant emittance is increased 16-fold.

Differentiating Planck's law with respect to λ, and setting

$$\frac{dW(\lambda)}{d\lambda} = 0$$

for a maximum, one finds that W is a maximum at

$$\lambda_{\max} = \frac{a}{T}$$

where $a = 2898 \, [\mu\text{mK}]$. This is Wien's law.

This law shows that the wavelength for maximum radiant emittance is inversely proportional to the temperature, as shown by the dotted line in Figure 2.95.

At a given temperature any body will emit radiation of all wavelengths, but not as much as a blackbody, which, as stated above, is the ideal radiator. The ratio of a body's radiant emittance, at a given temperature and wavelength, to that of a blackbody at the same temperature and wavelength is called the "radiant emissivity", or simply "emissivity"

$$\epsilon = \frac{W(\lambda)}{W_0(\lambda)}$$

Here ϵ is always less than unity.

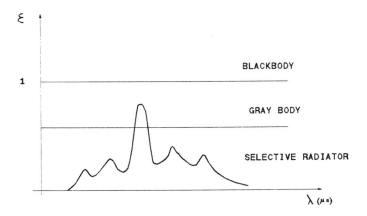

Figure 2.96 Emmissivity, that is, the ratio of the radiant emittance of a body to the radiant emmittance of the blackbody, is always less than unity for all bodies.

A body with constant λ is called a "gray body;" when ϵ changes with λ, the body is called a "selective radiator" (Fig. 2.96).

Radiation falling on a body is absorbed, reflected, or transmitted (Fig. 2.97).

Because of the law of conservation of energy,

$$\alpha + \rho + \tau = 1$$

where α is the absorptance, ρ is the reflectance, and τ is the transmittance.

If a body is opaque, there is no transmission, and $\tau = 0$. Therefore, for an opaque body,

$$\alpha + \rho = 1$$

For a blackbody, $\alpha = 1$, $\rho = 0$, and $\tau = 0$.

Kirchhoff observed that, at a given temperature, the ratio of a body's radiant emittance to its absorptance is constant,

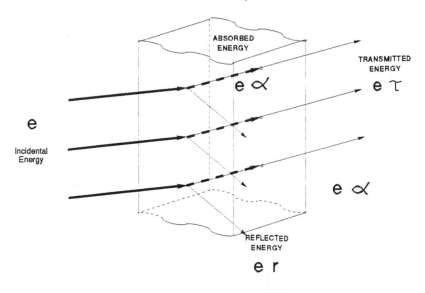

$$e = e\,r + e\,\alpha + e\,\tau$$
$$1 = r + \alpha + \tau$$

Figure 2.97 Absorption, reflection, and transmission of radiant energy.

and equal to the radiant emittance of a blackbody, W_0, at the same temperature:

$$\frac{W}{\alpha} = W_0$$

This is Kirchhoff's law.

Therefore, if a body is a good absorber of radiation (α close to unity), its emissive capability, as expressed by W, must be good too, since the ratio is constant. That is why a good absorber is also a good emitter.

Thus

$$\frac{\epsilon \sigma T^4}{\alpha} = \sigma T^4$$

and, therefore,

$$\epsilon = \alpha$$

(i.e., the emissivity of a body is equal to its absorptance).

ϵ , and therefore α, depends on:

- the nature of the body.
- its temperature.
- its surface finish.
- the wavelength.

2.3.1.2 Radiation from gases

Among emission spectra, *line*, *band* and *continuous* spectra may be distinguished. Generally speaking, line spectra are produced by freely vibrating atoms, for example by electric discharge in a gas, band spectra by gas molecules, and continuous spectra by heated solids and liquids.

For an analysis of radiation emission by gases, interested readers should consult the references. Here, it will suffice to recall that when an atom in its normal state acquires energy by, for example, collision with another particle, an electron which initially was in a low energy orbit will jump to a higher energy orbit and the atom will be raised to an excited state. Falling back to a lower energy level, the electron will emit a photon whose energy is equal to the energy difference between the two states

$$\Delta e = h\nu$$

where h is Planck's constant and ν is the frequency of the radiation.

Since the energy change Δe is not continuous but assumes only discrete values, it follows that this type of radiation will occur at discrete frequencies. What these frequencies are depends on the atoms concerned. According to Kirchhoff's law, emission and absorption take place at the same frequencies.

When a beam of radiation with a continuous spectrum passes through a gas, the transmitted spectrum will be discontinuous because the gas will have absorbed radiation at just those wavelengths which it can emit, namely at those wavelengths which raise atoms of the gas to an excited state. When an atom returns to its normal state, emitting radiation at the same frequency, it may emit in any direction, and the amount of radiation having exactly the same direction as the original beam will be negligible.

However, in gases, radiation is emitted not only by atoms, but also by molecules. Molecular energy may be electronic, translational, rotational, or vibrational.

Of special interest for IR emission are energy transitions of the vibrational type that produce spectra in the band from 2 to 30 m. These transitions characterize the emission spectra of gases produced, for example, by combustion in a jet engine.

In the combustion process, water vapor (H_2O) and carbon dioxide (CO_2) are usually produced; the resulting spectrum is of the type shown in Figure 2.98, which shows the flame spectrum of a Bunsen burner burning natural gas. The peak at 4.4 μm is due to the energy transitions in the CO_2 molecules, while the peak at 2.7 μm is the overlap of bands in CO_2 and H_2O. The larger the volume of gas, the closer its radiant emittance approaches that of the blackbody at the wavelength considered and at the given temperature.

2.3.2 Infrared radiation produced by targets of interest

Since IR weapon and detection systems are based on the detection and processing of signals produced by targets, in what follows, data concerning the IR radiant emittance of some platforms of interest, namely jet engines and missiles, will be discussed.

From the Stefan-Boltzmann law

$$W = \sigma T^4$$

it follows that the higher the temperature, the higher the radiant emittance.

Some parts of a jet aircraft are especially hot (Fig. 2.99). They are:

- the exhaust nozzle of the jet engine.
- the hot exhaust gas area, or plume.

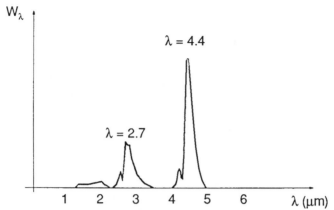

Figure 2.98 Radiant emittance spectrum of the flame produced by a Bunsen burner.

- the areas in which aerodynamic heating is highest.

The exhaust nozzle is usually the source of maximum radiant emittance, unless the aircraft uses an afterburner; in which case the plume will play this role.

A word of caution is needed here, before the approximate radiant emittance of these parts of the jet aircraft is calculated. In order to reach a sensor, radiation must pass through the atmosphere. However, the atmosphere is not transparent to all wavelengths (Fig. 2.102), and there will therefore be a strong selective attenuation of the radiated signals.

2.3.2.1 Nozzle

The exhaust nozzle can be regarded as a gray body with $\epsilon = 0.9$ whose radiant emittance is given by (in W/cm^2)

$$W = 0.9 \times \sigma T^4$$

Assuming for the exhaust nozzle a temperature of 500 degrees C (=773 degrees K), and recalling that $\sigma = 5.67 \times 10^{12}$ W cm^{-2} T^{-4}, the radiant emittance may be written (in W/cm^2)

$$W = 1.822$$

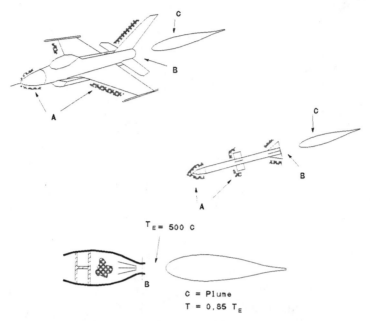

Figure 2.99 Aircraft and missile parts with strong IR radiant emittance.

Multiplying by the surface area S of the nozzle, measured in square centimeters, which is taken here to be 3500 cm^2, one obtains the radiant flux P (in W):

$$P = WS = 1.822 \times 3500 = 6377$$

The radiant intensity is given by

$$J = \frac{P}{\Omega}$$

Considering that a blackbody radiates over a hemisphere, of solid angle 2π steradians, according to the cosine law (Lambert's law), the value of Ω will be π, instead of 2π. Therefore, the radiant intensity will be (in W/sr)

$$J = \frac{P}{\Omega} = \frac{P}{\pi} = \frac{6377}{3.14} = 2030$$

2.3.2.2 Plume

The plume is characterized by the radiant emittance of the hot gases that are expanding into the atmosphere after passing through the exhaust nozzle. Calculating the radiant emittance in the small area in which the temperature is about 85 percent that of the exhaust nozzle, and recalling that the radiant emittance will strongly depend on the wavelength (as mentioned in section 2.3.1.2), one may write

$$W = \int_{\lambda 1}^{\lambda 2} \epsilon(\lambda) W_\lambda d\lambda$$

where W_λ is the radiant emittance of a blackbody at the temperature of interest, and $\epsilon(\lambda)$ is the emissivity of the gas at that temperature.

The approximation $\epsilon(\lambda) = 0.5$ in the region from 4.33 to 4.55 μm, and $\epsilon(\lambda) = 0$ elsewhere, yields

$$W = 0.5 \int_{4.4}^{4.55} W_\lambda d\lambda = 3.5 \times 10^{-2}$$

Applying Planck's law with $T=370$ degrees C and integrating, one obtains

$$\int_{4.3}^{4.55} W_\lambda d\lambda = 3.5 \times 10^{-2}$$

and therefore the radiant emittance is (in W/cm^2)

$$W = 0.5 \times 3.5 \times 10^{-2} = 1.75 \times 10^{-2}$$

The surface area of the gases in this region is not much greater than that of the exhaust nozzle, and it can be seen that the radiant emittance of the plume is much lower than that of the exhaust nozzle. With increasing distance from the exhaust nozzle, the temperature decreases rapidly, and, although the surface area of interest increases, its contribution can be ignored. For a surface area of 10,000 cm² the radiant flux is (in W)

$$P = W \times S = 175$$

so that (in W/sr)

$$J = \frac{P}{\Omega} = \frac{175}{3.14} = 27.8$$

That is, the radiant flux, and therefore the radiant intensity j (W/sr) of the plume region is about a tenth that of the nozzle. However this is no longer true when the aircraft employs an afterburner to increase its thrust. In this case, the radiant emittance of the plume can be several times higher than that of the nozzle, and the radiant intensity much larger.

2.3.2.3 Aerodynamic heating

When a body moves through space at high speed, it is heated by adiabatic compression of the air against its surface, and by friction. The temperature reached depends on the altitude, which determines the density of the air, on whether the boundary layer flow is laminar or turbulent, on the material and the geometry of the object, and on its speed.

It can be shown that a surface heated aerodynamically at speed Mach M, with a boundary layer in laminar flow, will reach an absolute temperature

$$T = T_0(1 + 0.164M^2)$$

where T_0 is the temperature of the air.

For example, the skin of an aircraft flying at Mach 2 at an altitude of 5000 m, with $T_0 = 250$ K, will reach the temperature

$$T = 250(1 + 0.164 \times 4) = 414 \, \text{K}$$

and therefore the radiant emittance will be (in W/cm^2)

$$W = \sigma T^4 = 5.67 \times 10^{-12} \times 414^4 = 0.16$$

Assume that in the hemisphere of interest the exposed surface is $20\,m^2 = 2 \times 10^4\,cm^2$. Then

$$P = 0.16 \times 20 \times 10^4 = 32000\,W$$

and (in W/sr)

$$J = \frac{P}{\pi} = 10188$$

Maximum emission will be at wavelength

$$\lambda_{max} = \frac{a}{T} = \frac{2898}{414} = 7\,\mu m$$

About 25 percent of the radiation, that is, 2500 W/sr, will be at shorter wavelengths than this.

2.3.2.4 IR Background

In the foregoing it has been shown very briefly, but with sufficient accuracy, how the IR signal originates and how much radiation is produced by targets of interest such as aircraft, missiles, stet turbojet engines, their nozzles, their plumes, and the radiation produced by aerodynamic heating of stet surfaces.

The interested reader should consult the specialized literature for a discussion of other IR sources, including:

- the turbofan engine, which gives a lower IR signature because it operates at lower temperatures,
- the ramjet engine, a jet engine with no compressor ahead of the combustion chambers, and nozzle temperatures on the order of 1600 degrees C, and
- the rocket engine, which produces signatures on the order of 1000 to 100,000 W/sr,

However, it should be noted that the background itself, that is, whatever is not a target, is also capable of producing strong IR signals; for example, the ground heated by the sun, the smokestack of a furnace, the engine of a vehicle, the blasts of bombs, or the firing of guns on the battlefield, and so forth. The main problem of IR sensors is precisely this: to discriminate the IR signal produced by the target of interest from the one produced by the background.

The background interfering with a sensor is the background seen in the elementary *field of view* (FOV) (Fig. 2.100). As shown in the figure, the background signal competing with the target signal is equal to the background radiance multiplied by the FOV and by the atmospheric transmittance.

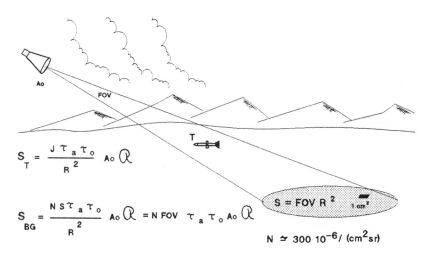

$$S_T = \frac{J \, \tau_a \, \tau_o}{R^2} \, A_o \, \mathcal{R}$$

$$S_{BG} = \frac{N S \tau_a \tau_o}{R^2} \, A_o \, \mathcal{R} = N \, FOV \, \tau_a \, \tau_o \, A_o \, \mathcal{R}$$

$$S = FOV \, R^2$$

$$N \simeq 300 \cdot 10^{-6} / (cm^2 sr)$$

Figure 2.100 IR signals produced by a target and by the background.

Below $3 \, \mu m$, the radiance of the background can, in general, be considered to be essentially that generated by the reflection of solar radiation. Above $3 \, \mu m$, this radiation becomes negligible because of the atmospheric attenuation at such wavelengths. Since the temperature of the background is on the order of $300 \, K$, the radiation generated directly by the background starts to be relevant above $5 \, \mu m$. In the band between 3 and 5 μm, the background tends towards a minimum, and its radiance integrated over the band varies between 20×10^{-6} Wsr^{-1}cm^{-2} (clear sky) and 300×10^{6} Wsr^{-1}cm^{-2} (mixed terrain).

2.3.3 IR *range equation*

In this section an equation determining the range of an IR system will be derived as was done for radar in Section 2.2.2. An IR system usually consists of (Fig. 2.101):

- an optical system, equivalent to the antenna of a radar system, which directs radiation onto the IR detector.
- an IR detector which converts the incident radiant energy into a useful electric signal, but also generates noise (*sic*); if high performance is desired, such a device may be very complex.
- a computer which will process the signal produced by the sensor in order to maximize the SNR and minimize unwanted signals produced by the background, and which will generate the appropriate information: images, warning signals, tracking signals, and so forth.

Figure 2.101 Block diagram of an IR system.

In the preceding section, IR signals from targets of interest, produced both by the heat of the engines and by aerodynamic heating, have been examined. In contrast with the RF case, for IR signals the influence of the atmosphere on transmission is very conspicuous. In practice, only radiation corresponding to special bands called atmospheric windows can pass through the earth's atmosphere (Fig. 2.102).

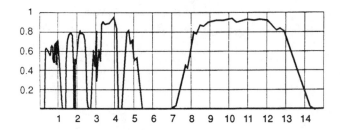

Atmospheric Transmittance
Standard Conditions,
Visibility 23Km,
Horizontal Path of 1 Km
at an Altitude of 1Km)

Attenuation Coefficient (dB/Km) (Visibility 23 Km)		
Quota	BANDA IR	
	35 µm	10 µm
0.2 Km	0.88* 0.6	0.57*0.33
2 Km	0.37	0.17
5 Km	0.16	0.05
10 Km	0.05	0.04

* = Visibility 5Km

Figure 2.102 Transmittance of the earth's atmosphere is strongly dependent on wavelength, because of absorption by gas molecules. Of special relevance are the two atmospheric windows in the 3 to 5 μm and the 8 to 12 μm bands.

In order to derive a range equation it is necessary to start by analyzing the performance of an IR detector, and by defining some parameters used by manufacturers to characterize detectors.

The responsivity R of an IR detector of area A_d is the ability of the detector to give a signal voltage V_s, when irradiated in its response band with an irradiance $H(\mathrm{W/cm}^2)$.

The higher the R, the higher the voltage produced. For a flat detector

$$V_s = R H A_d$$

whence

$$R = \frac{V_s}{H A_d}$$

Unfortunately, an IR detector produces noise that obscures the useful signal (Fig. 2.103). Among the many different types of noise, the so-called 1/f noise, whose intensity decreases with the increase in frequency, is the most disturbing in normal applications.

Noise equivalent power (NEP) is the power, or the radiant flux $H A_d$, which must be supplied to the sensor so that an rms voltage equal to the detector noise is generated:

$$R \times (NEP) = V_n$$

whence

$$(NEP) = \frac{V_n}{R} = \frac{V_n}{V_s/H A_d} = H A_d \frac{V_n}{V_s}$$

The detectivity D is a measure of a detector's ability to detect radiation. It is defined by

$$D = \frac{1}{(NEP)}$$

The smaller the NEP, the higher the detectivity.

Since the product $D \times A_d^{1/2}$ and the product $D \times \Delta f^{\frac{1}{2}}$ are constant, it is convenient to define a quantity much used by manufacturers of IR detectors, D^* (pronounced dee-star); that is, detectivity multiplied by $\Delta f^{\frac{1}{2}}$ and by $A_d^{\frac{1}{2}}$ (Fig. 2.104)

$$D^* = D(A_d \Delta f)^{\frac{1}{2}} = \frac{(A_d \Delta f)^{\frac{1}{2}}}{(NEP)}$$

whence

$$R = \frac{D^* V_n}{(A_d \Delta f)^{\frac{1}{2}}}$$

Consider an IR system (Fig. 2.104) with a lens of area A_0 and FOV ω and assume that at range R there is a target T

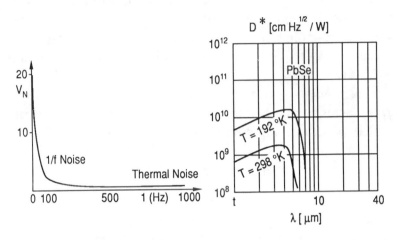

Figure 2.103 An IR detector is characterized by a quantity related to detectivity D^*. The figure shows also the characteristic pattern of $1/f$ noise.

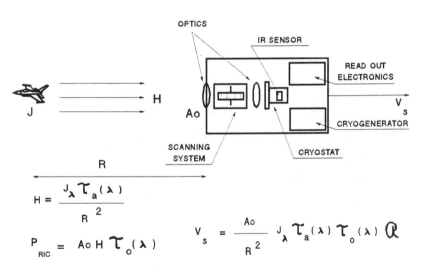

Figure 2.104 Diagram of an IR sensor head of scanning type.

emitting toward the system, at wavelength λ a radiant intensity J_λ (W/sr). Taking into account the atmospheric transmittance $\tau_a(\lambda)$, this target will produce at the optical system an irradiance

$$H_\lambda = \frac{J_\lambda \tau_a(\lambda)}{R^2}$$

Taking into account the transmittance $\tau_0(\lambda)$ of the optical system at that wavelength, one finds that at the detector the signal power will be

$$P_\lambda = H_\lambda A_0 \tau_0(\lambda)$$

If the detector is characterized by responsivity $R(\lambda)$, the signal voltage will be

$$V_s = P_\lambda R(\lambda) = \frac{J_\lambda \tau_a(\lambda)}{R^2} A_0 \tau_0(\lambda) R(\lambda)$$

and integrating over the band of wavelengths of interest,

$$V_s = \frac{A_0}{R^2} \int_{\lambda 1}^{\lambda 2} J_\lambda \tau_\alpha(\lambda) \tau_0(\lambda) R(\lambda) d\lambda$$

To simplify matters, replace the values that depend on λ by their averages over this band; then

$$V_s = \frac{A_0}{R^2} J \tau_a \tau_0 R$$

Recall that

$$R = \frac{V_n D^{*\frac{1}{2}}}{A_d \Delta f}$$

substitute

$$V_s = \frac{A_0}{R^2} J \tau_0 \tau_0 \frac{V_n D^*}{(A_d \Delta f)^{\frac{1}{2}}}$$

and solve with respect to R^2, to obtain

$$R^2 = \frac{A_0 J \tau_\alpha \tau_0 D^*}{(A_d \Delta f)^{\frac{1}{2}} V_s / V_n}$$

Thus, as in the radar case, the range is expressed in terms of known parameters of the system and of the SNR, which is here represented by the ratio of the voltages V_s and V_n.

However, it is more convenient to express the range in terms of the diameter of the optics D_0, the instantaneous FOV ω, and the numerical aperture NA. Recalling from the laws of optics that

$$(NA) = \frac{D_0}{2f}$$

where f is the equivalent focal length, and that

$$A_d = \omega \left[\frac{D_0}{2(NA)} \right]^2$$

substituting in the last expression for R^2, and remembering that

$$A_0 = \pi \left(\frac{D_0}{2} \right)^2$$

for a circular lens, one may write

$$R = \left[\frac{\pi(NA)D_0\, J\tau_\alpha\tau_0\, D^*}{2(\omega\Delta f)^{\frac{1}{2}} V_s/V_n} \right]^{\frac{1}{2}}$$

where R is in cm, D_0 is in cm, D^* is in cm $\mathrm{Hz}^{1/2}\mathrm{W}^{-1}$, J is in W/sr, J is in sr, Δf is in Hz, and NA, τ_a and τ_0 are dimensionless.

Grouping terms together, it is possible to exhibit the contributions to the range of various components of the problem:

$$R = (J)^{\frac{1}{2}}(\tau_\alpha)^{\frac{1}{2}} \left[\frac{\pi}{2}D_0(NA)\tau_0 \right]^{\frac{1}{2}} (D^*)^{\frac{1}{2}} \left[\frac{1}{(\omega\Delta f)^{\frac{1}{2}} V_s/V_n} \right]^{\frac{1}{2}}$$

where J depends on the target (a method for calculating it has already been given above), τ_a depends on the atmosphere (Fig.

2.102), $(\tau/2)D_0(NA)\tau_0$ depends on the optics, D^* depends on the sensor (Fig. 2.104), ω is the instantaneous FOV which depends on the system, Δf is the equivalent noise bandwidth which depends on the system, and V_s/V_n is the signal-to-noise ratio that depends on P_d and P_{fa} that can be accepted by the system (see Fig. 2.13).

2.3.4 *Suppression of background effects*

The amount of background radiation collected by an IR system depends on the responsivity of the sensor to that radiation, and on the elementary FOV of the system. Since the background is distributed over a great area, the smaller the aperture, the smaller the IR energy produced by the background that will be introduced into the system (Fig. 2.105). However, to ensure coverage of an angular sector of the right dimensions the instantaneous FOV cannot be made too small, so that the signal due to the background is usually stronger than the signal produced by targets of interest.

In order to reduce the background signal, IR systems usually exploit two types of filtering: temporal/frequency filtering and spatial filtering.

For temporal/frequential filtering, a filter matched to the duration of the phenomenon to be observed is used.

Spatial filtering exploits the characteristics of the signal of interest, which will in general come from positions different from those of the background objects, by suitable modulation with a rotating reticle.

Consider, for example, Figure 2.105, where a point target T is seen against a background consisting of a large cloud at a certain temperature.

If the reticle is rotated rapidly, the signals reaching the sensor from the background, which is uniform and widely distributed, will give a constant dc signal, while the point target will give a signal modulated in amplitude by the openings of the reticle. The background can be suppressed by inserting a filter at the modulation frequency of the signal.

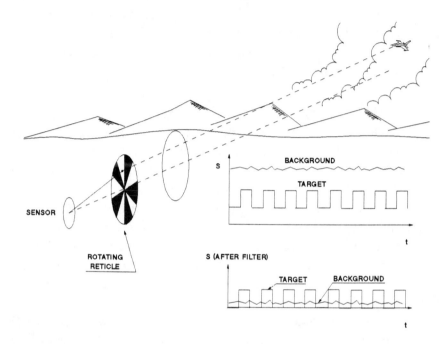

Figure 2.105 Spatial filtering of the IR background by means of a rotating reticle.

2.3.5 IR *systems*

The peculiarities of the IR emissions from targets of interest and the technologies for detection of such emissions have led to the development of many systems dedicated to the vision, to the search for and detection of targets (warning systems), and to target tracking.

The first category includes:

- devices for night vision, such as IR binoculars.
- *forward looking infrared (FLIR)* systems which present on a TV-like screen the image of a given angular sector; the sector is scanned in azimuth and elevation by an IR sensor, employing suitable rotations of small mirrors in a way similar

to that in which the TV scans: In systems of this type a decision about the presence or absence of a target of interest is left entirely to an operator.

- IR line scanners, which are systems capable of performing a line scan of a scene, for example scanning the vertical line from -60 a to +10 of elevation on one side of the aircraft; because of the motion of an aircraft in flight, the whole band of space seen from that side can be explored, and sent to memory for later analysis.

Warning systems are much more complex pieces of equipment. After detection of the IR signal, heavy processing is required to achieve discrimination of targets of interest against the IR background. While in a vision system all signals detected are shown to the operator, in search (warning) systems the output must be restricted to targets of interest, even if they are weaker than the background.

This category includes the following:

- *missile launch warning (MLW)* systems, capable of detecting the launch of a missile.
- *infrared search and track (IRST)* systems, capable of detecting the presence of a target, tracking it, and providing the right coordinates for a possible reaction.

The third and final category encompasses all systems that track targets by exploiting the IR emission produced by the targets themselves:

- IR pointers, which are devices for angular pointing; given the very low reflectivity of the sea and the narrow elementary FOV of these systems, pointers are often used as additional equipment for low altitude precision tracking over the sea; an example of application of these systems is shown in Figure 2.106, where an IR pointer is depicted, associated with a principal tracking radar.
- IR seekers, which are the guidance heads of heat-seeking missiles; these seekers exploit the IR emission of targets to track them and generate signals for missile guidance; and suitable

Figure 2.106 IR system for precision pointing at low altitude, MIRA, associated with a tracking radar.

techniques provide for the necessary background rejection and the extraction of pointing signals. These techniques will be discussed in more detail in the next Chapter.

REFERENCES

[1] J.W. Sherman, "Aperture-antenna Analysis," chapter 9 in M.I. Skolnik (ed.) *Radar Handbook*, New York: McGraw-Hill, 1970.

[2] R.E. Kell and R.A. Ross, "Radar Cross Section of Target," chapter 27 in M.I. Skolnik (ed.) *Radar Handbook*, New York: McGraw-Hill, 1970.

[3] C. Siegel, *Methods of Radar Cross Section Analysis*, New York: Academic Press, 1968.

[4] M.I. Skolnik, chapter 9, New York: McGraw-Hill, 1962.

[5] J.V. Di Franco and W.L. Rubin, *Radar Detection*, Norwood, MA: 1980.

[6] D.K. Barton (ed.), *Radars, Volume Two: The Radar Equation*, Norwood, MA: Artech House,1975.

[7] J.H. Dunn and D.D. Howard, "Target Noise," chapter 28 in M.I. Skolnik (ed.) *Radar Handbook*, New York; McGraw-Hill, 1970.

[8] M.I. Skolnik, "An Empirical Formula for the Radar Cross Section of Ships at Grazing Incidence", *IEEE Transaction AES*, March 1974, p. 292.

[9] M.I. Skolnik, *Introduction to Radar Systems*, chapter 8, New York: McGraw-Hill, 1962.

[10] F.E. Nathanson, *Radar Design Principles*, chapter 7, New York: McGraw-Hill, 1969.

[11] F.E. Nathanson, *Radar Design Principles*, chapter 6, New York: McGraw-Hill, 1969.

[12] L.V. Blake, "Prediction of Radar Range," chapter 2 in M.I. Skolnik (ed.) *Radar Handbook*, New York: McGraw-Hill, 1970.

[13] M.I. Skolnik, *Introduction to Radar Systems*, chapter 11, New York: McGraw-Hill, 1962.

[14] P. David and J. Voge, *Propagation of Waves*, Oxford: Pergamon Press, 1969.

[15] D.I. Kerr, *Propagation of Short Radio Waves*, MIT Radiation Laboratory Series, Vol. 13, New York: McGraw-Hill, 1963.

[16] F.E. Nathanson, *Radar Design Principles*, chapter 9, New York: McGraw-Hill, 1969.

[17] G. Galati, "Il circuito autogate nella rivelazione radar," *Rivista Tecnica Selenia*, Vol. 1, No. 3, 1973.

[18] F.E. Nathanson, *Radar Design Principles*, chapter 5, New York: McGraw-Hill, 1969.

[19] D.K. Barton, *Modern Radar System Analysis*, chapter 3, Norwood, MA: Artech House, 1988.

[20] F.E. Nathanson, *Radar Design Principles*, chapters 12 and 13, McGraw-Hill, New York, 1969.

[21] F. Marcoz and G. Galati, "A Sub-optimal Detection Technique: the Accumulator Detector," *Alta Frequenza*, Vol. XLI No. 2, February 1972, pp. 77-89.

[22] S. Rotella and F. Marcoz, "Analisi di un rivelatore a finestra mobile," *Alta Frequenza*, Vol. VI, No. 12, December 1967, pp. 1102-1110.

[23] T.C. Cheston and J. Frank, *Array Antennas*, chapter 11 in

M.I. Skolnik (ed.) *Radar Handbook*, New York: McGraw-Hill, 1970.

[24] R. Mailloux, "*Phased Array Theory and Technology*," *Proc. IEEE*, Vol. 70, No. 3, March 1982.

[25] J.W. Caspers, *Bistatic and Multistatic Radar*, chapter 36 in M.I. Skolnik (ed.) *Radar Handbook*, New York: McGraw-Hill, 1970.

[26] G. Picardi, *Elaborazione del segnale radar*, chapter 8, Franco Angeli Editore, Roma, 1988.

[27] M.I. Skolnik, *Introduction to Radar Systems*, chapter 5, New York: McGraw-Hill, 1962.

[28] Ibid.

[29] S.M. Sherman, *Monopulse Principles and Techniques*, Norwood, MA: Artech House, 1984.

[30] M.I. Skolnik, *Introduction to Radar Systems*, chapter 5, New York, McGraw-Hill, 1962.

[31] D.K. Barton, *Modern Radar Systems Analysis*, chapter 9, Norwood, MA: Artech House, 1988.

[32] J.H. Dunn, D.D. Howard and K.B. Pendleton, "Tracking Radar," chapter 21 in M.I. Skolnik (ed.) *Radar Handbook*, New York: McGraw-Hill, 1970.

[33] F. Neri, S. Pardini, S. Sabatini and M. Tarantino, *Analisi dell'effetto multipath per bersagli estesi a bassa quota*, 27th Congresso Scientifico Internazionale sull'Elettronica, Roma, 1980.

[34] G.W. Simpson, *Introduction to Airborne Radar*, Hughes Aircraft Company, El Segundo, CA: 1983.

[35] C.V. Morris, *Airborne Pulse Doppler Radar*, Norwood, MA: Artech House, 1988.

[36] Ferranti Defence Systems Ltd, "Pulse Doppler Airborne Radar," *Military Technology* No.6, 1987, pp. 182-200.

[37] R.D. Hudson, *Infrared Systems Engineering*, New York: John Wiley & Sons, 1968.

[38] I.J. Spiro and M. Schlessinger, *Infrared Technology Fundamentals*, New York: Marcel Dekker, 1968.

[39] W.L. Wolfe and G.J. Zissis, *The Infrared Handbook*, Wash-

ington D.C.: Environmental Research Institute of Michigan, 1978.

Chapter 3

Weapon Systems

3.1 Introduction

The main weapon systems employed by national armed forces were described in chapter 1. From the discussion there, it follows that for the development of effective *electronic countermeasures* (ECM) the operation of the following systems needs to be analyzed.

- Early warning systems, provide a general warning to all other defensive layers of a territory. To defeat these systems the best tactic is to avoid being detected by them. As the effectiveness of these systems is mainly based on the search radar performance, the most effective ECM against them are those discussed in chapter 2.

- Artillery systems of the following types:

(a) radar-guided *anti-aircraft artillery* (AAA) systems.

(b) radar-guided anti-ship fire control systems.

(c) anti-tank systems with optical guidance, laser range finders, and guided projectiles.

• Missile systems of the following types:

(a) anti-aircraft *surface-to-air missile* (SAM) and *air-to-air missile* (AAM) systems, radar or IR-guided.

(b) anti-ship systems: either missiles launched from another ship, *air-to-surface missiles* (ASM) launched from an airborne platform, or *surface-to-surface missiles* (SSM) launched by a coastal defense system; these are at present radar-guided missiles; in the future they may be either IR-guided or have hybrid guidance.

(c) anti-tank systems, at present wire or IR-guided, in the future guided by millimeter-wave radar.

(d) *anti-radiation missile* (ARM) systems.

Moreover, since communications play an important role in the performance of an armed force, communication systems should also be included among the systems which must be countered electronically.

In practice, all weapon systems are organized as in Figure 3.1. A warning center detects and analyzes the threats, and decides which of those within its area of competence should be destroyed. It further assigns and designates the threats to be destroyed either to missile or to artillery fire control centers. Usually, a fire control center consists of a tracking radar which, upon reception of the coordinates designating the threat, acquires and tracks it. The accurate position data generated by the tracking radar are fed to the fire control center computer, which will either aim the gun or launch and guide the missiles.

This type of organization is normally found in all land, sea, and airborne systems, although on-board aircraft, search, designation and tracking of targets are all usually performed by one single radar.

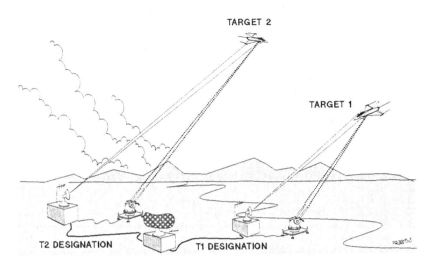

Figure 3.1 A weapon system normally consists of a search radar and a few fire control centers each guided by a tracking radar.

3.2 Artillery systems

To distinguish between land and sea artillery systems is beyond the scope of this book. Here it will suffice to recall that naval fire control systems are confronted with an additional problem: They have to compensate for platform motion, and therefore require stabilization in roll, pitch, and yaw [1, 2, 3].

Upon receiving a designation from the command and control center, the tracking radar will train its antenna and its range gate onto the designated azimuth and range, respectively. The time required for this change can take from three to five seconds. The radar will then start searching for the designated target, sweeping its pencil-beam antenna in azimuth and elevation (Fig. 3.2). This must be done because azimuth data are often inaccurate, and elevation data generally lacking, in the designation given by a 2-D radar. When the search radar is of the 3-D type, giving accurate information about all three coordinates, range, azimuth, and elevation, this search may be avoided.

Generally, the tracking radar keeps searching in azimuth and elevation until a detection circuit reveals the presence of the target in the acquisition gate. The antenna is then locked onto the detected azimuth and elevation data, and the first tracking phase is started, to determine the kinematic data of the target. This phase may last from three to ten seconds, depending on the altitude of the target. The reader will recall, from chapter 2, that radar range and radar tracking capabilities are not very good at low altitude, so that attacks at low elevation, exploiting radar limitations, are to be expected.

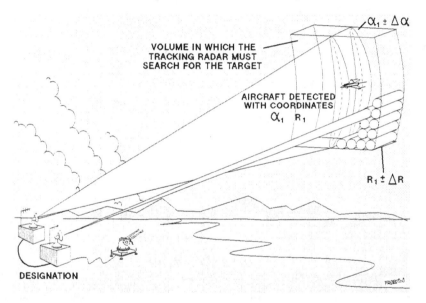

Figure 3.2 Acquisition search by a fire control system after designation by the search radar.

As soon as tracking starts, the data are fed to the computer of the fire control center, which computes the interception point, that is, the point at which target and weapon will meet, if the target keeps to the computed course. Wind, air temperature and pressure, gun powder temperature, gun ballistics, warhead dispersion pattern, and so forth must be taken into account in computing the interception point.

Up to this point, operation is automatic without any intervention on the part of the operator, whose task is merely to check that everything functions normally and to take action only in a few crucial cases, or when there is a malfunction. Usually, the operator must intervene, normally after an order from a superior, to associate the gun to the radar, that is, to enable the gun to aim onto the calculated interception point and to fire. In any case, a signal confirming that the threat is within the range of the associated weapon system must be received before firing.

At this point, the number of shots expected to achieve a given kill probability is fired, and immediately after this the operator performs a kill assessment, to check whether the target has been destroyed. If it has not, the burst of fire is repeated.

Sometimes, to improve firing accuracy, devices are used that automatically provide the real-time measurement of the miss distance. The miss distance is the minimum distance from the target to the projectile trajectories. This measurement, based on the data provided by the tracking radar, may be used to introduce, manually or automatically, a correction or "countermiss," which improves the aiming of the weapon.

3.2.1 Firing accuracy

The firing accuracy required of a weapon system depends on the type of ammunition used. If the ammunition is of small caliber and has no proximity fuze, it will have to hit the target, while if it is of medium-to-high caliber, with a proximity fuze, it will have only to get within lethality range.

The lethality of a projectile depends on the amount of explosive, the type and quantity of fragments it is capable of producing, the distance from the target at which it detonates, and the vulnerability of the target. A very simplified way to express lethality L [4, 5, 6] is

$$XL = \exp(-r/r_0)$$

where r is the projectile-to-target range at which the explosion happens and r_0 is a reference range, for example, the range at which the projectile is certainly lethal.

For medium caliber ammunition, for example, 76 mm caliber, r_0 can be on the order of 3 m. Thus, to destroy a threat at a 5-km range, the weapon delivery system must have a precision of less than one milliradian in angle and one meter in range!

Since the projectiles fired by a gun will take a time T_c, the *time of flight,* to reach the target, it will be necessary to aim the weapon at an interception point calculated by the fire control system on the basis of data provided by the tracking radar. For accurate processing of the data, narrowband filtering on the order of 0.1 to 1 Hz is required. However, when the target is making a rapid pass over the weapon system, such a band is not broad enough because it introduces too large a delay error (Fig. 3.3).

To solve this problem, a fire control center generally performs a coordinate transformation. In polar coordinates ρ, ϑ the situation is evolving quite rapidly, particularly if V is high and L is short, but in Cartesian coordinates x, y centered at 0 (0,0), the same situation is practically static, since $y = $ constant, and x is changing at a constant rate. That is why fire control systems often convert from polar to Cartesian coordinates, perform narrowband filtering and extrapolation in this reference frame, and finally, after conversion back into polars, guide the weapons and sometimes also close the tracking loop [1, 2, 7].

An example of a simplified interception point calculation will help to explain the importance of precision in calculating the estimated velocity of a target. Consider a target in uniform rectilinear motion toward the fire control center (Fig. 3.4). Neglecting the effects of altitude and of terrestrial gravity, one may write

$$X_F = R_0 - V_t T_c$$

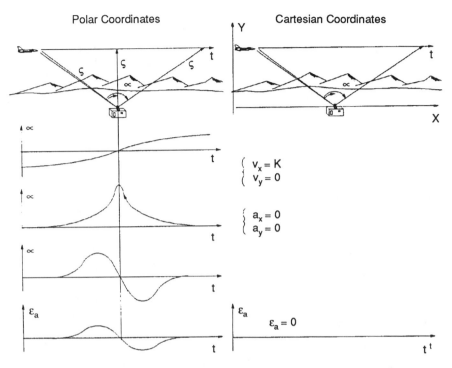

Figure 3.3 To reduce tracking errors when a target is making a fast pass over the weapon system, polar coordinates are converted into Cartesian coordinates and, after narrowband filtering, are converted back to polars.

$$X_F = \int_0^{T_c} V_p(t)dt$$

Ins are charged to the same voltage and the gate positioning error is zero.

If the gate position is advanced or delayed with respect to the radar center of the target, the late circuit will charge to a voltage higher or lower than the early circuit. This will generate an error signal at the discriminator output showing that the range measured by the radar, or the time elapsed from the transmission of the pulse, is lower or higher than the correct value.

In the most recent radars, the split gate is $\pm 5\,m$, so as to simulate a speed of $10\,m/s$; this small error multiplied by the

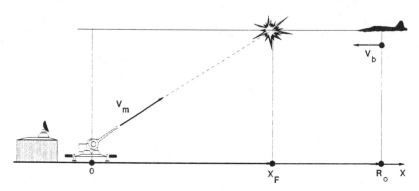

Figure 3.4 An artillery system must fire its projectiles toward the interception point hoping that the target does not change its trajectory

time of flight could become an error in range of 80 m, thus drastically reducing the effectiveness of the weapon.

3.2.2 Susceptibility to jamming of an artillery system

From the above it may be seen that a jammer can reduce the kill probability of an artillery system in the following ways:

- by jamming the search radar to prevent a quick designation to the weapon system.
- by jamming the tracking radar in its acquisition mode, thus preventing the determination of the data needed to extrapolate the interception point.
- by jamming the tracking radar in its tracking mode, thus generating errors which may yield an incorrect interception point.

This shows that protection does not necessarily require a "break-lock" situation in which the tracking radar loses track of its target altogether. However, if break lock is achieved, the effectiveness of the weapon system is reduced to zero, at least until the whole sequence of designation, acquisition, tracking, and weapon implementation has been repeated.

3.3 Missile systems

Artillery systems are very effective, but limited by the fact that their accuracy is high only if the range is short, if glint has been substantially reduced and, above all, if the target is not maneuvering.

At long range, the time of flight of projectiles is very long, which gives the target an opportunity to maneuver, thus invalidating the calculated interception point. The effect of the wind and the fact that errors in the system are angular, which implies that miss distance increases with range, make the problem worse.

At short range, on the other hand, the system can be saturated easily if the number of incoming threats is large. Resorting to missile systems may circumvent these problems. Missiles can be guided to the target despite potential evasive maneuvers by the target after the missile has been launched (Fig. 3.5) [8, 9, 10].

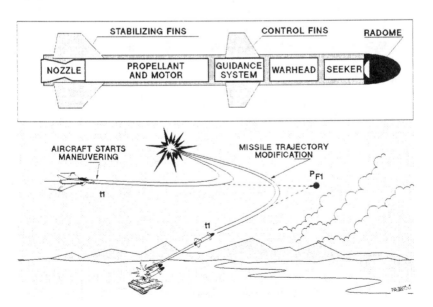

Figure 3.5 A missile modifies its trajectory to track a target that performs evasive maneuvers.

A missile usually consists of:

- an airframe, inside which are fitted.
- a seeker, protected by a radome, to detect the target and generate command signals.
- a warhead, consisting of an explosive charge and heavy metal material, prefragmented or not, to damage the target.
- a fuze, to assure detonation of the warhead explosive even in the absence of a direct impact.
- an autopilot, namely a guidance system which intercepts the signals produced by the seeker to position the control fins, and thus directs the missile toward the point of impact [11].
- a propellant motor to provide the correct thrust.
- a series of stabilizing fins.

Not all missiles follow this pattern exactly. For example, in command missiles the seeker is missing; in some missiles guidance is by the tail fins, and so forth.

Missile systems are organized in much the same way as artillery systems. A search radar reports to the command and control center, which evaluates the threats, and designates them to the various weapon system batteries, each comprising a tracking radar and a launcher usually able to launch more than one missile.

With missiles, the simple "within range" function of artillery systems becomes more complex, because of the different kinematics and the high cost of the missile. Each missile system covers a certain zone, depending also on target speed, within which it is almost certain that the target will be hit. Computations of coverage zones are usually made by the missile center computer after the tracking radar has started its tracking.

Missile systems can be medium-to-long range (50 to 150 km), to defend a relatively wide area (local area missile system), or medium-to-short range, to defend a target of great value such as an airfield or a railhead.

Missiles may be distinguished by their guidance systems:

(1) command (short-range) missiles.

(2) beam-riding (short-range) missiles.

(3) semiactive homing (medium-to-long range) missiles.

(4) active homing (medium-to-long range) missiles.

3.3.1 Command missiles

A command missile does not receive information directly from the target. It is guided by commands transmitted from the ground via a command link. There are two radars, one tracking the target, the other tracking the missile, which is usually equipped with a beacon in order to be seen more easily by the tracking radar (Fig. 3.6) [9, 10, 12].

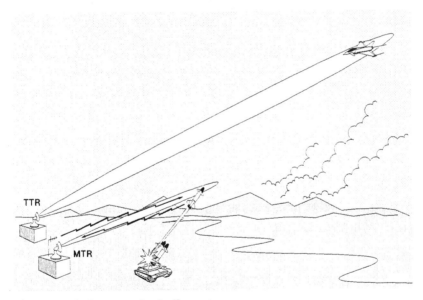

Figure 3.6 Command missile system.

Missile and target position data are sent to a computer, which processes them to generate command signals for missile guidance. The use of two independent systems for tracking missile and target ensures that the best possible trajectory for impact is chosen for the missile.

Alternatively, a single radar may be used to track both missile and target. In this case, the missile must be commanded to stay always within the radar beam. This type is called a *command-to-line-of-sight* (CLOS) missile. Often the missile is commanded in an "advanced" way; that is, it is sent part way toward the interception point so as to avoid excessive accelerations in the terminal phase of its trajectory.

Command missiles execute orders only; they do not have their own seekers. Their accuracy depends on the precision of the tracking radar, and their effectiveness normally decreases as the radar-to-target range increases.

As shown earlier, the angular accuracy of the radar is some fraction of the antenna lobe width, ϑ_B. If σ_t (milliradians) is the precision of the radar in tracking the target, and σ_m the precision of the radar in tracking the missile, then, neglecting other guidance errors, the rms target-to-missile miss distance at the range R (km) will be given in meters by

$$m_d = R\sqrt{\sigma_t^2 + \sigma_m^2}$$

The expression for the miss distance is in practice more complex, since all the other parameters of the missile guidance loop must be taken into account.

A command missile is therefore to be preferred for short ranges, as may easily be seen from Figure 3.7. The advantages of these systems are the simplicity of the missile and the power of the tracking system on the ground. An operator or a powerful computer can be of enormous help in tracking and in missile guidance, especially in difficult situations, such as in the presence of a jammer. A command link is also necessary. If the target descends to very low levels, disturbance of the radar by ground and sea clutter must be minimized.

For dynamic reasons, these missiles are generally not employed for air-to-air missions.

3.3.1.1 Susceptibility to jamming of command missiles

The susceptibility to jamming of a command missile is very similar to that of a fire control system, with the advantageous

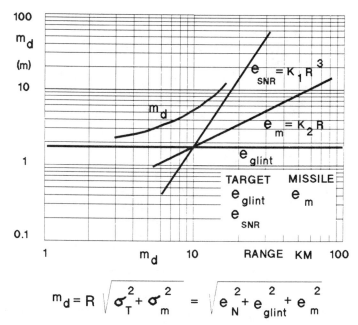

$$m_d = R \sqrt{\sigma_T^2 + \sigma_m^2} = \sqrt{e_N^2 + e_{glint}^2 + e_m^2}$$

Figure 3.7 Guidance errors, in meters, of a command missile as a function of range.

difference that the problem of sudden target maneuvers is very much mitigated.

3.3.2 Beam-riding missiles

A beam-riding missile has an on-board receiver capable of sensing whether it is centered within the radar beam that tracks the target, and can automatically correct its course to align itself with the radar boresight where, sooner or later, it will meet the tracked target.

With this type of missile there is no need for a command link to the center, as the necessary information is extracted directly from the radar beam. However, this means that the missile is forced to follow a trajectory that requires strong accelerations in the terminal stage, even in the absence of target maneuvers, as is shown in Figure 3.8. In compensation, this type of system can be relatively simple.

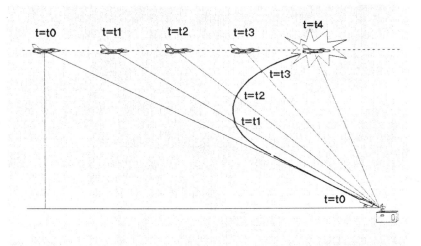

Figure 3.8 Typical trajectory of a beam-riding missile.

Some beam-riding missile systems derive the signals needed for missile guidance from an integrated closed-circuit TV system aligned with the tracking radar boresight.

3.3.3 Semiactive homing missiles

In a system of this type, the radar tracking the target illuminates it by means of a CW radio-frequency signal to highlight the target of interest (Fig. 3.9) [13, 14, 15].

The missile has a passive seeker capable of seeing the CW signal scattered by the target itself. The seeker can thus track the target by one of the methods listed in the preceding chapter (e.g., monopulse, conical scan, LORO, and so forth) without the need to carry on-board or transmitter. Sometimes an *interrupted continuous wave* (ICW) is used, which allows control of more missiles. As the signal of interest is in CW (or ICW), it is very easy for the seeker to discriminate target from clutter by using doppler filtering (Fig. 3.10).

The great advantage of CW is that the missile receiver may extract the angle-tracking data while operating in an extremely narrow band, i.e., on the order of one kilohertz. The system

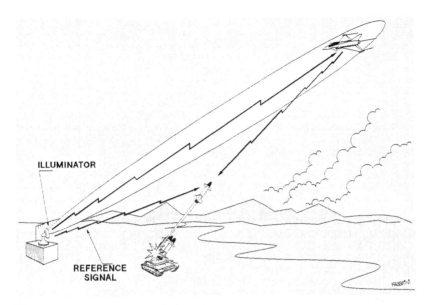

Figure 3.9 Semiactive missile system.

precision does not depend on range measurement, but on the quality of the seeker and the maneuverability of the missile.

The causes of miss are essentially the following:

- aiming errors of the launcher.
- target maneuvers.
- target glint.
- seeker noise.
- guidance loop parameters.

Since there is no need for the missile to remain within the radar beam, the principle of proportional guidance may be exploited as follows (Fig. 3.11) [16, 17, 18, 19].

The missile is launched toward the predicted interception point, while the seeker antenna tracks the target. The command correction to the missile velocity vector is proportional to the rotation rate of the seeker boresight, λ, with constant of proportionality N, called the navigation constant:

$$\dot{\gamma} = N\dot{\lambda}$$

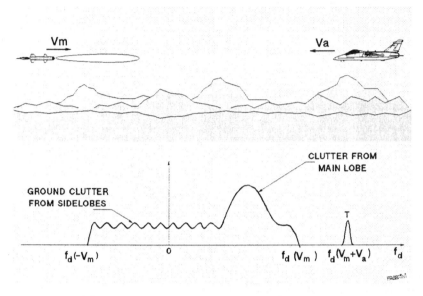

Figure 3.10 Spectrum of the signals seen by the seeker of a semiactive missile.

Or better, if given a new constant N', the effective navigation constant, is introduced so that

$$N = N' \left(\frac{V_c}{V_m} \right)$$

where V_c is the relative target-to-missile velocity and V_m is the missile velocity, then

$$\dot{\gamma} = N' \frac{V_c \dot{\lambda}}{V_m}$$

and therefore,

$$V_m \dot{\gamma} = N' V_c \dot{\lambda} = A_m$$

Here $V_m \dot{\lambda}$ is the lateral acceleration A_m to be given to the missile.

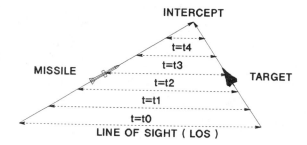

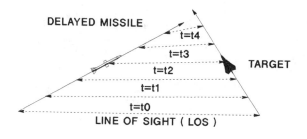

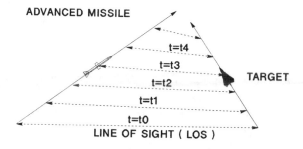

Figure 3.11 According to the kinematic conditions, a missile using proportional navigation can find itself on the correct impact trajectory, or on either an advanced or delayed trajectory.

With this type of guidance the missile has practically no need to accelerate in order to intercept a target on a constant course [20], in contrast to the situation with guidance systems of the beam-riding or CLOS type. The full attainable acceleration can be exploited to compensate for evasive target maneuvers.

The block diagram of a typical seeker is shown in Figure 3.12. The ground radar system [13, 15] illuminates the target and sends to the missile the stable reference frequency necessary to ensure the coherence of the local oscillator. The seeker makes a doppler search for the signal backscattered by the target. As soon as the target doppler shift is detected, the missile can begin tracking.

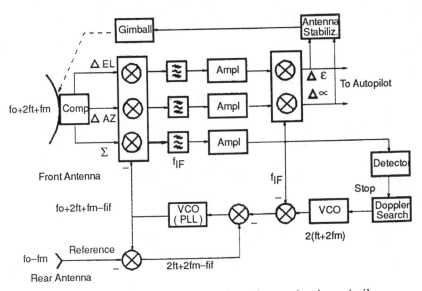

Figure 3.12 Block diagram of the seeker of a semiactive missile.

The intermediate-frequency band of the receiver is very narrow and centered on the target doppler. The thermal noise of this type of receiver is therefore very low. Recalling that $N = KTBF$, and assuming

$$F = 6\,\mathrm{dB}$$

$$B = -30\,\mathrm{dB/MHz}$$

while

$$KT = -114 \, \mathrm{dB \, m/MHz}$$

one obtains for noise

$$N = -138 \, \mathrm{dB \, m}$$

which shows that a semiactive missile is capable of tracking with signals on the order of -125 to $-130 \, \mathrm{dB \, m}$.

Missile systems with semiactive guidance are very effective, which is why today they constitute the majority of the medium- and long-range missile population. The only major disadvantage is that constant target illumination is required during the entire time of flight, and for an aircraft that has launched an air-to-air missile it is quite dangerous to keep approaching a target just for the sake of illuminating it; the enemy too can start to launch missiles!

3.3.3.1 Susceptibility to jamming of semiactive missiles

The susceptibility to jamming of a semiactive missile can be very low, given the extremely narrow processing band employed (circa 1 kHz on a carrier of many gigahertz). As in the case of an artillery systems to jam a semiactive missile it is necessary first to jam the search radar, and then the tracking radar, in both acquisition and tracking modes.

When the missile fire control center turns on its illuminator, this usually means that a missile launch is near. While the missile is in flight, either the tracking radar must be forced to break lock, or the missile must be jammed. A conical-scan missile seeker is more susceptible to jamming than a monopulse one, as will be seen in chapter 5.

3.3.4 Active homing missiles

To avoid the constant illumination of the target required by semiactive missiles, missiles with active seekers have been developed. These seekers are tracking radars equipped with transmitters which can independently engage targets after launch, needing no further assistance. These are "fire-and-forget" missiles.

A missile of this type usually exploits a dual guidance system. In the first part of its flight toward the target, it is under inertial guidance. Upon arrival in the target area, the missile's own seeker is activated, and, once the target has been detected and acquired, the missile starts its terminal phase under active homing guidance.

High cost and the lack of covert operation due to the transmitted pulses are two disadvantages of this type of missile. The indubitable advantage is that practically no assistance is required after launching.

3.3.4.1 Susceptibility to jamming of active homing missiles

It should be noted that this type of missile is very modern, and will therefore be equipped with a monopulse seeker using coherent waveforms (pulse doppler radar). Only a highly sophisticated ECM systems can succeed in jamming it. Here, as in the cases listed above, much can be done by jamming the search radar and the radar responsible for the launching of the missile.

3.3.5 Track-via-missile systems

Command missiles are very simple, but suitable only at short range. Active guided missiles are very good, but to avoid too great expense, they are relatively simple. Semiactive missiles appear to be the best compromise. To improve the missile performance, while at the same time keeping an optimum cost/effectiveness ratio, the *track-via-missile* (TVM) system has been devised (Fig. 3.13) [21].

This system requires a ground illuminator and a semiactive radar sensor on board the missile. However, the data for missile guidance are not processed locally, as is the case in semiactive systems. The signals received by the seeker are retransmitted

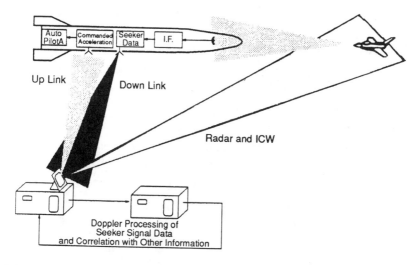

Figure 3.13 Diagram of a track-via-missile (TVM) system.

to the main system on the ground, where a powerful computer, too large for emplacement on the missile, processes the trajectory data of both target and missile and sends precision guidance commands to the missile.

In this type of system, there are one down-link and one up-link more than in the case of a semiactive missile, but, as remarked above, both the precision and the flexibility of the system are greatly improved.

The target is illuminated by an interrupted continuous wave. This means that a single system can launch missiles against several different targets. Therefore, radar and illuminator must both be capable of pointing the antenna beam successively and rapidly at the various targets. This is achieved by using tracking radars with phased-array antennas.

3.3.6 Passive IR-guided missiles

For several decades IR-guided missiles have enjoyed enormous popularity in the air-to-air role. In recent years, these missiles have achieved SAM notoriety, as they have been responsible for many target kills in international conflicts. As surface-to-air missiles, they are almost always of the short or very short range type ($R < 5\,\text{km}$), mostly portable and shoulder-launched, and used for the defense of troops and tanks. Usually, they are ill-coordinated, as they lack direct link with command and control centers.

The IR seeker (Fig. 3.14) is protected by an IR dome, transparent to IR and integrated with the missile. It comprises a telescope equivalent to the antenna of an RF seeker, a gimbal supporting a lens; a field stop; a reticle; a condensing lens; and an IR sensor sometimes cooled by a cold "finger,"

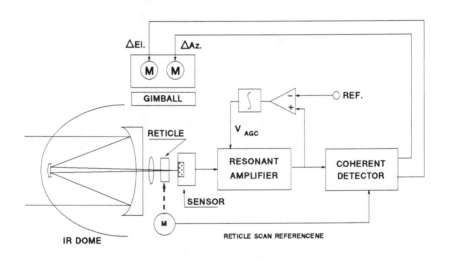

Figure 3.14 Block diagram of an IR seeker.

Given the limited lifetime of the missile after launch, the cooling system is often fitted in the launcher, outside the missile because following a launch, there is insufficient time for significant changes in temperature to take place.

Seekers of several different types can be distinguished by the reticle used to produce the angle-error signals needed for tracking [22]:

(1) seeker with rotating reticle.

(2) seeker with stationary reticle.

(3) seeker without reticle.

(4) seeker with focal plane array (FPA).

IR-guided missiles are very effective on account of their high maneuverability and of the absence of angular glint. They are well-liked because of their fire-and-forget mode. Normally, the pilot of an aircraft designates to the missile seeker the threat to be destroyed, and then waits for the "tell-back" signal which shows the readiness of the missile to track the target. At this point, missile release is activated by the pilot, and the missile flies toward its victim on a proportional navigation course.

3.3.6.1 Seeker with rotating reticles

Only an off-axis target (T_2) produces modulation in a seeker with a rotating reticle, as illustrated in Figure 3.15. A target on the axis (T_1) does not produce modulation. Therefore, this system does not allow space filtering, since an aligned target gives no modulation.

The reticle can be made to generate either amplitude modulation (equally spaced opaque and transparent segments) or frequency modulation (Fig. 3.16).

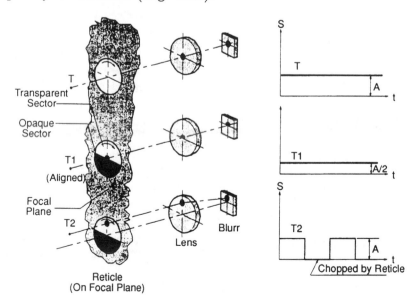

Figure 3.15 The primary function of the reticle of an IR seeker is to provide angular information for missile guidance.

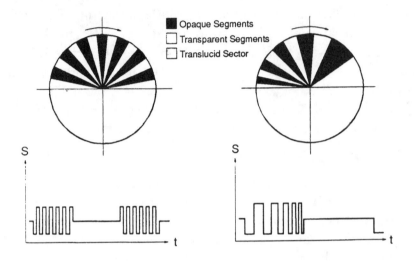

Figure 3.16 Amplitude and frequency modulation reticles.

3.3.6.2 Seeker with stationary reticles

To avoid losing the modulation deriving from space filtering, and to be able to "cancel" the background, seekers rotate
around the axis of the reticle. The optical axis of the lens is
parallel to the axis of the reticle, but displaced a distance d
(Fig. 3.17).

A target on the optical axis will project its image (blur circle) on the reticle at a point that does not coincide with the
center of the reticle, because of the displacement of the lens.
When the lens is rotated, the image of an aligned target will
describe on the reticle a circle with its center on the rotation
axis; if the target is not aligned, the center of the circle will be
displaced. Frequency modulation indicating the non-alignment
of the target will thus be generated (Fig. 3.18).

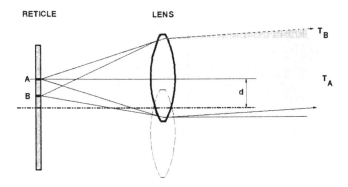

Figure 3.17 Seeker with stationary reticle.

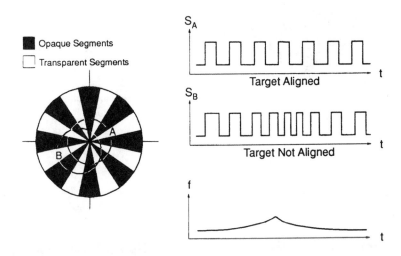

Figure 3.18 Signals generated by a stationary reticle.

Many types of reticles have been designed for missile IR seekers and manufactured to optimize the response and to minimize the effect of any potential jamming. Here, the variable transmission reticle, whose performance is analogous to that of a conical-scan radar, should be mentioned (Fig. 3.19). Variable transmission can be achieved by suitably segmenting the stationary reticle.

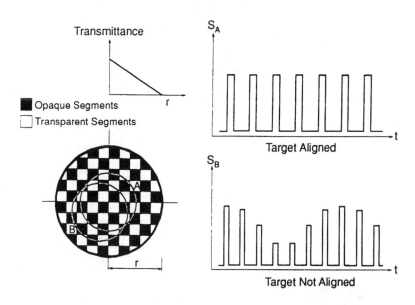

Figure 3.19 Reticles with transmission variable along a radius generate signals similar to those of a conical-scan radar.

3.3.6.3 Seekers without reticles

In this type of seeker, again a lens rotates around an axis offset with respect to the axis of the telescope. There is no reticle, but the sensor is positioned as shown in Figure 3.20, so that it is possible to determine from the signals generated whether the target is aligned with the axis of the seeker (A), or not (B).

3.3.6.4 Seekers with focal plane arrays

With progress in IR technology, an increasing number of focal plane arrays with more and more pixels (elementary sensors) will be available. In these arrays the instantaneous FOV

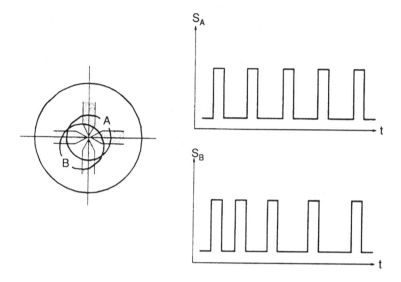

Figure 3.20 Seekers with no reticle resort to special detector geometry.

of the seeker, a few degrees in azimuth and elevation, will be divided into many elementary FOVs, with extremely high discrimination. Signal image processing will guarantee high accuracy and very high resistance to deception jamming, such as may be caused by the launching of flares.

3.3.6.5 Susceptibility to jamming of IR-guided missiles

IR-guided missiles are very resistant to countermeasures. Section 5.4 will discuss the very few methods and devices available for the jamming of IR missile systems.

3.3.7 Sea-skimming missiles

An active missile of a special type is the anti-ship missile flying at very low altitude, as if it were "skimming" the surface of the sea. Guidance in the vertical plane is by a small radar altimeter, which the enemy's ESM-ECM equipment can detect and jam only with difficulty. Guidance in the horizontal plane is of a dual type: In an initial inertial guidance phase, the missile exploits the data provided by a small on-board inertial platform to head for a computed waypoint in the target area. The missile's own seeker is then switched on, and an active homing phase begins, during which the missile heads for the target exploiting the data provided by its seeker (Fig. 3.21).

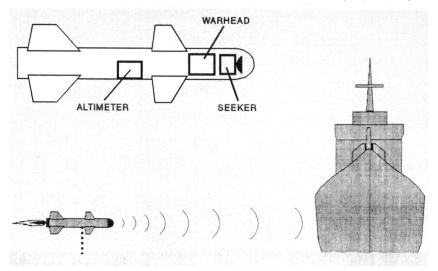

Figure 3.21 A sea-skimming, anti-ship missile heads for a target ship, guided in azimuth by its active seeker at an altitude controlled by the radio-altimeter. Some missiles resort to passive electro-optic seekers either as an alternative or as an aid.

The problems of these missiles are like those of the other types of missile. However, the angular glint of the ship is a special problem. A ship is a very extended target; therefore the error due to glint is very large:

$$\sigma_g = \frac{1}{3}\frac{L}{R}$$

The effect of glint may be reduced sufficiently by filtering the radar error with time constants that take into account the desired miss distance, the time available between the switching on of the seeker and the impact, possible maneuvers of the ship, and the noise of the seeker.

Usually, the error residues due to angular glint increase the penetrability of the anti-ship missile, as they do not allow the anti-missile artillery systems on board the ship to predict the actual interception point with great accuracy.

The need to calculate the interception point sometimes frustrates the use of electronic defense for anti-missile defense by naval artillery systems. To defend itself against an incoming missile, a ship usually begins by trying, with its ED devices, to prevent acquisition by the missile or even to break the missile's tracking mode. However, after break-lock, the missile can often shift immediately to memory and succeed in reacquiring its target, thus quickly returning to its active tracking mode. Under these conditions, the course of the incoming missile will not be very smooth because of its frequent transition states. Hence, under these conditions it will be almost impossible for an artillery system to succeed in predicting the interception point, and therefore in firing at the missile. For this reason, even if there are no problems of electromagnetic compatibility with continuous noise jammers, the use of ED devices may even be prohibited when the missile is within the weapon range of the ship.

This problem must be solved by using ED systems that are compatible, and possibly cooperating, with the "hard kill" systems.

In order to design an anti-sea-skimmer countermeasures system it is also necessary to remember that the RCS of a ship is generally extremely high. This means that the missile can count on a very strong signal; therefore, having no SNR problems, it can use pulses of very short duration. There are examples of missiles that use *pulsewidths* (PW) on the order of a hundred nanoseconds, equivalent to a radar range cell of little more than ten meters. But in this case if the processing on reception takes into account only the content of a single narrow range bin, it can happen that the missile will head for a very precise point on the ship without glint or, even worse, for a non-vital point of the ship. To track the central part of the ship, where signals are generally stronger and the ship is more vulnerable, the pulses used by the seeker must not be too short; in any case, with very short pulses, the range-tracking, early-late gate must be much wider than the pulsewidth (Fig. 3.22).

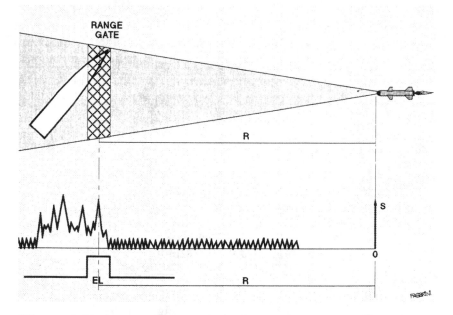

Figure 3.22 A very narrow radar pulse can cause the active seeker to lock onto a point of the ship which is not very vulnerable.

3.4 Passive anti-radiation missiles

The *anti-radiation missile* (ARM) is capable of homing with high precision onto a radar, guided by the radiation from that radar. This is accomplished by means of a passive seeker, in practice very similar to a small ESM system, capable of extracting the necessary angular data from the emissions of the victim radar [23, 24].

ARMs have great importance in the operational theater; their acknowledged presence will arouse many doubts about use of radar in the enemy's mind. ARMs are frequently air-to-surface missiles installed on board aircraft dedicated to the *suppression of enemy air defence* (SEAD) (Fig. 3.23).

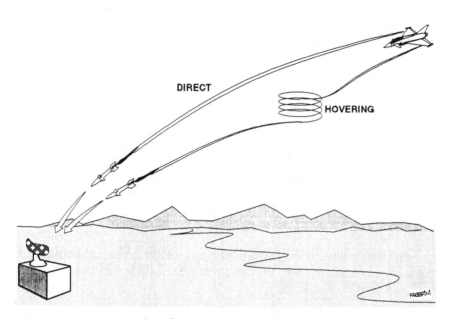

Figure 3.23 An *anti-radiation missile* (ARM) homes on to the radiation generated by the target radar. It can be of either direct or hovering type, to compensate for periods of suppressed transmission.

Usually the ESM system on board the aircraft intercepts, identifies and locates the victim radar, and designates it to the ARM seeker by means of parameters such as PW, *pulse repetition interval* (PRI), frequency, and possibly real-time gate, that is, the time gate in which the arrival of the radar pulses is expected. The ARM receiver "closes" onto the information received, and when the seeker senses the designated emission, neglecting all the rest, it locks onto the signals of the victim radar, sending the "tell-back" signal to the pilot who can now arrange for the launching.

From the signals, the ARM receiver deduces the angular errors needed for guidance, which takes place exactly as in the case of missiles with active guidance. Once launched, the missile does not require assistance. In order to protect itself, the victim radar can stop transmitting, but this does not mean assured survival since the missile can go on tracking on the basis of memorized coordinates. The radar may alternatively seek the help of a fire control system to have the missile destroyed.

Many ARMs have been developed; they now cover the range of bands from 0.5 to 18 GHz. The primary objective of an ARM is to prevent the use of radar; if a radar cannot operate, it becomes completely ineffective. The ARM can represent our ECM technique not only active, but also of the "hard kill" type, since it may lead to the radar being destroyed. The ARM is a typical example of a *standoff* weapon; it can be launched from a great distance, which means that the aircraft does not need to come dangerously close to the target. Figure 3.24 shows a block diagram of an ARM and of its seeker.

The angular precision required is on the order of one degree, and must in any case be compatible with the miss distance needed for the warhead to be effective. The data rate is on the order of a few tens of hertz.

The receiver can be tuned to a very broad band, thus covering most of the known types of radar. However, there are some limits to the choice of the band: downward (0.5 to 1 GHz), because the dimensions available for the antenna do not allow for good *direction finding* (DF) as the slope of the angular gradient is insufficient for the wide beamwidth resulting from the limited antenna aperture; upwards, the limit is fixed by the fact that, for the moment, there are no radars in the millimeter band which are of tactical significance.

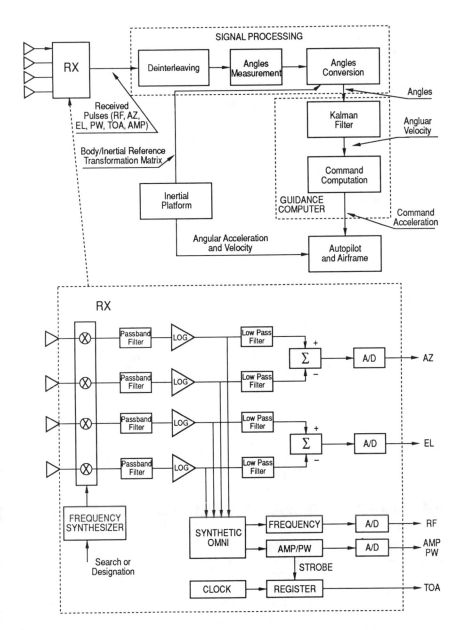

Figure 3.24 Block diagram of an ARM, and of its seeker.

3.5 Laser weapon systems

3.5.1 The laser

Weapon systems using a laser as the optical source for their aiming have been under development for the last twenty years.

LASER is an acronym for light amplification by stimulated emission of radiation, which describes concisely and effectively the operating principle on which this family of devices is based.

The characteristics which distinguish sources of this type from all other optical sources are:

- high coherence in space and time.
- high energy density, of the order of joules per square centimeter.
- high monochromaticity.
- high directionality.

These are based on the phenomenon known as "'stimulated emission," which will now be briefly described.

The energy in the outer electron bands of an atom of a gas can assume only discrete values. If an electron in an atom undergoes a transition to a higher energy state E_1, it will sooner or later return spontaneously to the ground or lower state E_0, emitting a photon whose energy is $E_f = h\nu = E_1 - E_0$ (radiative emission) (see section 2.3.1.2).

If several atoms are in the excited state E_1, the various emission events will, in the absence of any synchronization mechanism, be uncorrelated. This is the phenomenon of spontaneous emission on which traditional light sources operate.

Conversely, an atom in the ground state can absorb a photon of energy E_f, jumping to the excited state E_1; this phenomenon is called *resonance absorption*.

However, it can happen also that a photon interacts with an atom that is already in the excited state E_1, stimulating its decay; the stimulated atom will emit an additional photon with the same phase and, obviously, the same frequency as the stimulating photon. This is stimulated emission.

If there are N atoms, the ratio of probability of resonance absorption to probability of stimulated emission will be equal to the ratio of number n_0 of atoms in the ground state E_0 to number $n_1 = N - n_0$ of atoms in the excited state E_1.

In a thermal equilibrium situation, n_0 is higher than n_1, with

$$\frac{n_1}{n_0} = \exp[-(E_1 - E_0)/kT)]$$

and, therefore, resonance absorption is preferred to stimulated emission.

However it is possible, by providing energy to the system from the outside through a process known as "pumping," to create a condition, "population inversion," where n_1 is higher than n_0; that is, the number of photons produced is higher than the number of photons absorbed.

The amplification of the number of photons with the required frequency is attained by means of a resonant optical structure, called a "cavity": for example, the mirror cavity of a Fabry-Perot configuration, on which depends the optical feed-back capable of sustaining laser oscillations.

A laser consists of the following fundamental elements:

- The laser material, or active medium; for example, in solid-state laser, a rod made of ruby, of Nd:YAG (neodymium:yttrium aluminium garnet), or of doped glass. In gas laser, a volume of gas such as argon or carbon dioxide.
- The pump; for example a flashlamp, capable, when activated by a high energy pulse from a power supply, of emitting an intense light pulse that excites the atoms of the laser active medium.
- The resonant cavity; this can consist of two highly reflective mirrors, separated by a suitable distance, capable of selecting photons with the required energy, and of sustaining excitation.

Because of the properties of laser light, beams with minimum divergence, on the order of seconds of arc, and no sidelobes can be obtained. They are therefore able to preserve laser energy and to transport it over very long distances.

Laser energy can be controlled both in time of emission, with the attainment of high peak powers, up to a few megawatts, for times on the order of nanoseconds, and in repetition rate, up to a few kilohertz. It is also possible to construct CW lasers capable of emitting up to a few hundred kilowatts.

Finally it should be noted that, because of laser monochromaticity, it is possible to construct laser light receivers of high sensitivity and reduced size by the use of very narrowband optical filters.

3.5.2 The laser equation

The range of a laser system can be easily calculated by considering the signal power and noise power attainable at the output of the laser receiver. The same considerations as for radar range, concerning the number of integrated pulses, integration losses, SNR, P_d, and P_{fa} apply.

If the target is smaller than the cross section of the laser beam, the received laser signal power is given by

$$\varphi_0^2 R_1^2 \exp(-\gamma_1 R_1)\rho A_B \cos \vartheta_L \frac{1}{\pi R_2^2} \exp(-\gamma_2 R_2)A_r \tau_0$$

while if the target is larger than the cross section of the laser beam, it is given by

$$S = P_T \exp(-\gamma_1 R_1)\rho \cos \vartheta_L \frac{1}{\pi R_2^2} \exp(-\gamma_2 R_2)A_R \tau_0$$

where symbols with subscript 1 refer to the laser transmitter-to-target range and those with subscript 2 refer to the target-to-laser receiver range. If transmitter and receiver are colocated, the corresponding values are equal. The symbols have the following meanings: P_T is the transmitted peak power, φ_0 is the semiaperture of the laser beam, γ is the atmospheric attenuation coefficient, ρ is the reflectance of the target, A_B is the surface area of the target, ϑ_L is the Lambert angle, A_R is the surface area of the optics of the receiver, and τ_0 are the losses in the optics.

The useful signal and the SNR can be obtained, by analogy with the IR case, on multiplying the power of the received signal S by the sensor responsivity expressed in terms of D^*:

$$V_s = SR = \frac{SD^*V_n}{(A_d\Delta f)^{\frac{1}{2}}}$$

whence

$$\frac{V_s}{V_n} = \frac{SD^*V_n}{(A_d\Delta f)^{\frac{1}{2}}}$$

Once the value of V_s/V_n has been fixed to obtain the desired P_d and P_{fa}, it is possible to solve the equation with respect to R, and thus to find the laser range.

3.5.3 Laser applications

Laser devices have been employed mainly as essential components of some weapon systems, to perform one of the following functions:

Range measurement (laser range finder). As in a conventional radar, this measurement is based on the time elapsed between the emission of a laser pulse toward a target and the return of the reflected echo. Thanks to its high accuracy, use of a laser range finder permits a significant increase in the effectiveness of the fire control systems of tanks, anti-aircraft batteries and missile systems.

Lasers for this application typically operate at a wavelength of 1.06 m, transmitting pulses whose duration is a few nanoseconds and whose peak power is a few megawatts, with repetition frequency of 1 to 20 Hz.

Target illumination (laser target designator). In this application, the task of the laser is to illuminate a target so as to enable a missile, a bomb, or a projectile, equipped with a suitable laser receiver, to home onto it.

Typically, these lasers transmit pulses at a wavelength of 1.06 m, with peak power up to 10 MW, and with repetition frequency of a few tens of hertz, using sequential codes.

Among the systems using this technique are, laser-guided GBU 16 bombs, guided Copperhead projectiles (155 mm), the Hellfire modular missile (anti-tank missile), and the laser Maverick missile.

Illumination of an attack route. In this application the task of the laser is to provide a missile with guidance for a pre-established attack mode, exploiting techniques of the beam-riding type. A special receiver on board the missile ensures that the missile is constantly in the laser beam.

Directed energy weapon. A power laser can be used to damage directly the most vulnerable parts of a target, normally a sensor. This technique, which is not as yet a full-fledged one, is suitable for space applications such as killer satellites, and, in the future, for building powerful directed energy weapons.

The performance of laser equipment is strongly conditioned by the following factors:

- Atmospheric attenuation. The dependence of propagation on the atmosphere markedly constrains laser all-weather performance. Frequently atmospheric attenuation and scintillation cannot be countered by an increase in transmitted power.
- Low efficiency. Unfortunately, in most cases generation of laser light is a low efficiency process of only a few percent. Therefore the generation of high power entails enormous problems of heat dissipation.

Because of these limitations, lasers are normally employed in short-range (6 to 10 km) and very-short-range (3 to 6 km) weapon systems.

3.6 Stealth aircraft

A new weapon system, capable of remarkably degrading the performance of the radar air defense line, is the *stealth aircraft* (Fig. 3.26). Its manufacture depends on a series of technological devices that allow it to attain an extremely low RCS, so low that stealth aircraft are said to be invisible.

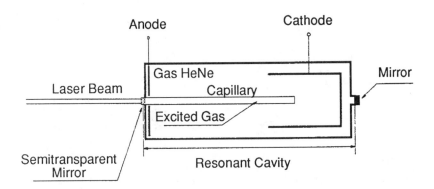

Figure 3.25 Schematic diagram of a laser. The energy transferred between cathode and anode causes a discharge in the gas in the capillary tube, raising electrons to a higher energy state. The resonant cavity, which is the space between the two mirrors, selects and sustains the desired radiation.

Stealth technology is important because the RCS reduction, compared to more traditional methods, can reach factors of 1000 or more, with a consequent reduction of radar range to 17% or less.

This technology is based on several fundamental principles:

(1) The use of *radar-absorbent material* (RAM) to coat those metal parts that would produce substantial scattering, such as the joining edges of turbine inlets.

(2) The use of synthetic materials transparent to microwaves $(\mu\epsilon \simeq \mu_0\epsilon_0)$, such as carbon fiber, to manufacture large surfaces like the wings.

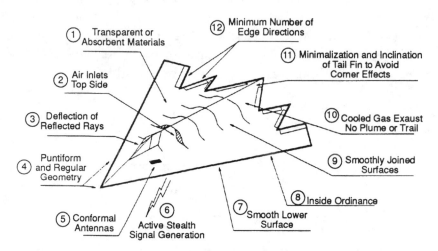

Figure 3.26 Stealth aircraft are practically undetectable by sensors. They exploit the techniques listed in the diagram to minimize scattered and reflected signals, and to focus the residuals in few directions, different from that of the sensors.

(3) The use of geometries producing minimum scattering, concentrated in a few directions, away from the direction of incidence. For example, minimally inclined edges, absence of dihedrals that could behave like corner reflectors, almost perfect joints between the various surfaces.

(4) The use of mirror-like materials to avoid corner reflector effects; for example, insertion of extremely thin metal threads in the transparent material of the canopy.

(5) Active systems for RCS reduction.

(6) Reduction of the IR signature and masking of gas exhausts by installing them on the top of the fuselage.

(7) Incorporation of ordnance and counter measures fit inside the airframe.

The drastic reduction of radar range resulting from stealth technology poses serious problems to air defense. The problem is not only that of the reduction of the signal, which can be solved by an increase in the sensitivity of the receiver, but, as the RCS is very small, it is less and less distinguishable from that of insects and birds.

Where there is also clutter, MTI devices must increase their improvement factor by 30 dB to detect so small a target, which may be impossible. Moreover, since lines of defense usually overlap, the range reduction resulting from a reduction of RCS leads to the creation of "holes" through which stealth aircraft can penetrate undisturbed until they are close to their targets (Fig. 3.27). To restore the continuity of the line of defense, the number of defensive radars would have to be increased by a factor of 4 or 5. To have the same coverage as before, the number would have to be increased by a factor of 16 or 25.

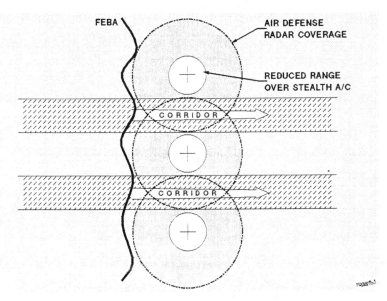

Figure 3.27 The low RCS presented by a stealth aircraft reduces the radar coverage, thus creating serious gaps in the air defense network.

The low RCS of stealth aircraft also makes their self defense easier as lower ERP jammers are required.

3.7 Communication systems

The very existence of military organizations and their need to perform missions in time of peace or war call immediately to mind the concepts of "command" and "control." Command and control functions require that orders, information, data, and so forth, be communicated.

Military operations often require the movement of a massive number of vehicles and units of the various armed forces. It is unthinkable that success could be achieved unless all actions are closely coordinated. Coordination is necessary not only before, but also during operations, when there invariably occur unforseen events which require changes of plan.

History has been the witness, on the one hand, to an impressive increase in the mobility of the forces engaged on the battlefield and, on the other, to a "globalization" of operations. Almost always modern battles involve several armed forces (ground, naval and air forces) of many nations acting in an extremely wide and interconnected scenario.

Military communications were bound to follow this development, moving from the simple communications networks of World War II, which consisted of a few nodes and some rather simple devices such as telegraph, teletype and radio, to the extremely complex, diversified and automated networks of today, almost always organized on an interforce and, at times, supranational basis, for example, the NATO communications network.

3.7.1 Networks

Communications networks enable the military apparatus to operate. Information is received at headquarters, and orders are issued to lower command levels and to operational units. Information, on which decisions can be based and orders originated, arrives in real time at centers for *command, control, and communications* (C^3 or, since *intelligence* operations are also often involved, $C^3 I$).

In the past, when the movements of armed forces were slow, communications networks could be organized in an axial or pyramid-like pattern (Fig. 3.28). Today it is unthinkable that networks should be connected only by means of single-link paths. Much more effective are the matrix or grid patterns in which each node can be reached by several different paths. Thus, overall connection between the users of the network is ensured, even if some of the redundant links fail or are severed.

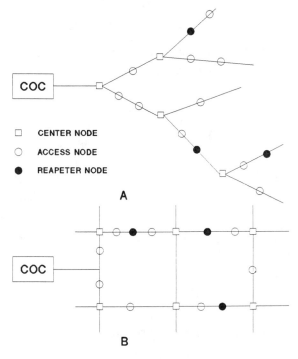

Figure 3.28 Unlike a pyramidal network, a matrix network allows access to each node by several different pathways, thus ensuring the necessary redundancy.

There are several types of networks:

- fixed multichannel networks for infrastructural or strategic use.
- fixed single-channel HF networks.
- mobile multichannel networks for tactical use.

- mobile single-channel networks for tactical use.
- time-sharing networks, shared by all users, for example, JTIDS (section 3.7.4).

Networks are by their nature very complex and generally very extended. From headquarters, through the various command levels, they must reach the different operational units, which may be located at short or long range, in space, on the ground, and even, in the case of submarines, under the surface of the sea.

According to the particular section of the network concerned, different techniques and solutions will be required in order to transmit the messages.

3.7.2 Types of transmission (links)

There are many types of communication systems. An effective classification rests on the basis of the type of link that has to be established, which almost automatically entails a precise choice of the electromagnetic frequency to be used [26].

In general, the possible links are the following:

- tactical, sometimes strategic, long-range links; the frequencies used are from 1.5 to 30 MHz (HF).
- tactical ground links; frequencies between 30 and 300 MHz (VHF), sometimes also HF.
- tactical ground-to-air and air-to-air links and radio relay systems; frequencies between 370 and 3000 MHz (V-UHF).
- microwave multichannel radio relay links.
- tropospheric scatter links.
- satellite links.
- fiber-optic links for local area networks.
- links with submerged platforms.

3.7.2.1 Transmissions in the hf band

These transmissions are used essentially for tactical and strategic long-range communications, often as an emergency reserve. The frequency range is normally between 1.5 and 30 MHz.

Messages can be transmitted either by analog voice (baseband 300 to 3100 Hz) or by digital data (up to 1.2 kbps) link. Message transmission by radio-teletype and Morse telegraphy are both still widely used.

In the HF band propagation occurs by direct wave and ground wave over short distances on the order of 100 km, and by sky wave reflected from the ionosphere, especially during daylight, over very long distances, from 100 to 1000 km and more (Fig. 3.29).

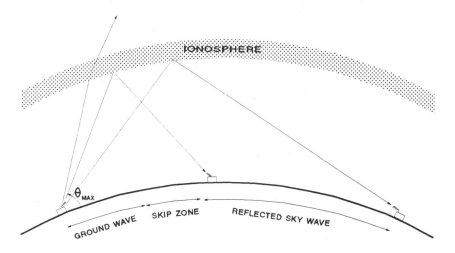

Figure 3.29 HF links can take place by ground waves or, over long distances, by sky-wave reflection from the ionosphere.

In order to exploit the ionosphere, the antenna must be tilted upward, with the reflection characteristics of the ionosphere creating an area close to the transmitter where no links are possible (skip zone). Reflection by the ionosphere takes place only below 30 MHz.

Modern radio link equipment is run by computers. These computers perform the analysis and optimized automatic selection of transmit-receive parameters (power, frequency, etc.), in order to ensure good link quality during daylight hours, a function previously performed by expert operators. Modern devices also digitize messages, which allows encryption, to render them unintelligible to the enemy, and speeding up of the transmission by suitable formatting and the use of a modem. The use of suitable codes also permits the automatic correction of errors.

3.7.2.2 Transmissions in the vhf band

These transmissions occupy the band between 30 and 300 MHz and are almost exclusively for tactical use, at distances up to about 50 km. The frequency range exploited in practice is between 30 and 88 MHz.

In the most recent equipment, a 25-kHz channel is used both for digital voice transmission (16 kbps), often by means of devices introducing pseudorandom codes (encryption), and for data transmission (up to 4.8 kbps). In data transmission, a *forward error correction* (FEC) code is often used. *Combat net radio* (CNR) is generally realized with this type of transmission. To reduce deleterious effects of intentional jammers, the following ECCM techniques are used:

- direct sequence: transmission of data embedded in noise-like signals.
- frequency hopping: switching from one to another of a great many channels, each occupied for an extremely short time, on the order of milliseconds.
- hybrid techniques, for example, combinations of the two preceding techniques.

3.7.2.3 Meteor-burst transmissions

A particular case of VHF transmission is *meteor-burst* transmission, so-called because it exploits the ionization of the upper atmosphere, at 85 to 120 km, produced by the billions of micrometeorites that cross it daily.

In this case, transmitter powers of up to 1000 W are needed. Because it is highly probable that the link will be maintained for up to 1.5 seconds, tranmission and reception are controlled by a computer which, by means of a "hand-shake," or exchange of data sequences, guarantees the security of the link. For this reason messages are divided into "packets" and transmitted at high speed. This type of transmission takes place between stations as far as 2000 km apart.

3.7.2.4 Transmissions in the uhf band

The frequency range occupied by the UHF band goes from 300 to 3000 MHz. This band is especially used for multichannel radio relay systems, up to 1850 MHz, ground-to-air links, and air-to-air links for tactical communications, from 225 to 400 MHz.

Radio relay systems can also be very-long-range links using various intermediate relay stations for transmission and reception. Their principal characteristic is the grouping of intermediate stations into "node centers" where channels can be switched from one line to another.

In this band there are also *time-division multiple-access* (TDMA) systems, which are time-division data-transmission networks where each user can enter the network at a precise moment, and there can be many users (JTIDS is one of these; see Section 3.7.4).

3.7.2.5 Microwave transmissions

Frequency ranges for microwave transmission [28] go mainly from 4.5 to 15.5 GHz, and will soon make much more use of millimetric wavelengths up to 70 GHz.

The links are of high capacity and are used essentially for territorial and interforce strategic networks, but also for tactical short-range links, up to a maximum of 30 km. Covert communications are also often centerd on regions of high atmospheric obsorption in the atmosphere.

3.7.2.6 Links by tropospheric scatter

Whenever for geographical or political reasons, or because of time limitations, intermediate relay stations cannot be installed, recourse is had to a particular mode of transmission

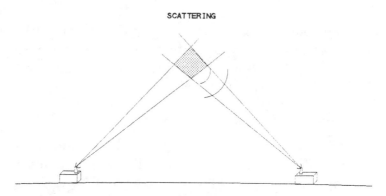

Figure 3.30 Tropospheric-scatter link.

that exploits upper atmosphere scattering. The scattering behaves as a source that can be received by a far-away receiver (Fig. 3.30).

Typical transmitter powers are on the order of 10 kW, and the frequency bands used are between 450 MHz and 5 or 6 GHz. These frequency bands permit the realization of links with high capacity, for example when differential *quaternary phase-shift keying* (QPSK) is used for modulation. The distance covered by a link depends on the altitude at which the scattering phenomenon occurs; it can be up to 1000 km. However, both the efficiency and the quality of the link are very low.

The equipment needed for this type of link is generally extremely cumbersome because of both the magnitude of the power required and the size of the antennas.

3.7.2.7 Satellite links

In the military world, as in the civilian, satellite links are being developed [29]. This type of link allows realization of redundant strategic and tactical networks with very long range and wide coverage, very resistant to countermeasures, and with *low probability of intercept* (LPI) characteristics. All this is achieved at very high frequencies, with antennas having particularly low sidelobes, and processing on board the satellites themselves.

An example of a satellite system for military telecommunications is MILSTAR (section 3.7.4), which uses a frequency band around 44 GHz for the up link and around 20 GHz for the down link.

3.7.2.8 Fiber-optic links

Optical fibers can be used when there is the need for local links of extremely high capacity, such as those forming local area networks, of which the data bus on board a ship or an aircraft is an example. They can also be used when it is necessary to lay a temporary line quickly in an operational environment [30]. Fiber optics allow good propagation even of incoherent light; in this case the attenuation is about 0.2 dB/km. Moreover, in view of the fact that it is also easy to obtain on-off light modulators with bands above a gigahertz, very broadband links (higher than 1 Gbps) have been realized.

Compared to other more traditional links, the fiber-optic link offers the advantage of being insensitive to the *electromagnetic pulse* (EMP) generated by nuclear explosions; it also has the characteristic of being a "safe" link, since it does not emit electromagnetic radiation that could be intercepted (TEMPEST).

The major disadvantage of the fiber-optic link is that a very thin cable has to be laid, which may be broken by tanks, bomb explosions, and so forth, and restricts the mobility of the units which it connects.

3.7.2.9 Links with submerged platforms

One of the greatest advantages of the submarine is that it is extremely difficult to detect when submerged. On the other hand, submarines need to exchange messages with their headquarters; therefore, they are confronted with the problem of establishing a link without the need to surface.

Electromagnetic waves propagate with difficulty in water. The signal power decreases exponentially with depth and is inversely proportional to the square root of the frequency. Therefore, to establish a link one has to resort to very low frequencies

(Fig. 3.31). At very low frequencies (VLF: 3 to 30 kHz) it is possible to establish a link with submarines by means of a loop antenna or a wire antenna, which is in practice a very long wire kept afloat by a buoy.

A loop antenna must generally be maintained near the surface, so that navigation becomes unsafe because of the risk of collision with a ship. A wire antenna requires a very long wire (more than 100 m) and must be kept more or less orthogonal to the direction of arrival of the electromagnetic waves.

The use of extremely low frequencies (ELF: 3 to 30 Hz, and below) allows the establishment of a link with submarines at greater depths; but at low frequencies the transmission of messages is very slow.

The depths which can be reached are 25–80 m at 100 Hz, 80–240 m at 10 Hz, and 280–780 m at 1 Hz.

Laser communications can overcome the problems presented by communications at very low frequencies [31]. Green-blue laser light ($f = 625$ THz) propagates to a depth comparable with that attainable at frequencies lower than 10 Hz. This makes it possible, for example, to establish a very high capacity link between an artificial satellite and a submarine.

3.7.3 The message

The information exchanged between the users of a communications network represents the message. The messages that can be exchanged are fundamentally of two types:

(a) Voice messages, which can be transmitted either in clear, or encoded so that they are intelligible only to the intended receiver.

(b) Data messages, which can be more or less complex according to the type of data to be transmitted. If the data concern services, they can be transmitted in a very narrow band; on the other hand, if a large amount of data has to be transmitted in a very short time, the band must be very broad.

The message is usually characterized by the type of modulation and the band occupied.

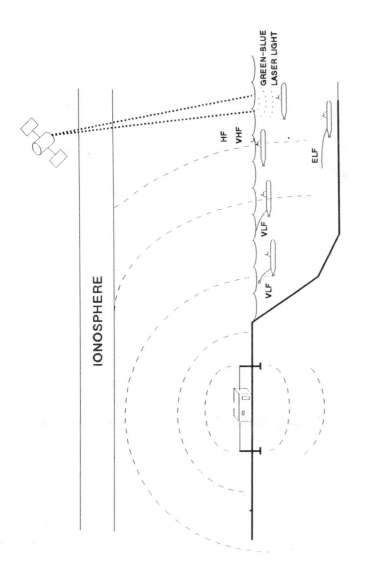

Figure 3.31 Links with submerged platforms.

In civilian usage, the bandwidth B_c occupied by a channel is as narrow as possible (according to the CCIR standard, a voice communication channel may occupy a 4000 Hz band), so that over a link characterized by a passband B a maximum number N of messages can be transmitted simultaneously, resulting

$$N = \frac{B}{B_c}$$

This does not in general apply to military communications, where in order to reduce the probability of interception, the tendency is to occupy a bandwidth B_c wider than the strictly necessary one.

Figure 3.32 shows two typical links for voice and digital data transmissions. The *modem* (modulator-demodulator) is responsible for the type of modulation impressed on the signal to adapt it to the transmission system.

The forms of modulation most used are:
- *amplitude-shift keying (ASK)*, in which the amplitude can assume m values; a special case is *on-off keying* (OOK), in which the amplitude is of the on-off type (Fig. 3.33).
- *frequency-shift keying (FSK)*, for transmissions up to 9600 bits/s (Fig. 3.33).
- *phase-shift keying (PSK)*, for transmissions up to $19,200$ bits/s (Fig. 3.33).

In FSK modulation, in order to transmit a 1 or a 0, the frequency is shifted from its nominal value in such a way that

$$f_1 = f_p - \Delta = 0$$

$$f_2 = f_p + \Delta = 1$$

In general, the frequencies used can assume m discrete values.

In the PSK system, on the other hand, the frequency is always the same; in order to transmit a 1 or a 0, the phase of

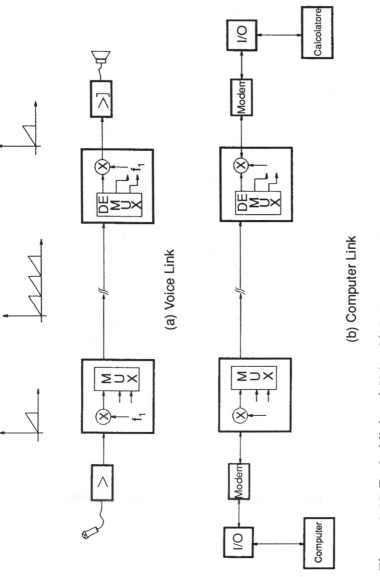

(a) Voice Link

(b) Computer Link

Figure 3.32 Typical links: a) Voice; b) Between computers.

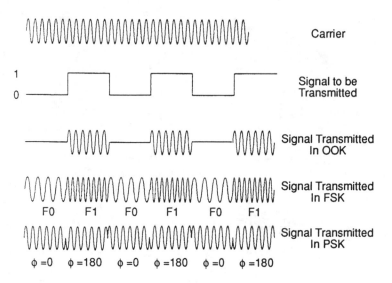

Figure 3.33 Principal modulations used in digital data transmission.

the transmitted signal is shifted. More than one phase shift can be used, namely, 0 to 180°, 0 to 90°, 90° to 180°, 180° to 270°, and so forth.

For more details on the preparation of messages, the reader is referred to the specialized literature [32].

3.7.4 Examples of communication systems

For ease of reference, a few communication systems are mentioned below.

Military strategic-tactical and relay satellite (MILSTAR). A satellite network for strategic communications, used by military command and control on a worldwide scale, which provides relay stations for early warning data, diplomatic communications and crisis management.

Eight geosynchronous satellites give the system very high survivability; the use of EHF, with an up link at 44 GHz and a down link at 20 GHz, makes it highly immune to intentional jamming.

Joint tactical information distribution system (JTIDS). JTIDS is a communications network for both voice and data transmission of the TDMA type; it is highly resistant to countermeasures. The capacity of the system is such that it permits command and control operations for many users, including ships, aircraft, and patrols, which may be widely distributed over an area with a radius of hundreds of kilometers.

A JTIDS network is characterized by a pseudorandom sequence that determines the transmission, that is, the hopping frequency and the spread-spectrum pseudonoise waveform. All users of the same network employ the same pseudorandom sequence.

A JTIDS system may consist of up to 128 networks. In each network a user can be both the sender and the receiver of messages and data, and can also operate as a relay for communication between other users. As a consequence, a JTIDS network remains active as long as there are at least two users.

A JTIDS network is organized in the following way (Fig. 3.34). The 24 hours of a day are subdivided into 112 cycles each 12.8 minutes long. Each cycle is divided into 64 frames of 12 s each and each frame consists of 1536 time slots whose duration is 7.8125 ms. Depending on the operational role, each user has available one or more slots in which a message can be transmitted, while for reception all the other slots of the frame are available. The user may select the information of interest.

Since a cycle is divided into 98,304 time slots, in the limiting case where each user were allowed only one slot, a JTIDS network could be used by over 98,000 users. In practice, however, the number of users will be much smaller.

Each message must include data concerning the identity, location, status, and so forth of the user, which have to be submitted to the network.

Each slot is in its turn divided in the following way: A synchronization sequence, an identification band, the message itself in coded form, and finally a guard band to allow time for the propagation of the message.

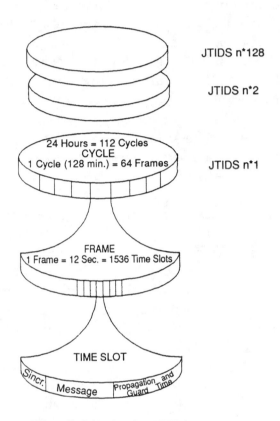

Figure 3.34 Time division in the JTIDS.

A message may contain 225 or 450 bits, according to whether detection and automatic correction of code errors is activated or not. Messages can be highly structured and formatted; in such a case, a few bits may carry a large quantity of information.

By measurement of the delay times between messages exchanged with other users, it is possible, with the help of the synchronization sequence, to obtain location data, so that the JTIDS can be used as a relative navigation system.

The operating frequencies of a JTIDS are between 965 and 1215 MHz, with the exception of the area around 1030 and

1090 MHz, where IFF operates.

Position location reporting system (PLRS). A UHF communications network to report automatically to a center the location of roughly 400 units. This system can be used by the army together with JTIDS.

Global positioning system (GPS). A satellite system (NAVSTAR) which enables users to determine their location with great accuracy.

To sum up, it appears that the electromagnetic spectrum is used by communication systems in the following way:

- ELF 3 to 30 Hz: communication with submarines at great depth.
- SLF 30 to 300 Hz: communication with submarines at great depth.
- ULF 0.3 to 3 kHz: communication with submarines.
- VLF 3 to 30 kHz: communication with submarines.
- LF 30 to 300 kHz: communication by reflection from the atmosphere.
- MF 0.3 to 3 MHz: long-range strategic communications.
- HF 3 to 30 MHz: off-line-of-sight tactical ground communications.
- VHF 30 to 300 MHz: ground-to-air and line-of-sight tactical ground communications.
- UHF 0.3 to 3 GHz: satellite and tactical ground-to-air communications.
- SHF 3 to 30 GHz: satellite communications.
- EHF 30 to 300 GHz: satellite communications.
- 625 T Hz (blue-green laser): communication with submarines.

REFERENCES

[1] W. Wrigley, *Encyclopedia of Fire Control*, Vol. 1, Fire Control Principles Instrumentation Laboratory, Cambridge, MA: MIT, 1957.

[2] J.L. Farrel and E.C. Quesimberry, "Track Mechanization Alternatives," *IEEE Proceedings*, Vol. 2, NAECON, 1981, pp. 596-602.

[3] W.E. Kolbl, "Fire Control Systems for Main Battle Tanks," *Miltech* 21.

[4] [Ed.] *Handbook of Weaponry*, Düsseldorf: Rheinmetall, 1982, pp. 59-60 and 199-227.

[5] R.E. Ball, *The Fundamentals of Aircraft Combat Survivability: Analysis and Design*, New York: AIAA Inc., AIAA Education Series, 1985, pp. 188-191.

[6] M. Held, "War-Heads for SAM Systems," AGARD Lectures Series, No 135: Advanced Technology for SAM Systems. Analysis Synthesis and Simulation.

[7] F. Berg Russel, "Estimation and Prediction for Maneuvering Target Trajectories," *IEEE Transactions on Automatic Control*, Vol. AC-38, No 3, March 1983.

[8] S.S. Chin, *Missile Configuration Design*, New York: McGraw-Hill, 1961.

[9] J.J. Jerger, *System Preliminary Design*, Princeton: D. Van Nostrand, 1960.

[10] A.S. Locke, *Guidance*, Princeton: D. Van Nostrand, 1955.

[11] J.H. Blakelock, *Automatic Control of Aircraft and Missiles*, New York: John Wiley & Sons, 1965.

[12] P. Garnell and D.J. East, *Guided Weapon Control Systems*, Oxford: Pergamon Press, 1977.

[13] A. Ivanov, "Improved Radar Design Outwits Complex Threats," *Microwave Journal*, Vol. 15, No. 4, April 1976, pp. 36-38, 40-42, 47-48, 50 and 53.

[14] M.W. Fossier, "The Development of Radar Homing Missiles," *Journal of Guidance, Control and Dynamics*, Vol. 7, No. 6, November-December 1984, p. 641.

[15] A. Ivanov, "Semi-Active Radar Guidance," *Microwave Journal*, September 1983.

[16] F.W. Nesline and P. Zarchan, "A New Look at Classical vs. Modern Homing Missile Guidance," AIAA 79-1727 R.

[17] P. Zarchan and F.W. Nesline, "Miss Distance Dynamics in Homing Missiles," Seattle, WA; AIAA Guidance and Control Conference, August 1984.

[18] F.W. Nesline, "Missile Guidance for Low Altitude Air Defense," AIAA, pp. 78-1317.

[19] F.W. Nesline and P. Zarchan, "Missile Guidance Design Trade-Off for High-Altitude Air Defense," *Journal of Guidance*, Vol. 6, No 3, May-June 1983.

[20] F.W. Nesline and M.L. Nesline, "An Analysis of Optimal Command Guidance vs Optimal Semi-active Homing Missile Guidance," Bedford, MA; Raytheon Company, Missile Systems Division, M3-55.

[21] D.R. Carey and W. Evans, "The Patriot Radar in Tactical Air Defense," *Microwave Journal*, May 1987.

[22] J.J. May and M.E. Van Lee, "Electro-Optic and Infrared Sensors," *Microwave Journal*, September 1983.

[23] R. McLendon and C. Turner, "Broad Band Sensors for Lethal Defense Suppression," *Microwave Journal*, September 1983.

[24] J.A. Mosko, "An Introduction to Wide Band Two Channel Direction-Finding Systems," *Microwave Journal*, February-March 1984.

[25] W.M. Mannel, "Future Communications Concepts in Support of U.S. Army Command and Control," *IEEE Transactions on Communications*, Vol. Com-28, No. 9, September 1980.

[26] P.G. Fontolliet, *Telecommunication Systems*, Chapter 3, Norwood, MA; Artech House, 1986.

[27] J.D. Oetting, "An Analysis of Meteor Burst Communications for Military Applications,"*IEEE Transactions on Communications*, Vol. Com-28, No. 9, September 1980.

[28] P.G. Fontolliet, *Telecommunication Systems*, Chapter 12,

Norwood, MA; Artech House, 1986.

[29] A.D. Dayton and P.C. Jain, "Milsatcom Architecture," *IEEE Transactions on Communications*, Vol. Com-28, No. 9, September 1980.

[30] P.G. Fontolliet, *Telecommunication Systems*, Chapter 14, Norwood, MA; Artech House, 1986.

[31] T.F. Wiener and S. Karp, "The Role of Blue/Green Laser Systems in Strategic Submarine Communications," *IEEE Transactions on Communications*, Vol. Com-28, No. 9, September 1980.

[32] P.G. Fontolliet, *Telecommunication Systems*, Chapters 4 and 8, Norwood, MA; Artech House, 1986.

[33] D.B. Brick and F.W. Ellersick, "Future Air Force Tactical Communications," *IEEE Transactions on Communications*, Vol. Com-28, No. 9, September 1980.

Chapter 4

Electronic Intercept Systems

4.1 Introduction

Electronic defense intercept systems [1] are used for timely detection of the presence in the operational scenario of one or more of the weapon systems described in the preceding chapter. In this chapter the following will be analyzed:

- *radar warning receivers (RWR)*, used to detect a threatening radar before it is able to give firing instructions to its associated weapons.
- *electronic support measures (ESM)* systems, used to detect the presence of an enemy platform in the intercepted electromagnetic scenario before it has had time to detect the defended platform.

- *electronic intelligence (ELINT)* systems, used to gather strategic data drawn from deep inside the territories of potentially hostile nations.
- *infrared warning receivers (IRWR)*, used to detect the presence of enemy platforms by their infrared emissions.
- *laser warning receivers*, used to detect the presence of a laser illuminator or laser range finder.
- *communications intercept* systems, used to intercept and locate the enemy emissions for tactical purposes and communication intelligence (COMINT) equipment used to discover strategic information concerning enemy communications systems.

4.2 The equation of a passive system

The first requirement of an intercept device is adequate sensitivity. In order to evaluate the degree of sensitivity needed by such a device [2, 3], the following equation, which expresses the power of a radar signal intercepted by an antenna with gain G, may be used

$$s = \frac{P_T G_T G \lambda^2}{(4\pi)^2 R^2 L_p} F_p^2$$

where F_p is the propagation factor, which in free space is equal to unity, and L_p are polarization losses arising from the fact that since the radar polarization is not known *a priori*, the intercept system will need antennas capable of receiving all polarizations, and is therefore not always perfectly matched to that of particular emitters.

Let the sensitivity of the intercept system be s_0. This is the operating sensitivity, that is, the sensitivity for which the signal-to-noise ratio (SNR) is such as to ensure signal detection and measurement with suitable precision. The range R_i of the intercept system can then be expressed [3]

$$R_i = \left[\frac{P_T G_T G \lambda^2}{(4\pi)^2 s_0 L_p} F_p^2 \right]^{\frac{1}{2}}$$

In general, the receivers of intercept systems are not matched to the pulse width of every received signal, but rather to the shortest pulse width to be detected. An exception is the detection of CW signals, which is performed in a narrow band.

In the case of a radar warning receiver, the requirement is that radar signal detection and identification occurs at a range in excess of the range of the associated weapon system radar.

In the case of ESM systems, the requirement is that the detection of the radar electromagnetic emissions must be achieved before the radar, in its turn, has detected the platform on which the ESM system is installed.

The ratio of the range at which the ESM can detect the radar to the range at which the radar detects the platform, is known as *range advance factor* (RAF), a parameter widely used for quick verification of the capability of the equipment to perform the desired role.

Often, and mistakenly, in computing RAFs only the radar *estimated range power* (ERP), that is, peak power multiplied by antenna gain, is considered, while the length of the transmitted pulse is neglected. However, it is essential to consider the radar waveform, because sometimes, especially in the case of pulse-compression radar, one might find that the RAF magnitude is less than unity.

It should be recalled that radar range is not determined by the peak power, but rather by the energy falling on the target during the time on target.

Recalling that the radar range is given by

$$R_R = \left[\frac{N_i P_T n\tau G_T G_R \sigma \lambda^2}{(4\pi)^3 KT F(S/N)_{\text{Pdfa}} L} F_{pt}^4 \right]^{\frac{1}{4}}$$

range and all other conditions being equal, a radar designer can act on the product $P_T n\tau$ and make P_T small enough so that RAF$\langle 1$ [2].

Here it should be noted that, since radar range depends on the parameter L, which incorporates the processing losses L_x, not being *a priori* known, it is usually difficult to determine the real value of the RAF.

4.3 Radar warning receivers

Relative simplicity and low cost characterize the *radar warning receiver* (RWR). It is generally installed on board platforms requiring protection against weapon systems guided by known radars [1, 4]. Data such as pulsewidth, frequency, pulse repetition interval, and so forth characterizing the threatening emission, are often preset in the RWR's memory. When the RWR detects an emission similar to one of those stored in its memory, it identifies the threat and gives the appropriate warning.

The lists of data in the memory, which are called *libraries*, may be very simple or very sophisticated.

When the RWR is installed on board aircraft, the dangerous or "alarming" threats are represented by the signals generated either by radars guiding *anti-aircraft artillery* (AAA) or SAMs in lock-on phase, or by airborne sensors in *track-while-scan* (TWS) mode: The most serious threats are however represented by the reception of stable CW or ICW emissions: A CW signal may in fact mean an incoming missile (Section 3.3.3).

The lock-on phase is generally recognized by the constancy of the detected signal amplitude, and by the persistence of the signal in time (Fig. 4.1). In fact, if the radar is still in search mode, the RWR will receive a signal, amplitude modulated according to the radar antenna pattern, that will be of maximum amplitude only when illuminated by the radar beam. In the lock-on phase, however, the radar beam is stably on the target, and fluctuations are due exclusively to the instability of antenna pointing and to variations in range.

Upon generation of the warning, with its associated information about the *angle of arrival* (AOA) of the threat, possible actions are:

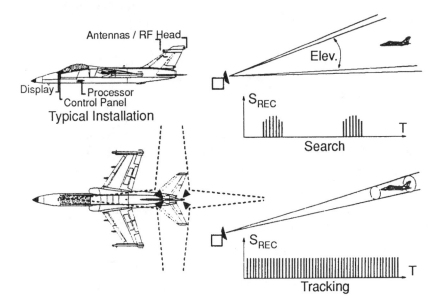

Figure 4.1 RWR and intercepted signals. An RWR normally deals only with tracking radar signals in lock-on or TWS phase.

- the pilot/operator maneuvers so as to reach a safer region, or makes evasive maneuvers with respect to the weapon system.
- passive countermeasures are activated to deceive the tracking radar, if the aircraft is equipped with them. An example would be the launching of chaff (section 5.5.1.2).
- an active countermeasures system is automatically activated to disable the weapon system (chapter 5).
- if the mission provides for it, an *anti-radiation missile* (ARM) is launched.

Clearly, an RWR must be characterized by great reliability, a very high *probability of intercept* (POI) of threats of interest, and a very low rate of false alarms [5, 6, 7, 8].

For good performance of an RWR, the following problems must have been solved satisfactorily:

- sensitivity; that is, the ability to detect hostile weapon systems before they reach their firing range.

- traffic; that is, the ability to provide correct information in the presence of a high volume of both pulsed and CW emissions.

4.3.1 RWR Sensitivity

To evaluate the order of magnitude of the sensitivity required of an RWR, some simple calculations will be carried out for two typical weapon systems, an AAA and a medium-range SAM system.

These two systems may be characterized typically by the following parameters:

AAA Tracking radar:
$$P_T = 100\,\text{kW} \ , \ G_T = 35\,\text{dB}$$
$$PW = 0.5\,\mu\text{s} \ , \ \lambda = 0.03\,\text{m}$$

AAA Gun:
$$R_{\text{max}} = 6\,\text{km}$$

SAM Illuminator:
$$P_{\text{ill}} = 200\,\text{W}$$
$$G_T = 37\,\text{dB}$$

SAM Missile:
$$R_{\text{max}} = 13\,\text{km}$$
(seeker lock-on at 20 km).

The RWR must be able to warn the pilot in good time of the possibility of being hit by such weapon systems, namely well before they reach their firing range.

Suppose that the radar cross section (RCS) of the aircraft is 10 m^2, and that the platform is heading for the artillery system. Let $V_{AC} = 300\,\text{m/s}$ be the speed of the aircraft, $V_m = 600\,\text{m/s}$ the projectile velocity, with $T_{\text{cmax}} = 300\,\text{m/s}$. (It is worth knowing that the kill probability of a fire control system decreases drastically after about 6 s of flight time.) From these data it is possible to calculate the range at which the weapon system can open fire, and therefore the minimum range at which the RWR must be capable of giving warning (Fig. 4.2).

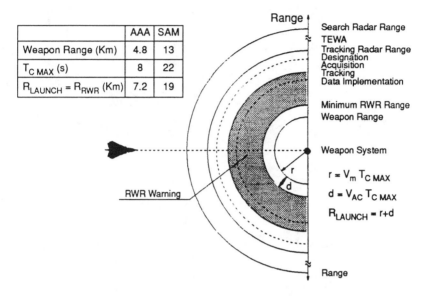

	AAA	SAM
Weapon Range (Km)	4.8	13
$T_{C\,MAX}$ (s)	8	22
$R_{LAUNCH} = R_{RWR}$ (Km)	7.2	19

Range

Search Radar Range
TEWA
Tracking Radar Range
Designation
Acquisition
Tracking
Data Implementation

Minimum RWR Range
Weapon Range

Weapon System

$r = V_m\,T_{C\,MAX}$
$d = V_{AC}\,T_{C\,MAX}$
$R_{LAUNCH} = r + d$

RWR Warning

Range

Figure 4.2 Ranges at which an RWR must give warning of the presence of AAA or SAM threats.

If the antenna gain G of the RWR is 3 dB, recalling that

$$s_r = \frac{P_T G_T G \lambda^2}{(4\pi)^2 R^2 L_p}$$

converting the parameters into decibels and calculating, one finds that the sensitivity required for correct detection at a range of 10 km is -17 dB m. Detection of the CW emission of the illuminator of a semiactive SAM system, which indicates the probable imminent launching of a missile, would, from the last equation, require detection of a CW signal of about -49 dB m (Fig. 4.3).

The sensitivity of crystal-video receivers is on the order of -40 dB m for pulsed signals, and -50 dB m for a CW signal, because of the convenient narrowing of the video bandwidth. That is why today the majority of RWR systems are of crystal-video type: their performance is adequate against both high-

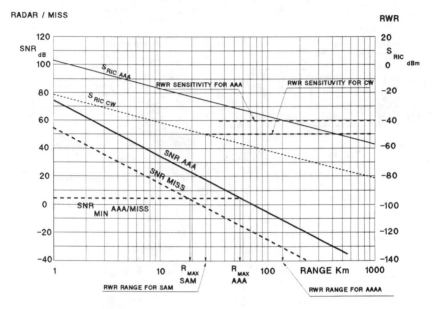

Figure 4.3 Sensitivity requirement of RWR systems for pulse radar and CW illuminators.

powered, low-repetition pulsed weapon systems and CW systems.

In the presence of a large number of radars with high pulse repetition rates, an RWR will have great difficulty in performing its sorting function, that is, the grouping of pulses with similar characteristics into trains, because of its structural simplicity. Generally an RWR can measure only the *pulsewidth* (PW), the *direction of arrival* (DOA), the *time of arrival* (TOA), and the amplitude. If a large number of pulses are to be sorted effectively, the pulses must be more highly defined; for example, not only their bands but also their frequencies will have to be determined. Better selectivity also could be required to avoid erroneous measurements caoused by overlapped fulses.

RWR systems can be classified as follows:

- wide-open crystal-video receivers.
- swept narrowband superheterodyne receivers.
- tuned radio-frequency (TRF) receivers.

- wideband superheterodyne receivers.

Figure 4.4 shows block diagrams of the above RWR configurations. The attainable performances are listed in the following table [1, 8], where also the wideband superheterodyne architecture with a channelized receiver is mentioned. However, this last configuration is more appropriate for an ESM system, and will therefore be discussed below.

Type	Sensitivity (dBm)	Probability of intercept (POI)	Traffic	Cost
Crystal video	-40 / 50	Very high	Low	Low
Narrowband superheterodyne	-70 / 80	Low	High	Medium
Tuned RF	-50	Medium	Medium	Medium
Wideband superheterodyne	-60	High	High	High
Wideband superheterodyne with channelized Rx	-70 / 80	High	Very high	Very high

4.4 Electronic support measures

The function of *electronic support measures* (ESM) systems is to provide those in charge of operations with information about the electromagnetic scenario or *electronic order of battle* (EOB), so that the correct decisions can be. It is therefore necessary to detect all emitters in operation, not only those which seem to present immediate threats. Much more is required here than in the case of RWR systems, which have to deal exclusively with "a priori" known tracking emitters. Moreover, the sensitivity of ESM systems must be sufficiently high for exploitation of the operational advantages inherent in timely knowledge of the electromagnetic scenario.

The ESM traffic can be therefore very dense; it may be on the order of millions of pulses per second. Since no *a priori* hypotheses should be made about the threats which may be detected, high precision is required in a great many parameters, including AOA, PW, frequency, and so forth, for adequate sorting and reconstruction of the electromagnetic scenario [9,10].

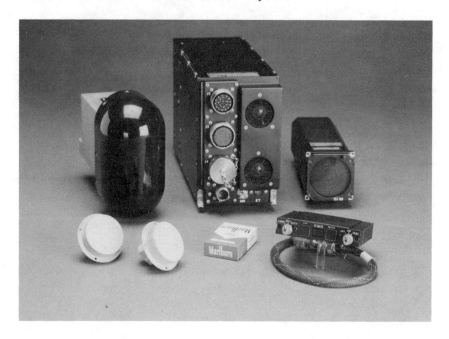

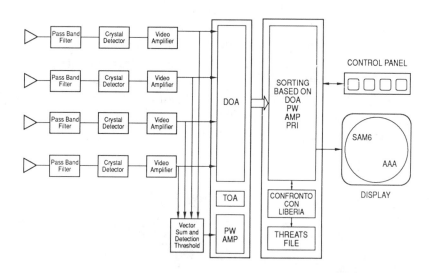

Figure 4.4a (a) Photograph of a crystal-video RWR system; (b) block diagram of wideband superheterodyne RWR.

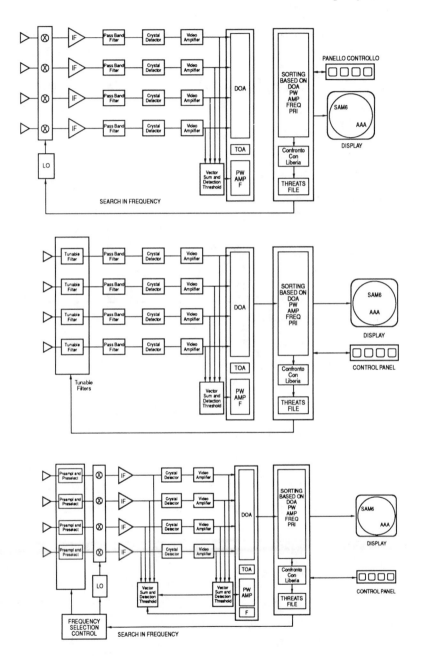

Figure 4.4b (c) Block diagrams of more complex RWRs; (d) narrow-band superheterodyne receiver; (e) RWR using TRF filters.

For this reason, ESM systems are much more sophisticated than simple RWRs and are equipped with many auxiliary circuits. A possible ESM configuration with blocks characterizing the various equipment modules is shown in Figure 4.5.

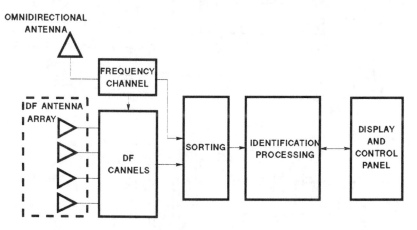

Figure 4.5 Block diagram of an ESM system. Unlike an RWR, an ESM system deals with all intercepted emitters, including scanning emitters.

Since frequency measurement is a costly process, it is performed only in an independent channel whose antenna is able to cover the whole area and frequency bands to be explored. The antenna is omnidirectional with a horizontal 360-degree beam; the latter is either a dedicated beam, or is synthesized from the *direction-finding* (DF) channels [1]. Detection, frequency measurements, and sometimes PWs are generally derived from this antenna channel.

The DF receiver [11, 3] generally consists of n simple channels each complete with a directional antenna. Sometimes, the gain of these antennas is exploited to avoid costly amplification before the main receiver. Comparison of the amplitudes received from the different channels determines the direction of origin of a signal. Pulse characteristics, translated into digital messages specifying TOA, PW, frequency, DOA, and amplitude, are sent to the sorting system to be segregated into trains.

A high-speed computer, or fast dedicated hardware, subdivides the pulses into groups whose similar characteristics indicate that they probably come from the same emitter (sorting or preprocessing). This computer performs a preliminary analysis of the data [9, 10, 11].

A second computer more slowly correlates the different emitters detected by the sorter, determines the operational mode of each emitter (scanning or tracking), computes the *antenna scan periods* (ASP), and possibly identifies the emitter by comparing its parameters with those memorized in the library. The data are then shown on a display.

ESM systems can be used to form a higher level system. In this case, the ESM systems form a passive surveillance network [12, 13]. They are installed on the ground, either along a frontier or in a zone to be protected, in well-defined positions, and are interconnected via a telecommunication link so as to be able to exchange data. The expected coverage of each ESM station is such that it overlaps with two contiguous stations.

When two stations intercept an identical emission, triangulation can be employed to locate the source of the emission, that is, to obtain also its range (Fig. 4.6). To obtain reliable location data, high DF precisions are required, on the order of a degree or tenths of a degree.

Passive location of emitters can be very useful to naval platforms where, once the presence of a hostile vessel has been detected by its emitted signals, it is desirable to launch anti-ship missiles without switching on the platform's own radars. In this case there must be a second naval platform linked to the first by a "protected" communication link with low probability of intercept by the enemy. Passive location of stationary emitters is easy when the ESM equipment is on board a helicopter. In this case, no other platform is required. The helicopter is able to make a DF measurement in one position (a fix), quickly move to a second position, and, after another fix, proceed with the triangulation.

Clearly it is even better if the location of an emitter can be

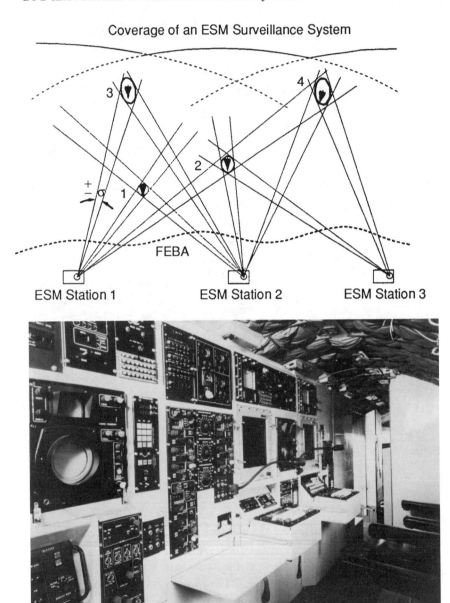

Figure 4.6 A network of ESM devices can form a substantial passive surveillance system capable of providing the location of targets by triangulation from various stations. The photograph shows a passive surveillance device installed on board a Falcon-20.

determined from an airborne platform; by correlating successive DF measurements with the successive positions taken by the aircraft in its flight, a computer can readily carry out the necessary calculations to locate the enemy radar.

Different types of ESM can be classified according to the methods used to implement the blocks in the diagram, as units of hardware. In the following sections, some of the most important constituents of an ESM will be discussed.

4.4.1 Omnidirectional antennas

An omnidirectional antenna is capable of providing 360 degrees coverage in the azimuth plane, and 30 to 40 degrees coverage in elevation, over a very wide frequency band (many octaves). It can be conical (Fig. 4.7, top) or biconical.

Certain designs require 45-degree polarizers. In general 45-degree polarization is preferred in ESM systems because it is capable of detecting all the possible polarizations used by radar: whether vertical, which is used especially by tracking devices, horizontal, which is used especially by search and acquisition devices, or right-hand or left-hand circular, often used by radars to reduce rain clutter.

4.4.2 Antennas for direction finding

The type of antenna used in a direction-finding ESM system [13] depends on the system characteristics (accuracy, security, bandwidth and so forth), in which it is installed. The most common are planar-spiral (Fig. 4.7) and log-periodic antennas. In general, direction finders can be of either the amplitude-comparison or the phase-comparison type. In the first, the DOA is determined by observing differences in amplitude at the output of the antenna system. In the second, the DOA is given by the measurement of phase shift between the signals of two adjacent antennas. A phase-comparison DF system gives better precision, but is much more complex because measurements depend on the signal frequency, and, for a wide angular coverage, direction finding is more subject to interference.

Figure 4.7 Above: omnidirectional ESM antennas. Below: planar-spiral DF antennas.

A particular type of amplitude direction finder is the multiple-beam-forming lens antenna shown in Figure 4.21, in which direction finding is performed by the comparison of 16 beams and is therefore very accurate.

4.4.3 Frequency receiver

Normally, the signal coming from the omnidirectional antenna is divided into several bands by a multiplexer [1, 14, 15, 16]. The various configurations possible for a frequency receiver depend on the performance requirement of the ESM device.

Wide-open structures cover the entire ESM spectrum instantaneously. Narrowband superheterodyne structures must be swept according to a given law to guarantee full spectrum coverage. The former are characterized by unit POI. The latter can encounter problems in trying to intercept scanning emitters within the desired reaction time with an adequate POI.

4.4.3.1 Wide-open frequency receivers

The most traditional configuration of a frequency receiver is shown in Figure 4.8. The signal coming from the omnidirectional antenna is *channelized* by means of a 5-plexer (a microwave filter capable of dividing the output into five adjacent RF bands) into the five traditional bands:

- 1 to 2 GHz (band L).
- 2 to 4 GHz (band S).
- 4 to 8 GHz (band C).
- 8 to 12 GHz (band X).
- 12 to 18 GHz (band K).

The signal is amplified by five amplifiers and sent to five interferometers for instantaneous frequency measurement.

The *interferometer* [17] (Fig. 4.9) is an instrument that causes the direct signal to interfere with the signal delayed by a delay line of length L, which is therefore out of phase by $\varphi = 2\pi L/\lambda$. If L has been suitably chosen, the two outputs I (in-phase) and Q (quadrature) represent the components of a vector whose phase shift is proportional to the frequency. This

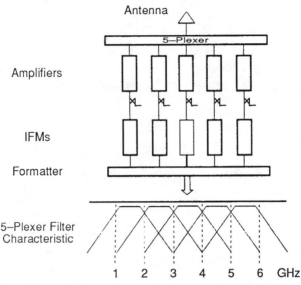

Figure 4.8 Subdivision in bands by a multiplexer in a frequency receiver.

phase shift is zero at the minimum frequency of the band of interest, and 360 degrees at the maximum frequency. If the bandwidth is Δf and the accuracy of the phase discriminator is α rms, the frequency precision of this interferometer will be

$$\sigma_f = \frac{\alpha \Delta f}{360}$$

When the delay line is such that the variation of frequency across the band causes a phase shift of 360 degrees, the interferometer is said to be simple. If the delay line caught is four times this, so that the frequency variation across the band causes a phase shift of four times 360 degrees (the vector makes four complete turns), the interferometer is said to be quadruple. In this case, the usual α error of the device causes an error in frequency which is one-fourth that of a simple interferometer:

$$\sigma_f = \frac{\alpha \Delta f}{4.360}$$

Normally, an interferometer works in amplitude-limited mode. In this case, an amplifier with strong gain in used such

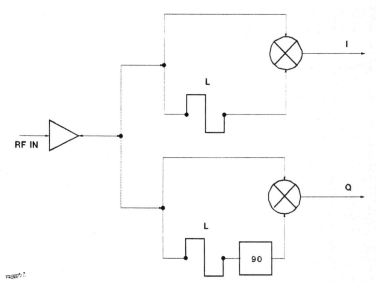

Figure 4.9 Diagram of a mixer interferometer for instantaneous frequency measurement.

that the output derived from a small signal has the same amplitude as that from the maximum signal. The amplifier is said to be *hard-limited*. When two signals are simultaneously present at the input of an amplifier of this type, at the output the weaker signal will appear with a much lower amplitude with respect to the stronger one (suppression of weaker signal) as shown in Fig. 4.10. However, when two signals of different frequencies and near equal amplitude reach the input of such an amplifier, it is generally impossible to predict which of the two will be the stronger at the output, because of the voltage standing-wave ratio (VSWR) and of the different behaviour of the amplifier stages with frequency.

The outputs from the five amplified channels to the circuits for detection of the incoming signals, and for the measurement of the TOA with respect to a reference point, the measurement of the PW, and the amplitude, are generally not kept in a linear range and do not reach the saturation. Measurements of these parameters are converted into digital form and sent, together

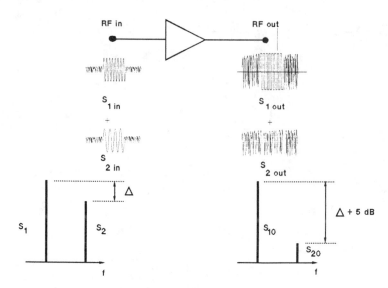

Figure 4.10 A hard-limited amplifier tends to suppress weaker signals.

with the measurement of the AOA, to a very fast digital processor which provides the deinterleaving (i.e., classification and segregation of the received pulses into homogenous trains).

Another wide-open receiver configuration is called the *band folded* receiver (Fig. 4.11). In this configuration, traffic from a number of bands is channeled into a single interferometer; so that, although fewer interferometers are needed, there is a danger that frequency measurements will be flawed by the increased probability of false overlapping.

Sensitivity of an Amplified Receiver

An amplified receiving chain (Fig. 4.12) is one in which the noise produced by the amplifier is much higher than the noise of the crystal video [18, 19, 20, 21, 22, 23]. In such a chain, when the RF bandwidth B_{rf} is tens of times larger than the video bandwidth B_V, the noise at the input is approximately

$$N = KTB_{eq}F$$

where

$$B_{eq} = \sqrt{2B_{RF}B_V}$$

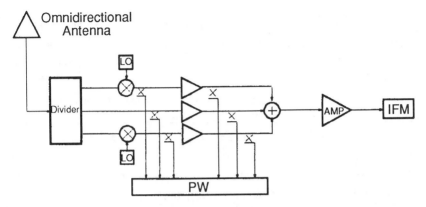

Figure 4.11 Diagram of a wide-open, band-folded frequency receiver.

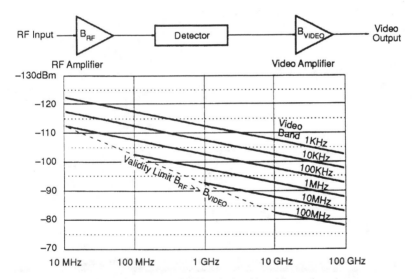

The sensitivity of a Preamplified ESM is calculated by adding to the values
in the graph the noise figure F (dB), the system Loss L (dB) and the SNR (dB)
required to measure the parameters, and subtracting the antenna gain (dB).

Figure 4.12 Sensitivity of a crystal-video ESM receiver with preamplification.

At the output a certain SNR will be needed to obtain the
required P_d and P_{fa} for detection from one single pulse (which

can be considered non-fluctuating), or to obtain the wanted precision in the measurement of the parameters. The signal at the input of the RF receiver must be higher than the noise by the required ratio.

For each passive element between the antenna and the amplifier, which introduces a loss L, the signal at the antenna output must be stronger by the same quantity L, to yield the required SNR at the receiver output.

For a more rigorous and detailed treatment, the interested reader should consult the literature mentioned.

4.4.3.2 Narrowband superheterodyne receivers

If a wide-open characteristic (POI=1) is not required, it is possible to obtain very good performance, in terms of sensitivity and of precision measurement, by using a *superheterodyne* receiver (Fig. 4.13) [24].

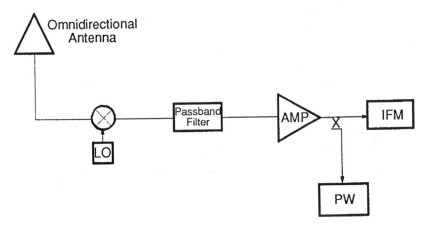

Figure 4.13 Diagram of a superheterodyne frequency receiver.

In this case, only a single intermediate-frequency (IF) channel of bandwidth B_{IF} is needed, to which all the possible n channels of the radio-frequency band are converted:

$$n = \frac{B_{RF}}{B_I F}$$

This is achieved either by the positioning of one or more *voltage-controlled oscillators* (VCO), or by frequency synthesis.

The disadvantage with this configuration is that, while tuned to one band, it can fail to detect a threat operating in another and transmitting for a very short time. Suppose, for example, that an airborne *track-while-scan* (TWS) radar (Fig. 4.14) illuminates the ESM receiver while this is tuned to another frequency, in accordance with the superheterodyne search strategy. Such a threat may go totally unnoticed, unless the TWS radar is operating continuously and the ESM sensitivity is such that the TWS radar can be detected via its sidelobe transmitted signal.

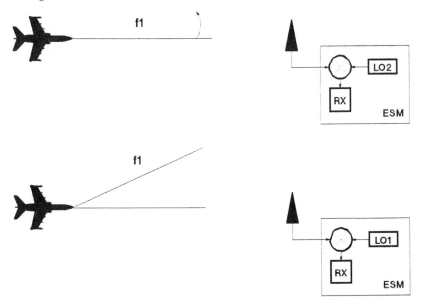

Figure 4.14 The probability of detection of a narrowband superheterodyne ESM receiver can be low. When the receiver is correctly tuned, the radar antenna may show a null from the sidelobes.

4.4.3.3 Wideband superheterodyne receivers

To overcome the disadvantage of the low POI of narrowband superheterodyne receivers, ultra-wideband superheterodyne devices, with bandwidths of 1 to 2 GHz, tunable over the full ESM band, are used. In this case the POI is increased but, to reduce the consequential effects of the increased bandwidth, namely higher susceptibility to interference, a receiver of this type must be complemented by a second conversion in a channel with a narrower bandwidth. This second receiver can be employed on a command from the operator or the processing system.

The more sophisticated wideband superheterodyne ESM systems, instead of using a second conversion, often employ a device, sometimes called a *panoramic receiver*, which is capable of analyzing the full IF band instantaneously (Fig. 4.15).

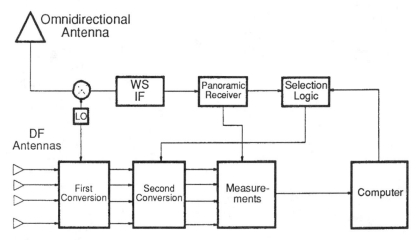

Figure 4.15 Block diagram of a wideband superheterodyne ESM with panoramic frequency receiver for spectral analysis and control of the DF receiver.

The most widely used advanced ESM receivers are:

- the channelized receiver.
- the Bragg-cell receiver.
- the microscan (compressive) receiver.

4.4.3.4 The channelized receiver
When the traffic is expected to be very intense, and it is desired to avoid a high rate of incorrect measurements from pulse overlap, channelized receivers should be used.

The basic idea of this receiver is very simple: By means of suitable contiguous filters, the total coverage band of the receiver is divided into many channels each equipped for measurement of signal parameters [16, 25, 26, 27, 28, 29].

If the bandwidth of the elementary channel is required to be very narrow, the total band covered cannot be very wide, due to the problems of volume, weight, cost and power consumption. A receiver of this type is often employed at the IF output of a wideband superheterodyne receiver.

Cost-effective channelized receivers, capable of covering the whole of the useful spectrum, will probably be feasible with the new technologies of the *microwave integrated circuit* (MIC), and even more the *microwave millimeter-wave integrated circuit* (MMIC). These technologies may be used for appropriate RF filtering, or to supply banks of mixers, filters and amplifiers for frequency conversion. For filtering and channelization, *surface acoustic wave* (SAW) filters (Fig. 4.16), which are compact and very stable, may be used. In this case, particular care must be exercised to avoid reflection echoes in the acoustic-wave device.

All the channels must have circuits for detection, for frequency measurement, and, if necessary, for PW and amplitude measurements, and all the data in all the channels must be converted into digital form and processed in real time.

The ensuing processing may be a bottleneck for the channelized receiver. Moreover, notwithstanding its high performance in precision measurements and in protection against interference, it tends to be very expensive, at least for the time being. In compensation, this type of receiver offers optimal performance in respect of:

- sensitivity.
- dynamics.
- ability to process signals overlapping in time but in different channels.
- ability to measure the amplitude and PW of signals.

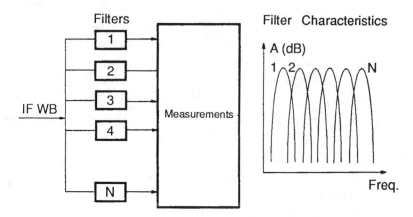

Figure 4.16 Panoramic receiver channelized with *surface acoustic wave* (SAW) filters.

4.4.3.5 The acousto-optic Bragg-cell receiver

This type of receiver is based on a special device called the "Bragg cell" (after its inventor) [30 to 42]. The Bragg cell is an electronic device that, by exploiting the acousto-optical characteristics of certain materials, may be used as a frequency analyzer in several applications, for example, EW, telecommunications, and radar (Fig. 4.17).

The operating principle of the Bragg cell is based on an interaction, in a suitable acousto-optical medium, (such as lithium niobate), between a beam of coherent light generated by a laser source and a traveling sound wave generated by a piezoelectric transducer fed by an RF signal.

The piezoelectric transducer thereby transforms the RF signal into waves of compression and rarefaction, which propagate in the acousto-optical medium with a wavelength

$$\lambda_a = \frac{v_p}{F_{RF}}$$

where v_p is the propagation rate of the acoustic waves in the acousto-optical medium, and F_{RF} is the frequency of the RF input signal.

The wavelength in the acousto-optical medium will be much shorter than the wavelength of the electromagnetic signal -

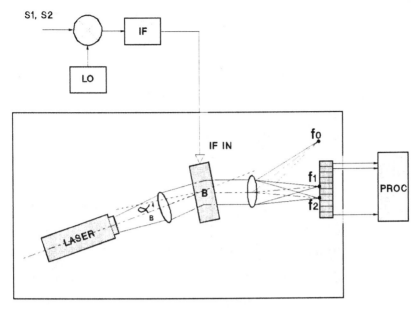

Figure 4.17 Bragg-cell panoramic receiver.

because the propagation rate of the acoustic wave inside the medium is much lower than the speed of light.

Since the index of refraction of the medium is a function of the density of the material, the periodical compressions and rarefactions will induce a modulation of the index of refraction distributed in the medium, throughout the propagation zone of the acoustic wave.

This region is thereby transformed into a diffraction grating for the incident coherent light beam, and for incidence normal to the direction of the acoustic wave, the diffracted laser beam will consist of a very large number of component beams spread out in space (diffraction fringes).

When the laser beam is inclined to the normal at a certain angle, called the "Bragg angle," another diffraction mode, called "Bragg scattering" is established. In practice, under these circumstances, the diffracted light beam consists of only two components. All the other components interfere destructively and are strongly attenuated.

The first of the two beams (order zero) is not deflected by diffraction and therefore cannot be used. The second, however, is deflected with respect to the incident laser beam by a quantity α (which, calculated in the center of the band, is equal to twice the Bragg angle α_B) given by

$$\alpha = \frac{\lambda_0}{n_0}\frac{f_{RF}}{v_p}$$

where λ_0 is the wavelength of the laser light, n_0 is the index of refraction of the acousto-optical medium, v_p is the propagation speed of the sound wave, and f_{RF} is the sound wave frequency, which is equal to the frequency of the RF signal at the input of the piezoelectric transducer. The angle of deflection of the laser beam is therefore proportional to the frequency of the received RF signal.

The Bragg cell can therefore be used as a spectral analyzer. If a suitable lens is placed after the acousto-optic interaction zone, it is possible to focus the primary light beam diffracted by the cell at a point in the focal plane. Since the image point depends on the direction of arrival of the light beam, and since, in this specific case, the direction is a function of the carrier frequency of the cell, a one-to-one relationship is established between the ratio frequency and the points in the focal plane of the lens.

If the piezoelectric transducer is fed by a signal whose spectrum has a large number of components, the spectrum will be seen in the focal plane as a series of bright points whose intensities are proportional to the power of the corresponding spectral lines. The spectrum of the RF signal can thus be analyzed by putting a series of photodetectors in the focal plane and reading out the signal levels at their outputs, either in series or in parallel.

The choice of the part of the spectrum to be collected by a single photodetector, or by a single channel B_c, is generally fixed so as to collect the power in the main lobe of the spectrum of the shortest pulses that have to be processed. For example, if pulses whose minimum duration is 50 ns have to be processed, the width of the main lobe will be 40 MHz; this value can be used to determine the frequency quantization (single channel bandwidth) for the array of photodetectors.

The above provides an overview of the way in which the system operates. For more details, the interested reader should consult the bibliography.

What is of interest here is the fact that the Bragg-cell receiver is fundamentally characterized by the following parameters, of which typical values are given:

- center frequency 1.5 to 2.5 GHz.
- maximum bandwidth B 500 to 1000 MHz.
- transit time 100 to 300 ns.
- number of elementary channels $n = B/B_c$.
- width of the channel function of B_c.
- instantaneous dynamic range $\simeq 40$ dB.
- number of bits per channel adequate to the dynamic range.
- readout time 50 to 500 ns.

Since, in order to operate correctly the Bragg cell needs an RF signal power on the order of -10 to 0 dBm, the input signal will need to be suitably amplified.

A very noticeable problem of these devices is their dynamic range. Here it is necessary to distinguish between the dynamics of a single correctly processed input signal and the dynamics of two correctly processed signals, overlapping in time but in separate channels. The first is in practice the dynamics of the ESM system in which the Bragg cell happens to operate. The second, limited by the spectra of the RF pulses and by the quality of processing, expresses the ability of the device to measure more than one signal at a time.

4.4.3.6 The microscan (compressive) receiver

Another type of receiver capable of analyzing a large portion of the spectrum, namely the IF band of a wideband super-heterodyne receiver, is the microscan or compressive receiver.

Its name comes from the fact that the analysis is carried out performing like a frequency scan in a very short time (less than one microsecond), of the operating band.

The microscan receiver [43 to 47] is based on the use of *dispersive delay lines* (DDL) with linear characteristics, that is, lines so designed that the delay introduced into the signal is proportional to its frequency.

These delay lines are easily realized by means of *surface acoustic wave* (SAW) devices. Their characteristic is shown in Fig. 4.18.

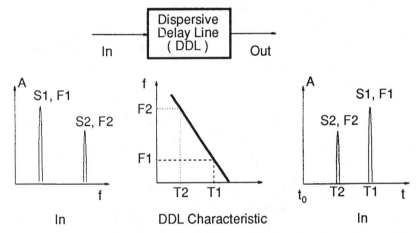

Figure 4.18 The characteristic of a dispersive line is to introduce a delay dependent on the signal frequency.

Suppose that signals present in a band of width B have to be analyzed (Fig. 4.19). It would be possible to scan the band with a swept superheterodyne receiver whose bandwidth is adapted to the duration of the expected shortest signal to be detected. In this way, however, some short signals might arrive at such times as to fall outside the IF filter band, and in any case the losses introduced because of the scanning dwell time would be high.

A better alternative is to exploit the full bandwidth at the mixer output, expanded by multiplication with the swept local oscillator signal, followed by a dispersive delay line which,

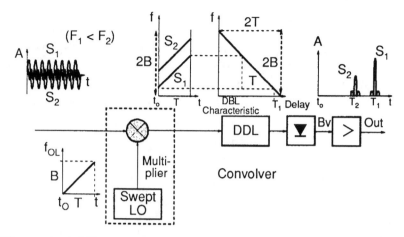

Figure 4.19 Microscan panoramic receiver.

thanks to its linear delay characteristic, will compress the expanded signal at a proper time.

In this way a fixed-frequency signal, for example a CW signal present at the input of the receiver for the whole of the local oscillator sweep time, T will produce at the mixer output a signal whose frequency varies linearly between $f_1 + f_{\text{LOmin}}$ and $f_1 + f_{\text{LOmax}}$. If the delay characteristic of the dispersive line is such that a delay T_1 is introduced at the lower frequency, while a delay equal to T_1 minus a delay equal to the duration of the sweep T is introduced at the higher frequency, a compressed signal centered at T_1 will be obtained at the output of the line.

The frequency information is thus transformed into time delay information: the spectrum of the input signal appears at the output of the dispersive line, within the limitations imposed by the fact that the sweep takes place for a limited time. Thus a continuous wave at the input is seen as a long pulse, of duration equal to the sweep time T. This pulse is not, however, a Dirac delta function (a spectral line), but a pulse with $(\sin x)/x$ dependence and its first zeros at $\pm 1/T$.

Using a detector after the dispersive line, it is then possible to analyse the spectrum of the signals at the input of the device. The delay of the detected signals with respect to the starting time of the sweep will show the frequency, while the signal intensities will be proportional to the amplitudes of the spectral components. This device has good sensitivity because the noise of the receiver is not the noise relative to the wide-band input, but the noise relative to the output filter of the dispersive line matched to the minimum duration of the pulse to be measured. The probability of intercept is 100%.

The device has the ability to discriminate and analyze signals overlapping in time, but at different frequencies, and performs a spectral analysis limited only by resolution. The dynamic range is good if care is taken to eliminate the sidelobes arising from the truncation, by means of a weighting function.

It is not possible to measure the PW if it is longer than the sweep duration. For very short pulses, namely much shorter than the duration of the sweep, sensitivity is reduced.

4.4.4 Amplitude-comparison direction finders

In an ESM system, the objective of a DF receiver is to provide information about the direction of arrival of each received pulse. In general, a DF receiver consists of a certain number of channels connected to directive antennas to form a monopulse network (Fig. 4.20).

A signal arriving from the direction α will be differently weighted by the gains of the different antennas. This will allow measurement of the angle of arrival [3, 11]. The antennas may be $4, 6, 8, \ldots$ in number.

If the antenna patterns are known, good precision will be obtained over a very wide RF band, especially when operating digitally, which allows the use of *programmable read only memory* (PROM) correction systems where correction data determined during calibration have been stored.

The problem with these devices is amplitude matching between channels, both between antennas and between subsequent amplifiers, detectors, and so forth.

Direction finders with four antennas give precisions between 10 and 15 rms in the 2 to 18 GHz band. Those with eight

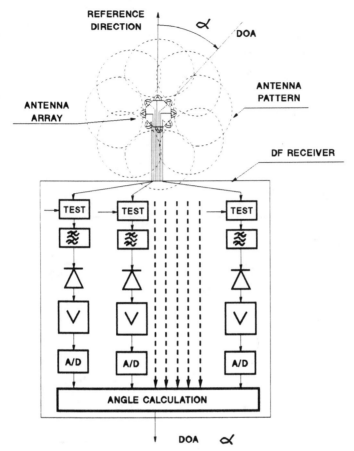

Figure 4.20 Direction finder.

antennas can reach 4° to 6° rms. Another severe problem is that, when the DF receiver is wide open, the presence of a strong CW can alter the levels in the relevant channels, thus yielding erroneous DF measurements for pulse signals that are not strong enough. If there are two or more CW emissions, then their DF measurement becomes a problem.

In order to reduce the intensity of this phenomenon, one can resort to channelization, or to superheterodyne channels, or the number of antennas can be greatly increased, so that the interference is confined to angles near its direction of arrival and could be cancelled by switching off the relevant DF channels. An example with many antennas and many channels is a direction finder that makes use of a multibeam antenna (Fig. 4.21). This antenna is capable of realizing 16+16 beams, thus obtaining extremely high preformance from the point of view of DF precision.

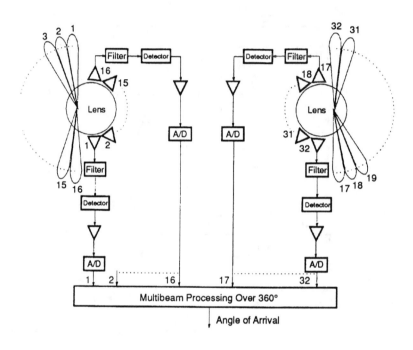

Figure 4.21 Multibeam direction-finding antenna.

A rotating antenna is another possibililty for DF measurements (Fig. 4.22). It is generally connected to a superheterodyne receiver tuned by an operator or by a suitable processor. The system receives a certain number of pulses from the emitter of interest. It examines the amplitude of the detected signals, correlates them with the antenna lobe, and finally calculates the angle of arrival of the signals. With this method, accurate measurement of fast scanning emitters is a problem, since the amplitude of the pulses can be altered by the scanning of the victim radar.

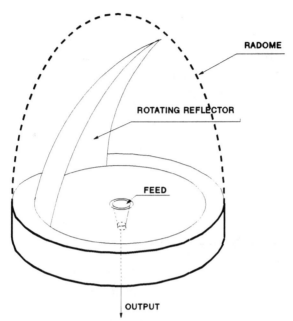

Figure 4.22 Rotation direction-finding antenna.

An improvement of this system is the use of a monopulse antenna capable of generating both and Δ channels, as in a tracking radar. In this case, the amplitude modulation generated by the scanning emitter has no effects.

Direction finding with a rotating antenna does not allow for exploitation of the DF data in the deinterleaving process, so that this method is little used in ESM equipment. However it is used extensively for ELINT purposes.

The DF method based on the measurement of the differential time of pulse arrival should also be mentioned. This method entails the use of widely spaced receiving antennas, some tens of meters apart, in such a way that by channel to channel comparison of arrival time, which can require an accuracy on the order of a few nanoseconds, it is possible to determine the angle of arrival of the signal.

4.4.4.1 Phase-comparison direction finders

Some DF equipment exploits phase information instead of amplitude information. Since a phase-comparison direction finder is capable of achieving high accuracy only over a limited angular sector, it may be used as an add-on kit or a new option to give a precise angular measurement.

The principle on which it is based is shown in Figure 4.23. A signal arriving from an off-axis direction α causes a phase shift

$$\varphi = \frac{2\pi}{\lambda} L \sin \alpha$$

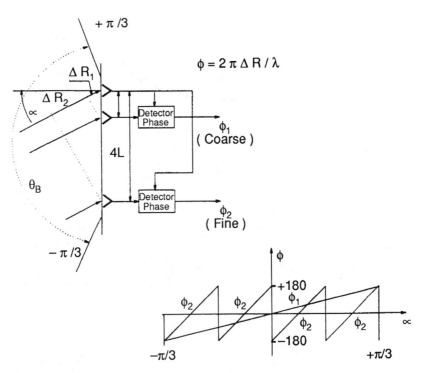

Figure 4.23 Phase-comparison direction finder.

If L is sufficiently large, the precision is very high. However it can happen that, even for relatively small angles, the phase shift is higher than 360°, so that the measurement becomes ambiguous. A third channel is then introduced, with a smaller

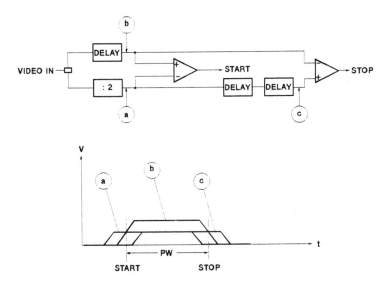

Figure 4.24 Pulsewidth measurement.

base, capable of giving a coarse measurement, but with higher angular dynamics, to remove the ambiguity in the more precise channel. The problems are similar to those encountered with interferometers.

The phase shift depends on wavelength; for this reason frequency data must be fed to the measuring apparatus if it is to be able to compute the AOA correctly.

4.4.5 Pulsewidth measurement

The measurement of pulsewidth (Fig. 4.24) has to be made linearly, that is, avoiding operation in non-linear dynamic areas of the receiver, for otherwise reception from the sidelobes and the main lobe of the same emitter would generate different tracks, because of the different responses of a receiver when saturated and when in the presence of weak signals. To keep the signals within the dynamic range, either a receiver controlled by instantaneous *automatic gain control* (AGC) (these receivers are at present very complex, requiring a delay line and so forth), or a receiver with a logarithmic video amplifier may be used.

When the receiver employs an RF amplified, the configuration shown in Figure 4.25 can be used in order to avoid saturation. Although it is more complex, it has the advantage of very large instantaneous dynamics.

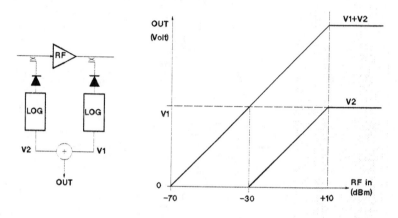

Figure 4.25 Very large, instantaneous, dynamic range receiver, suitable for PW and amplitude measurements.

4.4.6 Automatic detection

For each pulse received, the *front end*, that is, the set of frequency and DF antennas and receivers, will generate and pass on a *pulse descriptor word* (PDW) consisting of the following information:

- pulsewidth in general, from 0.1 to 100 μs.
- angle of arrival 0 to 360°.
- frequency from less than 1 to 18 GHz.
- amplitude −60 to 0 dB m.
- time of arrival.

The number of pulses received per unit time, which can be very high (up to a few millions per second), will depend on the traffic in the operational environment. From this enormous quantity of totally uncorrelated pulses, the subsequent digital

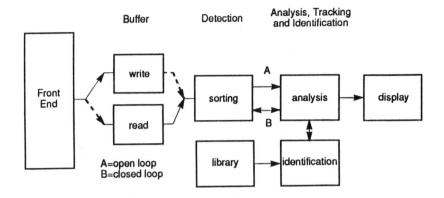

Figure 4.26 An automatic ESM detector extracts from the received signals the emitter present in the scenario and potentially identifies them.

processing, or automatic detection, must reconstruct the electromagnetic scenario in the theater of operations (Fig. 4.26) [1, 9, 10].

A correlation, called "sorting" or "deinterleaving," is first performed. This entails the grouping together of all the pulses that potentially come from the same emitter. This is not an easy task, if one thinks of the decorrelating effect arising from the imprecision of the receiver, or of the effects introduced by the parameters agility of the emitter, or the effects of the environment, causing interference and reflections.

After this first associative processing to detect probable emitters, complex software performs analysis to determine agility, stagger, mode and scanning period of these emitters, and finally identifies them. The results are presented on a display in both tabular and graphic form, together with a warning, whenever a high priority threat is identified.

There are essentially two types of automatic detectors: open-loop and closed-loop automatic detectors.

4.4.6.1 Open-loop automatic detectors

In an open-loop automatic detector (Fig. 4.27), a certain number of many cells are available for incoming pulses. These cells are labeled with information about the instantaneous parameters of the emission, for example PWs and DOAs. When a pulse arrives, a free cell will declare itself available to accept other pulses with the same PW and DOA as those of the first pulse, within a certain tolerance. Appropriate criteria may be used to take into account the agility of the parameters.

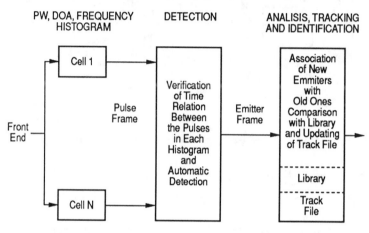

Figure 4.27 Open loop automatic detector.

For example, the following criterion serves to detect both fixed-frequency and frequency-agile emissions, if this is required. Upon arrival of other pulses of equal PW and DOA, the cell can adjust to accomodate more or less range in frequency, according to the frequency of the pulses which follow. If the first pulses are of a fixed frequency, the cell polarizes itself to accept only pulses with those frequencies, those PWs, and those DOAs, always within certain tolerances that must take into account the spreading of the measured parameters. If, on the contrary, the first pulses cover a wide frequency range, the cell will accept pulses of those PW and DOA, but with a spread of frequency up to 10 per cent of the bandwidth.

Each cell is open for an aperture time of a few tens of milliseconds, during which time it can collect only a certain maximum number of pulses. As soon as it is full, the cell will replace old pulses by new ones only if their amplitude is higher than the amplitude of the stored pulses. In this manner, given the limited memory capacity, some sorting is done, selecting pulses which are of higher intensity and therefore cleaner, and rejecting the information contained in pulses with a lower SNR.

This labeling process avoids saturation during the subsequent processing, preventing the processing of pulses which add no information, such as happens, for example, with pulse-Doppler radars which emit a very high number of pulses per second, all equal from the ESM point of view.

Once the aperture time has passed, the content of the cells is sent to the sorter, which tries to establish whether among the pulses stored in the memory cell there are any time relations typical of PRF radars. The search for and measurement of the PRF, or better, the PRI, is based on the analysis of the differential time of arrival, ΔTOA, betweeen these pulses. The process is usually performed by a search for the minimum ΔTOA existing between two successive pulses, and by verifying that other pulses fall in congruent time windows, that is, gates allocated in times equal to multiples of the minimum ΔTOA, plus or minus an adequate tolerance. If in the gates there is a number of pulses higher than a certain threshold (5 to 10), it is considered that an emitter has been detected and the next processing stage has begun. Otherwise, the process is repeated, considering other minimum ΔTOAs. Pulses that do not belong to any group are discarded. The process of measuring the PRI generally requires many operations in quasi-real time, and several algorithms have been proposed to speed up operation as much as possible.

In the end, a picture of the possible emitters is built up, and the process of sorting, that is, of detecting emitters in that framework, may be considered to be finished. From then onwards the processing is conducted by powerful computers, but of "normal" speed. This picture is compared with the file of emitters already established to decide whether new emitters have been detected or whether the new data only update known emitters.

As soon as the emitter file has been updated, it is possible to start computing to determine the scanning mode of the emitter under examination: lock-on, circular scan, sector scan, or track-while-scan.

Finally, the parameters are compared with those in the libraries to permit identification.

4.4.6.2 Closed-loop automatic detectors

In a closed-loop automatic detector (Fig. 4.28), whatever the detector has already learned is deleted from the flow of pulses to be used for new detections, but is used to improve or complete information about emitters already detected. As in the open-loop case, one begins to charge a pulse buffer (this time a much larger one) and proceeds to form histograms based on PW, DOA, or frequency. Once an emitter has been detected, a tracking channel is opened.

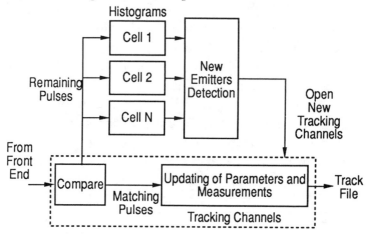

Figure 4.28 Closed-loop automatic detector.

The difference in this structure is in the fact that the incoming pulses are also directly compared with the tracking channels and, if they belong to them, are deleted. Therefore, after the switching-on transient, the detection process becomes safer and less stressful, and all pulses are used to consolidate and improve the data about emitters already detected. Guard circuits avoid possible saturation by very high PRF emitters.

It can safely be stated that design and realization of a reliable automatic ESM detector is one of the most difficult tasks of electronic engineering, and its implementation requires years and years of work.

4.4.7 Identification and data processing

Independently of the type of automatic detector used, the information stored in the files is later used for threat evaluation, identification and presentation.

Threat evaluation means either the detection of emitters that are declared a high priority threat, those of a hostile anti-ship missile for a shipborne ESM, for example, leading to an immediate warning, or the effective evaluation of the mode of the emitter.

If an emitter is detected in its search mode, and is not of the "alarming" type (that is, is not of a type associated with an enemy weapon system), it is scarcely threatening. But if the scanning is switched from search to lock on, that is, if the radar is tracking the platform, then the operator's attention must be drawn to it, even though the emission is not of a known threatening type. Since in time of war radars can change their operating parameters, frequency, so that their characteristics do not necessarily coincide with library data about threats, this is necessary in order to avoid the possibility that the platform comes under enemy fire without a threat warning being generated.

Normally, in ESM equipment there is a much larger second library in which the characteristics of many thousands of radar modes are stored. The ESM processing is such that, by applying an appropriate search and comparison strategy, it is possible to identify the detected emitters in a reasonably short time, on the order of a few seconds. The result is an identification to which a number expressing the reliability of the identification (the *confidence level*) is associated. Often the ESM can be programmed to give a *threat level* for an emitter which depends on whether it is locked-on, on the waveform characteristic and on library data.

4.4.8 Presentation

The information extracted is sent to a display and presented to the operator in a form chosen from among several possibilities. Three types of presentation are most usual (Fig. 4.29):

- tabular.
- Cartesian, in which the abscissa represents the AOA of the emission and the ordinate the RF (this type of display is also called $f - \alpha$).
- situational.

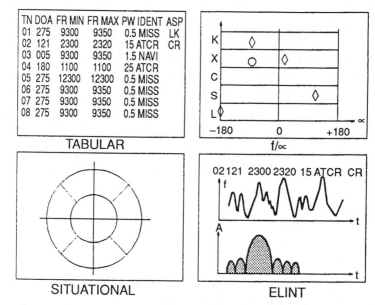

Figure 4.29 Typical presentation of an ESM system.

In a tabular presentation, tables are provided in which each emitter is shown by a *track number* (TN), followed by information on DOA, frequency, agility band, PW, PRF, PRF jitter or stagger, scan type, and so forth.

In the f-alfa presentation, the AOA can be either with respect to the platform, or absolute, that is, with respect to geographic North.

If the operator requires it, all data identifying each threat can be displayed; the tracks can be ordered by threat level, and so forth. Representation of the emitters may be synthetic, showing only an identification number, a TN, or, more coarsely, just the detected amplitude.

In the situational, or tactical, presentation, emitters are shown on a circular screen, like a *plan position indicator* (PPI), as a function of their AOA. The radial distance can represent either emitter class, or the amplitude of the emitter, thus giving a hint about the range. In this connection, it is appropriate to explain what is meant by a "hint." The range of an emitter is

$$R = \left[\frac{P_T G_T G \lambda^2}{(4\pi)^2 s_r L_p} \right]^{\frac{1}{2}}$$

where $P_T G_T$ is the radar ERP; although this is known *a priori* from intelligence, it may vary by ± 2 dB according to the condition of the equipment

G is the antenna gain of the ESM receiver, which can be known to within ± 2 dB for broad elevation sections.

L_p are polarization losses, which can be estimated with a 1-dB accuracy.

Combining the errors, one concludes that, once the amplitude of the detected signal has been measured, the precision in range could be on the order of $\pm 40\%$. However, the presence of the factor F_p^2, which takes into account reflection from the surface of the sea or from the ground, introduces large ambiguities in the amplitude/range relationship, worsened even further by the ducting effect. Therefore measurements of signal amplitude may be considered totally unreliable for estimates of range.

4.4.9 Problem areas in ESM

Besides the problem of accurately measuring the parameters and thereby accurately deducing the electromagnetic scenario, or EOB, of the theater of operations, ESM equipments, both shipboard and airborne, are confronted by another severe problem, namely the probability of intercept POI.

An emitter might be, for example, a radar installed on a submarine whose antenna emerges for only a few seconds to locate a target with a single scan. To intercept such an emitter with 100% probability, a wide-open structure is required. However, this structure may be incompatible with other onboard equipment.

For example, a ship needs to transmit strong CW signals, both for satellite telecommunications and for missile guidance. Given the power levels involved, the interference thus produced could hinder the ESM performance, even completely blinding the ESM device.

The insertion of filters in ESM channels is complicated and costly: both the frequency receiver and all the DF channels have to be filtered, and if their function is not to be degraded, severe matching restrictions must be met. The use of a channelized architecture would appear optimal, if it did not entail high cost when the number of channels is high. Another possibility is to renounce a 100% POI, and to resort to superheterodyne configurations, followed by receivers of the Bragg-cell or the microscan type. However, these problems have been mitigated by the development of new techonologies which allow the realization of hybrid configurations, as will be illustrated in Chapter 7.

4.4.10 Typical characteristics of a naval ESM system

For information, typical performances attainable by present-day ESM systems are outlined below:

- Coverage in frequency 0.5 to 18 GHz in angle 360° Az and 40° El.

- polarization linear at 45° , or circular.
- operating sensitivity (pulse) −50 to −65 dB m.
- operating sensitivity (CW) −60 to −75 dB m.
- PW (minimum; maximum) 0.1 μs; 100 μ s.
- PRF (minimum; maximum) 50 Hz; 300, 000 Hz.
- input traffic (maximum) $\simeq$ 300, 000 pps.
- precision
- Pulsewidth 50 to 200 ns.
- Frequency 0.1 to 0.5%.
- amplitude $\pm$ 1 to 2 dB.
- direction of arrival 3° to 5° rms.
- detectable emitters.
- pulsed; CW; interrupted CW.
- fixed.
- frequency agile.
- PRF agile.
- coded (without detection or extraction of the code).

4.4.11 Range advance factor in the operational environment

It is important that an ESM system detects enemy radars before they detect the platform on which the ESM is installed. Often ESM equipment is installed on-board ship; accordingly, the calculation of the (RAF) *range advance factor* will be for equipment (Fig. 4.31).

Suppose that the radars to be detected are an airborne early warning radar, an airborne interceptor, and a navigation radar, with the following characteristics:

- Early Warning Radar.
$$P_T = 100 \, \text{kW}$$
$$G_T = G_r = 30 \, \text{dB}$$
$$\lambda = 0.1 \, \text{m}$$
$$\tau = 0.5 \, \mu\text{s}$$
$$n = 52$$
$$h_a = 4000 \, \text{m}$$

- Airborne interceptor.
$$P_T = 200 \, \text{kW}$$

Figure 4.30 Receiving antenna of a naval ESM system.

$$G_T = G_r = 33\,\text{dB}$$
$$\lambda = 0.03\,\text{m}$$
$$\tau = 0.5\,\mu\text{s}$$
$$n = 1$$
$$h_a = 100\,\text{m}$$

- Navigation radar.

$$P_T = 200\,\text{kW}$$
$$G_T = G_r = 33\,\text{dB}$$
$$\lambda = 0.03\,\text{m}$$
$$\tau = 0.5\,\mu\text{s}$$

Figure 4.31 Typical situation of a naval ESM system.

$$n = 1$$
$$h_a = 25\,\text{m}$$

Make the following assumptions:

(a) RCS of the ship $= 10,000\,\text{m}^2$.
(b) Number of pulses integrated by the radar, $N_i = 20$.
(c) $L =$ total radar losses $= 10\,\text{dB}$ for all radars.
(d) $F = 6\,\text{dB}$ for all radars.
(e) P_d of 50% and P_{fa} of 10^{-6} (corresponding to a 14 dB $(SNR)_{\text{Pdfa}}$.
(f) $G = 0\,\text{dB}$ (effective ESM antenna and receiver gain).
(g) $s_0 = -62\,\text{dB m}$ (operating sensitivity of the ESM corresponding to a 16 dB SNR, that is an 8 dB RF).

Then, from the equations for radar range and ESM range

$$R_R = \left[\frac{N_i P_T n G_T G_R \sigma \lambda^2}{(4\pi)^3 KTBF(S/N)_{\text{Pdfa}} L} F_{pt}^2 \right]^{\frac{1}{4}}$$

one can plot a graph showing the dependence of the SNR on the radar and ESM, obtain the respective ranges, and therefore the attainable RAFs (Fig. 4.32). Note that the radar signal behaviours considered are for a point target at 7 m height. Since a ship is instead an extended target, the radar signals will in practice be just fluctuating points lacking the full range-related additional attenuation arising from a single multipath effect.

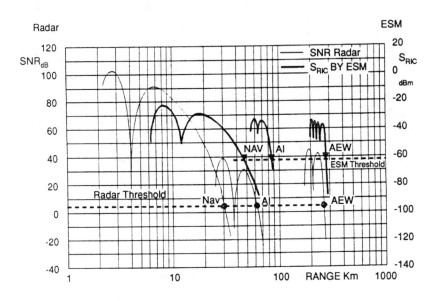

Figure 4.32 Range dependence of radar SNR and ESM signals, for RAF computation.

Before the advent of coded radar, a sensitivity on the order of $-40\,dB\,m$ sufficed to obtain RAFs much higher than unity. However, ESM is forced to operate with a very wide video band, matched to the shortest pulses expected, namely a bandwidth on the order of $10\,MHz$ when $0.1\,mus$ pulses are to be detected. Therefore if the radar operates with a very low peak power and uses code, it may have an advantage over the ESM. For equal average power, and therefore equal radar range, the use of an n element code reduces the radar peak power by a factor n, so that the ESM range will decrease by $n^{\frac{1}{2}}$.

Since signals undergo very strong attenuation immediately beyond the horizon, it is useless to attempt to improve the

RAF against radar surface emissions by increasing ESM sensitivity. What matters is that the sensitivity should be capable of correctly detecting an emission, without giving the radar a significant advantage in range.

When the ducting effect intervenes, the actual visibility of the ESM and the radar increase unpredictably and no calculation is at all reliable.

If the ESM device is installed on land, both the topography of the area and the attenuation due to trees and must be taken into account (see chapter 2). The topography has to be considered so that the attenuation due to diffraction, as well as the attenuation due to non-transparent obstacles such as mountains, hills, buildings, and so forth, may be calculated.

4.5 Electronic intelligence (ELINT) systems

The mission of this class of equipment is mainly strategic, but also tactical, as will be seen below [3, 23]. In their strategic role, ELINT systems are capable of providing information about the technological status of a potentially hostile country and on its military activity. This information will have to be translated into plans that can have an impact on the political, military, and industrial sphere.

Detection of signals generated by new equipment of higher quality or by the use of a new frequency band must lead to the initiation of military and industrial programs for the neutralisation, if necessary, of these new threats. Detection of unusual electromagnetic activity, or of the movement toward borders of a quantity of radar equipment, may suggest some political moves are required to clarify the situation.

However, ELINT systems have an important tactical objective, as well as a strategic one. Given the enormous amount of precise and detailed information known about emitters, an ELINT system is capable of following the displacement of one single piece of equipment and of providing a very accurate

EOB. Furthermore, the available sensors are so precise that an ELINT system can be the principal supplier of information for the libraries to be loaded into the ESM, RWR, and ECM devices.

4.5.1 ELINT sensors

Figure 4.33 shows the block diagram of an ELINT sensor. Since its task is not to give immediate warning, but rather to give very precise measurements, well protected from interference, an ELINT system is usually of the superheterodyne type, with a directive, rotating, receiving antenna, which can be trained in directions of interest by means of servomechanisms. This configuration provides ELINT systems with a high sensitivity that allows detection ranges of a few hundred kilometers. In order to minimise the range restriction resulting from the curvature of the earth, this equipment is often installed at elevated sites. Thus the receiver has the capability of measuring many parameters with great precision, to the point of being able to identify a specific device ("fingerprinting"), and, after integration with other data, may be able to identify the platform on which the device is installed. An ELINT system generally has several IF bands at its disposal and is capable of analyzing the spectra of single emissions. It can load very long pulse buffers stores to enable accurate analyses to be performed in delayed time. An operator may see the results on a display in both tabular and graphic form. The enormous mass of data is extracted by the ELINT sensor can be recorded and sent to ELINT processing centers, either by a protected link or on recorded tapes and discs.

4.5.2 The ELINT processing center

An ELINT processing center (Fig. 4.34) has the following functions:

- gathering of data and information.
- generation of a data base.

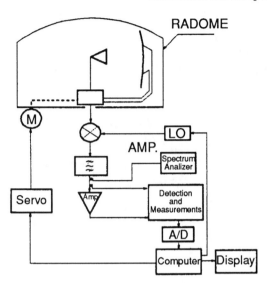

Figure 4.33 Block diagram of an ELINT system.

- generation of strategic information.
- generation of tactical information (libraries for ED systems).

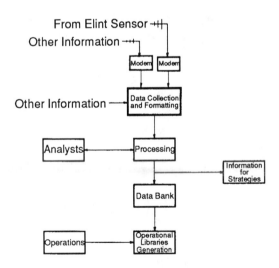

Figure 4.34 Diagram of an ELINT processing center.

To perform these functions, the ELINT processing center

must be able to receive information from both ELINT and other ESM sensors, and from other intelligence organizations. All information is formatted, that is, rendered homogenous to allow subsequent digital processing. It must be correlated, filtered, and analyzed by experts to extract potential strategic information and to add to the data base memory. Once the data base has been formed, depending on the political-military situation, operations experts will prepare libraries for the various operational departments that will insert them into the memories of ED equipment.

4.6 Infrared intercept systems

Many passive surveillance or warning systems exploiting the radiation emitted by aircraft, missiles, ships, tanks, and other targets in the IR part of the electromagnetic spectrum (see chapter 2) have been developed.

The most important systems are:

- passive IR surveillance and warning systems used at air defense radar sites as back-up, when there is the danger of ARM attacks.
- IR systems used as sensors for the coordination centers of short-range weapon systems, such as shoulder-portable IR-guided missiles.
- missile launch warning systems for air platforms.
- surveillance and warning systems for naval platforms.
- airborne surveillance systems, called line scanners.
- satellite detection systems.
- *forward looking infrared* (FLIR) systems for night navigation.

IR systems can use either scanning sensor heads, or fixed sensor heads called "staring heads."

Scanning heads use a system of mobile reflecting surfaces to ensure that the elementary *field of view* (FOV) scans the whole of the angular sector of interest. When the sensor element is fixed, scanning can cover up to 180 degrees in azimuth. For wider coverages, it is necessary to rotate the sensor set as well.

Often an array of sensors is used in order to cover a broad sector in elevation, while keeping the elementary FOV very small. In this way the background IR signal can be maintained, after suitable filtering, at a level lower than that of the signal of interest.

A staring sensor head consists of a mosaic of detectors, each realizing an elementary FOV, which together cover a very wide, instantaneous FOV. Coverages on the order of many tens of degrees can be achieved in both azimuth and elevation, depending on the width of the elementary FOV and on the number of sensors. The greater the number of sensors, the greater the complexity and the cost of the IR head. In compensation, the staring technique allows continuous parallel processing of the signal from each pixel, in contrast to scan systems, which can give information about the target only when they are looking at it.

4.6.1 Missile launch warner/Missile approach warner

Missile launch warners (MLW) are systems capable of detecting the launch of a missile, if it occurs within their FOV. These systems are designed to be installed on board aircraft that may be subject to sudden attack by IR-guided missiles. Since the latter can be launched without any radio frequency emission, the only means of detecting their launch with any reliability is to exploit devices capable of detecting their IR emissions or, sometimes, UV emissions. With this equipment an aircraft can launch its limited supply of flares in an optimized way, without risking lack of cartridges when they are really necessary.

IR systems capable of providing information about approaching missiles are under study, but, at present, radar techniques remain the best tools for performing this function. Devices specialized in detecting the presence of an approaching missile are called *missile approach warners* (MAW). They are small CW or pulse-Doppler radars able to detect fast approaching targets with small RCS.

The principal disadvantage of these systems is that they are active, and therefore do not allow the carrier aircraft to approach its target covertly. Also they do not usually give precise angular coordinates. To avoid informing the enemy of the approach of the air platform on which MAWs are installed, they can be switched on after the missile launch warner has given a warning. This means however, that the aircraft will have to be equipped with both systems, thus increasing both cost and weight.

4.6.2 Forward looking infrared systems

Forward looking infrared (FLIR) systems are scanning systems, like surveillance devices, but covering a limited angular sector and generate a TV-like picture based on IR signals. The video output is even compatible with standard CCIR.

On-board an aircraft, FLIR makes possible both night navigation and night landing. It displays on a TV screen whatever has been detected by the sensor during its scan of the full FOV, with no need for any special processing. The extraction of information from the display is left to the operator.

Infrared search and track (IRST) systems are more advanced IR systems. In fact, they are able to extract automatically targets of interest from the background, while keeping the false-alarm rate at an acceptable level. An advanced sensor head, generally able to monitor more than one IR band (multi-color sensor), and a sophisticated space-time filtering system provide a signal that is compared with an adaptive threshold and "libraries" so as to detect targets and, if possible, identify them.

The problems of IRST systems all stem from their low background filtering capability. These systems do not yet offer adequate overall performance.

4.7 Communications ESM and communication intelligence

The role of communications intercept devices (COM-ESM) is support those responsible for military operations by providing them with the following capabilities:

(1) Exploitation of enemy communications to obtain information about their content; about operational modes and C^3 activity peaks that can indicate an attack; troop or vehicle movements; and, whenever possible, the enemy's intentions. All this has both tactical ESM and strategic COMINT objectives.

(2) Location of communications centers and, possibly, C^3 centers.

(3) Designation of jammers.

Moreover, in peacetime, these systems constitute the best means for training personnel and for identifying weaknesses in one's own communications network.

4.7.1 Communications ESM

COM-ESM systems differ widely according to:

- the links which they have to intercept.
- the platforms on which they are installed.
- the operational functions of the communications networks which they have to intercept.

The characteristics of these systems will depend on the frequencies and techniques employed, from ELF to MF, from HF to UHF, with conventional, spread-spectrum or burst transmissions, radio relay systems, tropospheric scatter and satellite transmissions, and so forth.

For example, consider two types of battlefield communication systems, one based on V/UHF devices using frequency hopping, the other based on radio relay systems.

These two communication systems, often used together with conventional combat net radio, require different approaches to

ESM. V/UHF frequency-hopping communications are characterized by the extreme brevity of the emissions in each channel, and the fact that frequency information cannot be used as a discriminating parameter to catalog and locate the emitter.

On the other hand, radio relay systems require high sensitivity of the ESM system, while an extremely short reaction time is not essential. Moreover, radio relay systems work over a wide range of frequencies, from VHF to the D band, and use wideband multi-channel *frequency-division multiplexing* (FDM) and *time-division multiplexing* (TDM) modulation techniques, while hoppers are normally confined to the V/UHF band and use modified *frequency-shift keying* (FSK) modulation.

All these differences imply that the receivers, the devices for analysis, and the antennas must be of different types. Also the diverse platforms on which COM-ESM systems are installed require different equipment. For example, the propagation pattern varies according to whether the platform is airborne, shipborne, or installed on land, and consequently the required sensitivity and the electromagnetic scenario changes . The technique to be used for fixes varies too. While stationary emitters can be located either by a single airborne direction finder or by several cooperating ground-based DF stations, fixes of aircraft emissions can be performed only by cooperation of synchronized stations.

Moreover, ESM functions become complicated when real-time discrimination of emissions from fast-approaching air platforms (close air support) is needed. In such a case, a time analysis of the parameters of the intercepted signal, capable of rejecting fluctuations due to propagation anomalies, is required.

The operational functions of the communication system to be intercepted can also imply differences in the COM-ESM systems. Communication systems can be of the following types:

- divisional C^3 systems.
- artillery C^3 systems.

- *time-division multiple-access* (TDMA) data distribution network.
- support communications for second echelon attacks.
- close air support.
- combat net radio.
- support networks for intelligence and targeting.
- tactical air force integrated information systems.

The differences in traffic intensity, single emission duration, network organization, network mobility, and so forth, require that techniques appropriate to each case be used, in order to guarantee the desired POIs and the reliability of the measured parameters.

The general functional structure of a COM-ESM system is shown in Figure 4.36.

Figure 4.35 IR tube containing a focal plane array that follows realization of string an IR system with wide FOV.

The main functions of a COM-ESM system are:

- *continuous scan* (CS) of the band.
- *discrete scan* (DS) of the band.
- CS + DS.

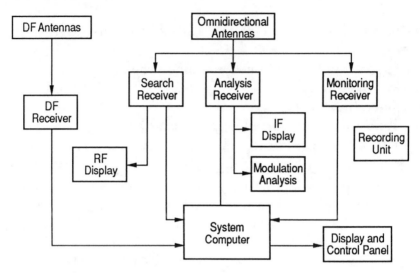

Figure 4.36 Block diagram of an ESM-COM system.

- designation for DF measurements.
- designation for ECM operations.

Automatic CS over one or more assigned frequency sub-bands presents the operator with the electromagnetic spectrum displayed in both tabular and graphic form (a synthetic RF display).

The different sub-bands will be characterized by such parameters as:

- priority.
- search filter and sweep rate of the receiver.
- detection threshold.
- channels to be suppressed because they are on friendly frequencies or of no interest.

During the scan, intercepted emissions are correlated in frequency according to suitable data base memorization criteria. An identification number and the essential parameters, namely track number, frequency, times of the first and last interception, modulation, signal band, priority, and so forth, are presented to the operator on an alphanumeric display. In general, emissions that are intercepted for the first time, which need to be listened to and analyzed by an operator, are emphasized. The synthetic RF display gives some indication of electromagnetic activity and of the duration of the emissions. It allows the operator to zoom over frequency areas as wide as 1 or 2 MHz with a resolution lower than that of the channelization.

Correct performance of the system is conditioned by the sensitivity of the superheterodyne receiver. Sensitivity values attainable with good receivers and active wideband antennas are between 0.5 and $5\,\mu V/m$.

The probability of detection, that is, the probability that all the emitters in an electromagnetic scenario, consisting of multiple emitters, are correctly detected and processed, depends not only on the scenario but also on the sweep rate of the receiver relative to the average duration of the emission. It also depends upon the processing capability of the computer, in particular, its coding management technique. For example, in the case of airborne radar, the duration of a communication in the UHF band is of a few seconds, so that sweep rates of 25 to 50 MHz/s are acceptable. These can be achieved with narrowband superheterodyne receivers with rates from 1000 to 2000 channels per second, which also give the best sensitivity, fidelity and frequency measurement precision.

In the search phase, the receiver should achieve good resolution and precise frequency measurement. These two parameters are, respectively, important for discrimination between the different emitters, and to allow the correlating algorithm to operate correctly without generating false tracks. Acceptable values are resolutions around 25 kHz and measurement precision of a few kilohertz.

For correct sorting of emissions it is important that modulation be identified automatically by the system. Such identification may also be performed manually by the operator, but automation improves the characteristics of the system by decreasing its reaction time.

A good COM-ESM system may be able to perform the pause and suppress functions. The pause function stops the search receiver at active channels, allowing the operator to select the parameters most suitable for listening to and analyzing an intercepted emission by sending the related measurements to the system computer.

On the basis of the analysis, the operator can decide whether to suppress the channel, by using the suppress function, or insert it in the discrete scan table, or whether to send it to the monitoring or locating subsystem for the COMINT functions, or to the jamming subsystem to generate the appropriate ECM. Suppressed emitters are no longer intercepted by the scan receiver, thus saving precious time. The DS mode allows scan only over a certain number of priority emissions preselected by the operator to check frequently for analysis, listening or surveillance.

Another possible performance mode is the hybrid CS + DS. In this case the system performs the CS and DS functions cyclically by time sharing, thus allowing the operator both surveillance of the electromagnetic spectrum and continuous monitoring of priority channels.

If it is necessary to intercept communications which exploit frequency hopping, the frequency parameter cannot be used for identification or classification of emissions. In this case it is of much greater importance to recognize that a frequency-hopping emission is present, to intercept as many channels as possible among those used by the hopper, and to determine the time length of channel occupancy and if possible the way in which the frequency varies. Then it is possible to preset the direction finder on the channels most used by the hopper and await the arrival of samples for DF measurement purposes. Usually, the content of a frequency hopping communication cannot be analyzed.

4.7.2 COMINT

The functions performed by COMINT systems are essentially to monitor, locate and record, for subsequent analyses, the emissions intercepted by the COM-ESM system.

The monitoring function entails:

- demodulating a certain number of channels of interest to make them available to the operator.
- allowing the operator to listen to the channels of interest.
- recording all data related to intercepted emissions together with the messages themselves.

The locating function entails:

- correlation between centers or, if the platform is mobile, between the various fixing measurements.
- recording the location data together with all the other intercepted parameters.

All intercepted data are classified and arranged in a data base and sent to the centers in charge of subsequent analysis and processing, for determination of possible strategies

REFERENCES

[1] D.C. Schleher, *Introduction to Electronic Warfare*, Norwood, MA: Artech House, 1986.

[2] D.K. Barton, *Radar Systems Analysis*, Prentice Hall, 1976.

[3] R.G. Wiley, *Electronic Intelligence: The Interception of Radar Signals*, Norwood, MA: Artech House, 1985.

[4] A. Brann, "RWRs Face New Threats," *Microwave Systems News*, November 1986.

[5] F.M. Herschell, "100 Percent Probability of Intercept?" *Defense Electronics*, February 1988.

[6] B.R. Hatcher, "Intercept Probability and Intercept Time," *EW*, March-April 1976.

[7] A.G. Self, "Intercept Time and its Prediction," *Journal of Electronic Defense*, August 1983.

[8] H.J. Belk, J.D. Rhodes, and M.J. Thornton, "Radar Warning Receiver Subsystems," *Microwave Journal*, September 1984.

[9] N.J. Whittall, "Signal Sorting in ESM Systems," *IEE Proceedings,* Vol. 132, Pt. F, No. 4, July 1985.

[10] C.L. Davies and P. Hollands, "Automatic Processing for ESM," *IEE Proceedings,* Vol. 129, Pt. F, No. 3, June 1982.

[11] L.G. Bullock, G.R. Oem, and J.J. Sparagna, "An Analysis of Wide Band Microwave Monopulse Direction-Finding Techniques," *IEEE Transactions on Aerospace and Electronic Systems,* Vol. AES-7, No. 1, January 1971.

[12] A.R. Baron, K.P. Davis, and C.P. Hofman, "Passive Direction Finding and Signal Location," *Microwave Journal,* September 1982.

[13] T.E. Morgan, "Spiral Antennas for ESM," *IEE Proceedings,* Vol. 132, Pt. F, No. 4, July 1985.

[14] J.B.-Y. Tsui, P.S. Madorn, and R.L. Davis, "Advanced Electronic Warfare Receiver Forecast," *SPIE,* Vol. 789, Optical Technology for Microwave Applications III, 1978.

[15] D.L. Lochead, "Receivers and Receiver Technology for EW Systems," *Microwave Journal,* February 1986.

[16] J.B.-Y. Tsui, *Microwaves Receivers with Electronic Warfare Applications,* New York: John Wiley & Sons, 1986.

[17] P.W. East, "Design Techniques and Performance of Digital IFM," *IEE Proceedings,* Vol. 129, Pt. F, No. 3, June 1982.

[18] J. Edwards, "Sensitivity of Crystal Video Receivers," *IEE Proceedings,* Vol. 132, Pt. F, No. 4, July 1985.

[19] H. Klipper, "Sensitivity of Crystal Video Receivers with RF Preamplification," *Microwave Journal,* Vol. 8, No. 8, July 1965, pp. 85-52.

[20] W.J. Lucas, "Tangential Sensitivity of a Detector Video System with RF Preamplification," *IEE Proceedings,* 1966, 113 (8).

[21] J.B.-Y. Tsui, "Tangential Sensitivity of EW Receivers," *Microwave Journal,* October 1981.

[22] W.B. Davenport and W.L. Root, *An Introduction to the Theory of Random Signals and Noise,* New York: McGraw-Hill, 1958.

[23] R.G. Wiley, *Electronic Intelligence: The Interception of Radar Signals*, Norwood, MA: Artech House, 1985.

[24] S.J. Erst, *Receiving System Design*, Norwood, MA: Artech House, 1984.

[25] J.E. Dean, "Suspended Substrate Stripline Filters for ESM Applications," *IEE Proceedings*, Vol. 132, Pt. F, No 4, July 1985.

[26] D.C. Webb, "AO, SAW, BAW, and MSW Technology for Frequency Sorting," IEEE Ultrasonic Symposium, 1986.

[27] T. Higgins, "Channelized Receivers Come of Age," *MSN*, August 1981.

[28] D.E. Allen, "Channelized Receivers: A Viable Solution for EW and ESM Systems," *IEE Proceedings*, Vol. 129, Pt. F, No 3, June 1982.

[29] J.B.-Y. Tsui, "Channelizers and Frequency Encoders," *Microwave Journal*, September 1989.

[30] A.E. Spezio, J. Lee, and G.W. Anderson, "Acousto-Optics for Systems Applications," *Microwave Journal*, February 1985.

[31] P.V. Gatenby, "Broadband Integrating Bragg Cell Receiver for Electronic Support Measures," *IEE Proceedings*, Vol. 136, Pt. F, No. 1, February 1989.

[32] D. Mergeria and E.C. Malarkey, "Integrated Optical RF Spectrum Analyser," *Microwave Journal*, September 1980.

[33] C.L. Grasse and D.L. Brubaker, "Acousto-Optic Bragg Cell Speeds EW Signal Processing," *MSN*, January 1983.

[34] T.R. Joseph, "Integrated Optic Spectrum Analyser," *IEE Proceedings*, Vol. 129, Pt. F, No 3, June 1982.

[35] M.C. Hamilton, "Wideband Acousto-Optic Receiver Technology," *Journal of Electronic Defense*, January-February 1981.

[36] N.J. Berg, M.W. Caneday, and I.J. Abramowitz, "Acousto-Optic Processing Increases EW Capabilities," *MSN*, February 1982.

[37] J. Yarbourough, "Second Generation Bragg Cell Receiver," *Journal of Electronic Defense*, October 1985.

[38] R. Adler, "Interaction between Light and Sound," IEEE

Spectrum, May 1967.

[39] E.H. Young and Shi Kay Yao, "Design Consideration for Acousto-Optic Devices," *IEEE Proceedings*, Vol. 69, No. 1, January 1981.

[40] E.I Gordon, "A Review of Acousto-Optical Deflection and Modulation Devices," *IEEE Proceedings*, Vol. 54, No. 10, October 1966.

[41] C.F. Quate, C.D. Wilkinso,n and D.W. Winslow, 'Interaction of Light and Microwave Sound', *IEEE Proceedings*, Vol. 53, No 10, October 1965.

[42] B.K. Harms and D.R. Hummels, "Analysis of Detection Probability for the Acousto-Optic Receiver,"*IEEE Transactions on Aerospace and Electronic Systems*, Vol. AES-22, No. 4, July 1986.

[43] M.A. Jack, P.M. Grant, and J.H. Collins, "The Theory, Design and Application of SAW Fourier-Transform Processors," *IEEE Proceedings*, Vol. 68, No. 4, April 1980.

[44] H. Matthews, *Surface Wave Filters Design, Construction and Use*, New York : John Wiley & Sons, 1977.

[45] M.A. Jack, G.F. Manes, P.M. Grant, C. Atzeni, L. Masotti, and J.H. Collins, "Real Time Network Analyser Based on SAW Chirp Transform Processor," Ultrasonic Symposium Proceedings, IEEE Cat. 76, September 1976.

[46] C. Lardat, "Improved SAW Chirp Spectrum Analyser with 80 db Dynamic Range," Ultrasonic Symposium Proceedings, IEEE Cat. 78, September 1978.

[47] J.B. Harrington and R.B. Nelson, "Compressive Intercept Receiver Uses SAW Devices," *Microwave Journal*, September 1974.

[48] D.L. Adamy, "Trends in Tactical Communication ESM," *International Countermeasures Handbook*, Palo Alto: 1987, pp. 53–57.

Chapter 5

Electronic Countermeasures Systems

5.1 Introduction

The objective of *electronic countermeasures* (ECM) systems is to prevent an enemy's weapon systems from operating correctly, without resorting to conventional weapons [1, 2, 3]. ECM directed against IR-guided weapon systems are distinguished by the name *infrared countermeasures* (IRCM). The table in Figure 5.1 shows a simple classification of ECM equipment. The description which follows will be based on this table.

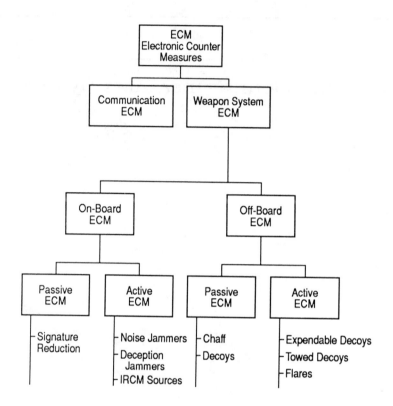

Figure 5.1 Classification of an ECM system.

5.2 On-board ECM systems

5.2.1 Passive systems

5.2.1.1 Signature reduction Following the well-known principle that the best defense is to avoid detection, the protected airborne or naval platform must in the first place try to minimize its visibility to enemy search systems, which may be of the *radio-frequency* (RF), the *infrared* (IR), or the optical type.

To reduce radar signature [4, 5], the following techniques can be exploited (Fig. 5.2):

• use of RF-absorbing materials.

- use of RF-transparent composite materials.
- reduction of edges, of surface inhomogeneity and of corner reflections.
- deflection of radiation in directions other that of the radar.

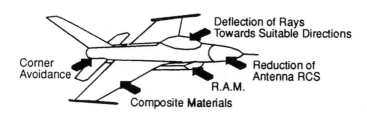

Figure 5.2 Some techniques used for stealth aircrafts can be used to reduce the signature of more conventional platforms.

To reduce the IR signature it is necessary:

- to shield the hottest parts of the platform, such as the engine exhaust nozzles of an aircraft, or the funnels of a ship.
- to minimize the output temperature of exhaust gases.
- to include additives in fuels so that emissions are centered in spectral areas in which atmospheric transmission is low.

Mimetic paints of low reflectivity, suitable coloring, and so forth, are employed to reduce the signature in the optical domain.

5.2.1.2 Reduction of the radar cross section of an antenna

The *radar cross section* (RCS) of a platform depends not only on its surface area, its geometry and the materials which compose it, but also on the antennas of the sensors installed on it. On a stealthy platform, the RCS of the antennas could be the major part of the radar signature.

The RCS of an antenna is, according to the general definition of RCS, the ratio of the signal power reradiated in the direction of observation to the incident power density.

Reradiated signal power has essentially two components: the power P_s scattered by the antenna structure, which is not easily predictable, and the power P_{tr} arising from reflection of the received signal because of the imperfect match between the antenna and the receiving line. This will be reradiated with transmission gain G_T. If the incident power density is p, and the antenna gain on reception is G_R, the power at the input of the receiving line will be

$$P_i = p \frac{G_R \lambda^2}{4\pi}$$

The power reflected by the line, for a *voltage standing-wave ratio* (VSWR) r, will be

$$P_r = P_i \left(\frac{1-r}{1+r} \right)^2 = \rho^2 P_i$$

where ρ is the reflection coefficient deriving from r. One may therefore write

$$(RCS)_A = \frac{P_r G_T + P_s}{p}$$

Suppose it were possible to realize an antenna with a negligible P_s, then the RCS would be

$$(RCS)_A = \frac{P_r G_T}{p} = \frac{\rho^2 G_T G_R \lambda^2}{4\pi}$$

If $G_R = G_T = G$, then

$$(RCS)_A = \frac{G^2 \lambda^2 \rho^2}{4\pi}$$

It should be remembered that G is the antenna gain in the direction of observation, and not the maximum antenna gain.

In order to minimize the RCS antenna one can only alter ρ, that is, rectify the mismatch of the receiving line. So minimization of antenna RCS is generally achieved by careful design and manufacture.

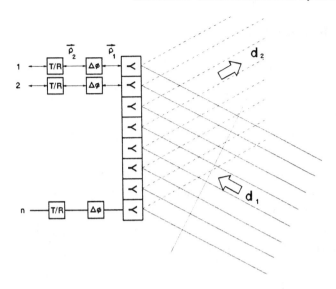

Figure 5.3 Phased array antenna.

If a particularly low antenna RCS is required, phased-array antennas are to be preferred, because of their flexible performance. The following considerations apply to phased-array antennas as shown in Figure 5.3:

- suppose that the significant ρ is due to a mismatch between the radiating element and the phase shifter, and assume that the signal reradiated by each element of the array undergoes the same phase shift. Then radiation arriving from a certain direction will be reradiated, on average, in the direction of specular reflection. Thus the RCS of a phased-array antenna will be particularly low in every direction, with the exception of directions of observation orthogonal to the surface of the array.

- suppose that the most significant ρ is due to a mismatch beyond the phase shifter. Then it will be necessary to analyze the behavior of the phase shifter. This may be frequency-independent, or can be realized by means of lines of different length (true time delay phase shifter) with frequency dependence of the form

$$\Delta\varphi = \frac{2\pi}{\lambda}L$$

where L is the length of a delay line with time delay $\Delta t = L/c$.

The phase shifter may be of reciprocal or non-reciprocal type. In the latter case, the phase shift depends on the direction of propagation. Some non-reciprocal phase shifters insertion loss only in a selected direction.

It is now easy to discuss G_R and G_T for this case. If the phase shifter is non-reciprocal, the reradiated signal will be the sum of randomly phase-shifted signals, so that G_T will on average be equal to the gain of the sidelobes, except in directions of observation orthogonal to the array. G_R will depend on the pointing direction of the antenna; as far as phase-scanning phase shifters are concerned, it will depend also on frequency.

For reciprocal phase shifters, in general $G_T = G_R$, the relation above is still valid, and the antenna RCS will be maximum when the antenna is pointed in the direction of observation. It should be noted that while for reciprocal frequency-scanning phase shifters the RCS will be maximum at all frequencies, for phase-scanning phase shifters this will be the case only at the antenna operating frequency. To minimize the RCS of a frequency-scanning phase shifter, it is possible to exploit the fact that the beam will be pointing in the direction of interest only for the time strictly necessary to perform the assigned mission, which could be of the order of microseconds, and then scanned elsewhere, thus presenting an extremely low average RCS to potential observers.

This can happen with an airborne interceptor that could for example track a target by pointing its beam at it only to transmit a pulse in that direction, and then point the beam again for the minimal time necessary to receive the echo (Fig. 5.4).

To sum up, the RCS of the phased-array antenna can be very low even when there is significant ρ due to mismatches after the phase shifter.

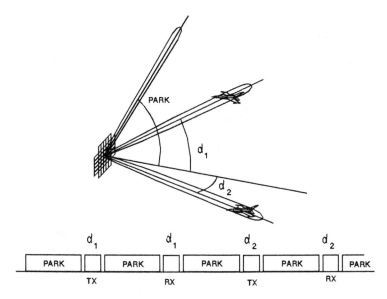

Figure 5.4 A phased-array radar can point its beam momentarily toward the targets, keeping it parked in a suitable direction at other times.

5.2.2 Active systems: noise jammers

A noise jamming system is an ECM device whose objective is to generate a disturbance in the radar receiver in such a way as to prevent detection of a target.

For the jamming to be effective, it is necessary that at the output of the radar receiver, the signal J produced by the jammer be of such intensity as to mask the radar signal S, which can be computed by means of the radar equation; that is, the jamming-to-signal ratio must be adequate [1, 7].

Ideal jamming is achieved by the generation of noise very similar to the victim radar's thermal noise, so that detection of neither target nor jamming signal is possible.

Generally a jammer (Fig. 5.5) consists of a receiver, a generator of jamming signals, and a transmitter. The receiver is needed to identify the signal to be jammed and to tune the jamming signal generator to the correct frequency.

The generated signal is a noise of a given bandwidth centered on the frequency of the victim signal. If the receiving

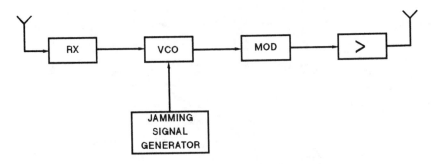

Figure 5.5 Block diagram of a noise jammer.

and transmitting antennas are not isolated from one another, tuning is carried out during "look-through" periods, when the transmission of jamming signals is interrupted in order that the victim radar's signal may be received correctly. This entails a loss in efficiency, so the look-through period must be carefully determined.

The fundamental characterstics of a jammer are:

- spatial coverage.
- frequency coverage.
- receiver sensitivity.
- dynamic range of the parameters acceptable to the receiver, including minimum and maximum *pulsewidth* (PW), minimum and maximum *pulse repetition frequency* (PRF), and so forth.
- tuning precision and speed of jamming signals generator.
- noise bandwidth.
- noise quality.
- effective radiated power (ERP), that is, transmitted power really multiplied by antenna gain.
- polarization.

The space coverage of a noise jammer is generally 360 degrees in azimuth and 20 degrees to 40 degrees in elevation. A coverage of 120 degrees in azimuth is considered acceptable in the forward and aft sectors of an air platform.

Since the *continuous wave* (CW) power of a *traveling-wave tube* (TWT) is not very high (a few hundred watts in the X and K bands), it is necessary to adjust the antenna gain to obtain the desired ERP. However, the antenna beam will thereby be narrowed; to achieve full coverage it will be necessary to provide the antenna with a servo-driven pedestal in order to position the jammer in the azimuth and elevation directions designated, for example, by the *electronic support measures* (ESM) system.

If the beam is very narrow (a few degrees), normal designation by the ESM, unless extremely accurate, is no longer sufficient; the victim emission must be tracked directly, which remarkably complicates the whole system. It should be recalled that if the ESM has a precision of σ degrees (rms), a beam of aperture $\pm\sigma$ degrees will only guarantee that in 67% of the cases the victim emitter is within the 3 dB beam of the jammer's antenna. Normally, a beam with an aperture $\vartheta_{BJ} = \pm 2\sigma$ is considered acceptable.

5.2.2.1 Noise generation

To jam a radar receiver effectively with noise, one has to generate noise that as far as possible emulates the thermal noise of the radar receiver. In this way, the radar operator cannot be sure that the radar is being jammed, especially if it has a *constant false-alarm rate* (CFAR) receiver that adapts its threshold to noise.

Thermal noise is effectively white noise, which means that its spectrum is uniform: All the frequencies in the band of interest have the same probability of being employed, and its amplitude has a Gaussian distribution. It is a signal with these characteristics that should therefore be introduced into the radar (Fig. 5.6).

One method of generating a signal of this type is as follows [1]: The noise band of interest from a noise source, for example a highly amplified diode, is filtered and directly amplified to the maximum power that can be generated by the transmitter, which consists of a power amplifier. This method, called *direct noise amplification* (DINA), is nowadays little used because, clearly linear wideband power amplification is not a very effi-

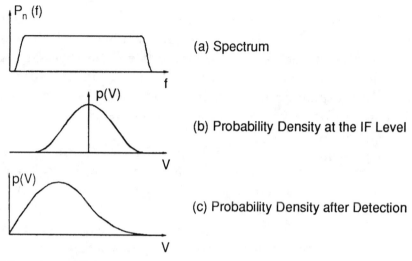

(a) Spectrum

(b) Probability Density at the IF Level

(c) Probability Density after Detection

Figure 5.6 White noise.

cient process. A saturated power amplifier of the TWT type, capable of operating in a wide band (more than an octave) with relatively high efficiency and high power is usually employed instead [3].

With this choice, however, the transmitted power is constant, and further steps must be taken to generate the noise. During the look-through mode, an *automatic frequency control* (AFC) device keeps a *voltage-controlled oscillator* (VCO) tuned to the frequency of the victim radar (Fig. 5.7). Later, noise is added to the tuning voltage of the VCO to obtain a random modulation of its frequency. The signal thus obtained is sent to a TWT power transmitter and radiated at constant power toward the victim radar. This signal, randomly entering and exiting the radar tuning frequency, produces at the output of the radar receiver a voltage whose amplitude varies randomly (Fig. 5.8) like a noise. The spikes thus produced in the radar will have a duration and amplitude that depend on how the jammer frequency passes through the radar band.

In the above process the jammer's noise power will be distributed over a bandwidth wider than the radar bandwidth

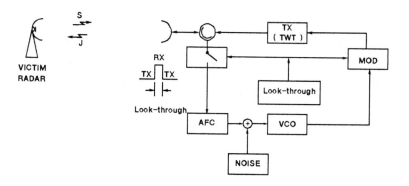

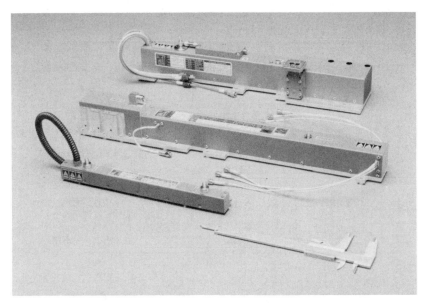

Figure 5.7 Generation of the noise jamming signal. The photograph shows TWTs used in ECM equipment.

(Fig. 5.9); this causes a loss

$$L_n = \frac{B_j}{B_R}$$

that generally includes the loss arising from the generation process of a noise of good quality.

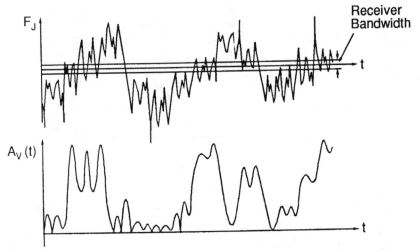

Figure 5.8 Conversion of swept CW into noise in the radar receiver.

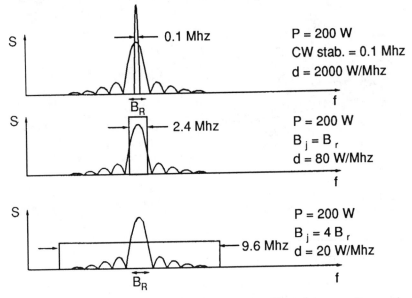

Figure 5.9 Power density and noise power collected in a radar receiver.

5.2.2.2 Types of noise

The forms of noise jamming used in practice are the following:

- CW is used to saturate radar receivers. It is generally ineffective against modern receivers.

- swept CW is a narrowband signal rapidly swept over the band of a radar receiver which generates high-intensity spikes that cause oscillations in the first receiver stages.
- spot noise is noise over a bandwidth limited to cover the spectrum of the pulse signal radiated by the radar and the potential small unwanted shifts of the RF radar carrier.
- barrage noise is wideband noise that covers the full bandwidth used by a frequency agile victim radar. It can be used against a naval platform to jam several radars of the same type simultaneously.
- gated noise is obtained when noise can be generated only in a portion of the radar range. This form of noise is very important because it permits jamming of several radars simultaneously. The technique is quite complex because it requires that the victim radar signal be received while jamming and that the jammer gate remain open to the radar PRF.

All these jamming systems may be amplitude modulated if it is thought that AM will improve their effectiveness. This is the case when jamming a conical-scan tracking radar is required.

Figure 5.10 shows the effects of the different types of noise against a search radar lacking ECCM.

5.2.2.3 The jammer equation

A jammer may be employed either to defend the platform on which it is installed (self-screening), to defend other platforms which are in the same radar beam (mutual screening), or to impede the performance of a search radar, to the advantage of other platforms, while the jammer remains at a distance from the zone of operations (stand-off) [6, 7].

The parameter measuring the effectiveness of a jammer is the ratio of the jamming signal power to the target signal power, J/S(or JSR); if J is higher than S, the radar performance could be compromised. But beware! The ratio must be measured at the output of the radar receiver, to take into account the signal processing gain.

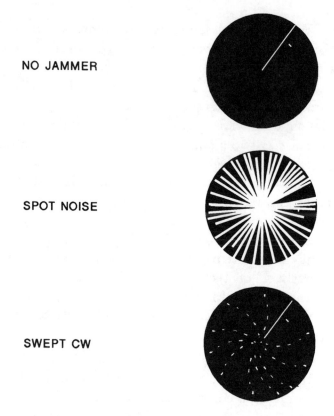

NO JAMMER

SPOT NOISE

SWEPT CW

Figure 5.10 Effects of various types of jamming on radar.

In the self-screening case, J/S is calculated starting from the signal-to-noise ratio (SNR) equation that expresses the radar performance:

$$\frac{S}{N} = \frac{N_i P_T n G_T G_R \sigma \lambda^2}{(4\pi)^3 KTFBR^4 L_i L_m L_x L_{Tx} L_{Rx} L_b}$$

Analogously, when a jamming signal of power P_j occupying a band B_j is transmitted, the *jamming-to-noise ratio* (JNR) that can be obtained at the output of the receiver will be

$$\frac{J}{N} = \frac{(P_j/B_j)BG_j G_R \lambda^2}{(4\pi)^2 KTFB}$$

where L_p are polarization losses arising from the fact that generally the polarization used by the jammer does not coincide with the polarization used by the radar. Here other the jammer transmission losses. The signal losses L_x and L_m are not losses for a jamming signal which resembles thermal noise. Either L_b is incorporated into the G_j, or it is negligible.

Considering that (JNR)/(SNR) is equal to J/S, one obtains, in free space (see Fig. 5.11),

$$J/S = \frac{(ERP)_j}{n(ERP)_R \sigma} \frac{B}{B_j} \frac{L_i L_m L_x}{N_i} \frac{4\pi R^2}{L_p}$$

where $(ERP)_j$ is the effective power transmitted by the jammer in the direction of the victim radar (i.e., it includes the transmission losses), and $(ERP)_R$ is the effective power transmitted by the radar.

Once again it should be noted that the radar performance depends on the energy reflected by the target and therefore on the number of pulses N_i effectively used by the integrator circuit.

It was shown in section 2.2.3.3 that, in an operational environment, propagation depends on F_p. The propagation should be considered two-way for the radar, and one-way for the jammer. Moreover, since the radar signal is the sum of contributions from all the elementary scatterers of the target, F_p^4 to be considered is the weighted average F_{pi}^4, of the propagation function. For this reason, if the target is an extended target (for example a ship in a calm sea), the typical multipath pattern tends to disappear (Fig. 5.12). The expression for J/S becomes

$$\frac{J}{S} = \frac{(ERP)_j}{n(ERP)_R \sigma} \frac{B}{B_j} \frac{L_i L_M L_x}{N_i} \frac{4\pi R^2}{L_p} \frac{F_p^2}{F_{pt}^4}$$

The same considerations apply to the case of mutual screening, with the difference that now two different geometries must

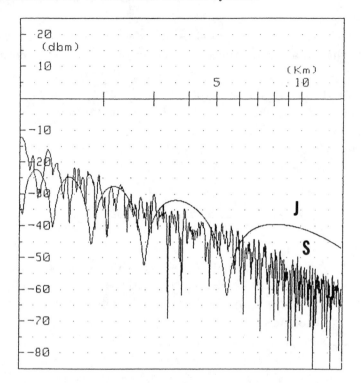

Figure 5.11 Power of the useful signal and jamming received by the seeker of an anti-ship missile in the presence of multipath.

be taken into account, that of the jammer and that of the protected target:

$$\frac{J}{S} = \frac{(ERP)_j}{n(ERP)_R \sigma} \frac{B}{B_j} \frac{L_i L_M L_x}{N_i} \frac{4\pi R_T^2}{R_j^2 L_p}$$

In the case of a standoff jammer, one obtains

$$\frac{J}{S} = \frac{(ERP)_j}{n(ERP)_R \sigma} \frac{B}{B_j} \frac{L_i L_M L_x}{N_i} \frac{4\pi R_T^4}{R_j^2 L_p} \frac{G_{SL}}{G_R}$$

where G_{SL} is the radar antenna gain in the direction of the sidelobes.

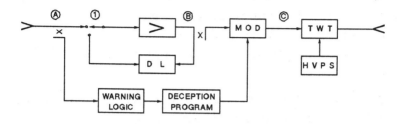

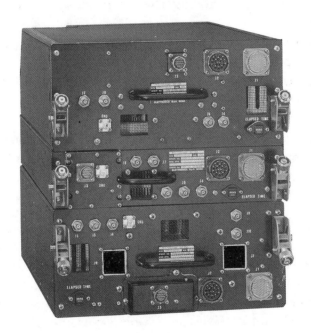

Figure 5.12 Block diagram of a deception jammer. The photograph shows an avionic deception jammer.

5.2.3 Deception jammers

The main objective of a deception jammer is to provide the victim radar with erroneous information by the generation of signals that are similar to the signals that the radar expects, but of much higher power, that is, of high JSR. This type of equipment is able to receive and memorize a radar signal, and retransmit it at the appropriate time, with suitable amplitude, phase and polarization modulations.

Basically, simple deception jamming can take the following forms:

- generation of multiple false to counter for a search radar or a tracking radar in the acquisition phase.
- *range gate pull-off (RGPO)* against a tracking radar, with the objective of moving the range gate onto an erroneous range.
- *velocity gate pull-off (VGPO)* against a tracking radar exploiting the Doppler effect.
- generation of targets with superimposed amplitude modulation to generate false angular data in tracking radar of the sequential type.

A discussion of the various attainable deception techniques will follow a discussion of the principal types of deception equipment.

5.2.3.1 Pulse deception jammer

Figure 5.13 shows the block diagram of a deception jammer capable of retransmitting pulses toward the victim radar. Signals received by the antenna are sent to a circuit where they are detected and processed to check for the presence of threatening radars. Often the logic employed comprises verification of pulsewidth and PRF. If these two measurements are within the limits for "alarming" samples, deception programs are implemented. These programs generally include RGPO, to deceive tracking radars in range with or without amplitude modulation to deceive in angles conical-scan, lobe-switching, *conical-scan on receive only* (COSRO), *lobe-switching on receive only* (LORO), and sometimes *track-while-scan* (TWS) radars. A *frequency memory loop* (FML) is the basic circuit employed to create replicas of the received pulse.

The operating principle of this circuit is very simple. The received signal passes through the switch 1, is amplified while

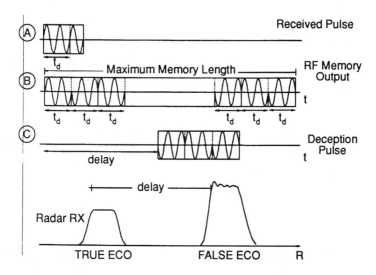

Figure 5.13 Generation of deceptive carrier and signal.

simultaneously being detected in the warning logic circuit, and is sent to a delay line (generally 100 to 200 ns). As soon as the delay line is full, switch 1 changes its position to interrupt reception and collect the signal at the output of the delay line. The signal is then sent back to the amplifier and the loop is repeated. At the output there will be a signal of frequency equal to that of the input signal, consisting of a series of adjacent "slices," each of duration equal to the time delay of the line (Fig. 5.14). Finally, a modulation circuit generates the deceptive pulse to be transmitted at a suitable time with suitable duration.

Since the amplifier has a noise figure F, and the signal is attenuated while passing through the delay line, at each cycle the noise present at the receiver output is increased. Obviously, after a certain number of cycles, the noise will be stronger than the signal. The loop must be interrupted before the signal is excessively corrupted by noise. This generally sets a severe limit on the memory duration [8].

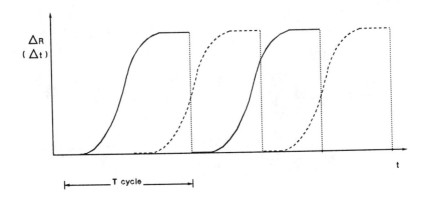

Figure 5.14 Interlaced RGPO.

The RF signal produced by the FML is sent to a modulator controlled by the generator of the deception program. Here the signal is modulated in time, duration, and amplitude to generate the best possible deception signal. Sometimes the duration of the transmitted signal does not coincide with that of the incoming pulse, but is kept at a constant reference value.

The delay law of the deception signal, that is, the movement of the false echo relative to the real one (skin return), is generally arranged as follows: First the deception signal is transmitted with minimum delay for a length of time such that the *automatic gain control* (AGC) of the victim radar is captured. In this way the "real" radar signal is attenuated in the same ratio as the generated J/S. Then, with an acceleration credible to the radar, the false echo is detached from the real one. Usually, the tracking radar's range gate follows the false echo. Once the maximum delay has been reached, it is possible to interrupt transmission of the jamming signal and leave the radar without a target, forcing it to start afresh the process of search and acquisition, and thus preventing initiation of firing by the weapon system. Alternatively, it is possible to create a false target remote from the real one, so that the enemy weapon system will fire at the wrong place. In any case, the deception system is iterative; that is, at the end of the cycle, if the warning conditions persist, it restarts the delay sequence automatically.

With this type of deception, because of the short memory duration (3 to 10 ms, corresponding to 450 to 1500 m of radar range) it is practically impossible to realize *range gate pull-in*, i.e., the creation of a false target approaching the weapon system (see section 5.3.6). RGPI can be achieved only against fixed PRF and fixed-frequency radars (against the latter, only if in addition an AFC system is able to maintain tuning of a VCO or other source of microwaves on the victim radar frequency).

The delay law may be simple or interlaced (Fig. 5.15), and in general may have various acceleration values: a high value, suitable for protection of airborne platforms, and a much lower one, suitable for protection of naval platforms.

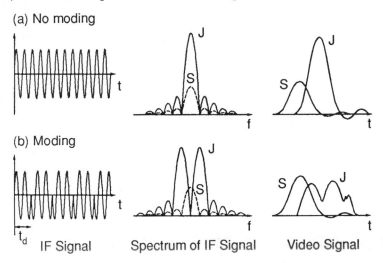

Figure 5.15 Moding can cause a significant loss of deception power.

If the antenna installation permits sufficient isolation between transmitter and receiver, it is possible to transmit while receiving [1, 3]. Otherwise, transmission is not possible until the delay line has been filled. For this reason, delay lines should be as short as the required delay allows.

Typical values for the parameters of deception jammers are as follows:

- Coverage ±50° Az ±20° El (airborne application) 360° Az ±20° El (naval application).
- band sub-bands between 2 and 18 GHz.
- sensitivity −40 dB m.
- PW (in) 0.1 to 2 μs.
- PW (out) 0.3 to 1 μs.
- delay (minimum) 0.15 to 0.25 μs.
- delay (maximum) 3 to 15 μs.
- power peak 1 kW.
- duty 1 %.
- antenna gain 1 to 5 dB (airborne application) 15 to 20 dB (naval application).

Deception jammers of the FML type suffer from a special shortcoming known as "moding." The FML is basically a "slice repeater," since it works by memorizing a slice of the received radar signal and transmitting a jamming signal prepared by repeating the slice M times. According to the duration and frequency of the slice, a phase discontinuity producing an attenuation of the deception signal appears between adjacent slices in the generated signal. This is the moding phenomenon.

Because of the phase modulation introduced, the energy of the jamming signal tends to be centered at particular points of the RF spectrum, at distances $1/\tau$ from one another, that depend on the frequency. The risk arises that the victim radar's bandwidth will not be covered adequately, which may result in significant reduction of the JSR [10].

Understandably, the worst condition will be when successive slices are 180° out of phase with one another. In this case the spectrum of the generated signal will be like the one shown in Fig. 5.16b.

There are at least two ways to mitigate the moding problem. The first is to spread the energy of the jammer artfully in order to be sure that no "hole" is formed corresponding to the victim radar frequency. In this way the worst case is mitigated, at the

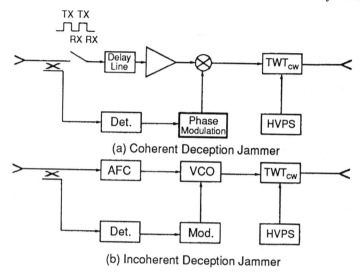

(a) Coherent Deception Jammer

(b) Incoherent Deception Jammer

Figure 5.16 Block diagrams of CW deception jammers a) coherent b) incoherent.

expense, however, of the best case, and on average a loss equal to the ratio of the band occupied by the jamming signal to the band occupied by the radar signal has to be accepted.

The second possibility is to attempt correction of the phase shifts between adjacent slices, with the objective of reducing them and of rendering the phenomenon negligible. In this way, almost all the energy of the jammer enters the victim radar, thus maximizing the ratio J/S. This second possibility gives better performance, but its realization is more difficult.

Figure 5.11 shows J/S as a function of range for various values of the ERP generated by the system.

5.2.3.2 CW Deception jamming

Two types of deception jammers belong to this category: coherent and incoherent [3]. Block diagrams are shown in Figure 5.17. The second diagram is practically identical to the diagram of a noise jammer; it is shown here only to recall that with such a system it is possible to achieve RGPI, provided that the victim radar has a fixed frequency and a fixed PRF. In order to generate a false target that looks as if it were approaching the victim, it is enough, in principle, to tune to the radar pulse on reception, and to use a modulator to generate pulses which are at first delayed by the exact PRI, and therefore coincide with the true echo, but which later are delayed by shorter times.

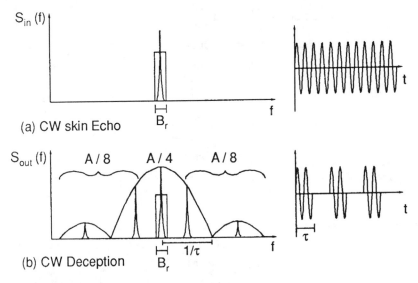

Figure 5.17 Spectrum of a CW deception signal.

In practice, however, all radars used for military objectives have either a jittered or a staggered PRF, and are often frequency agile, so that the field of application of the RGPI is very limited.

More interesting is coherent CW jamming. At present it is the most effective device against semiactive homing, which is widely used by radar-guided AAM and SAM systems.

This type of deception jammer is used for self-protection of air platforms, where lack of space makes it very difficult to achieve good isolation between transmitting and receiving antennas. The isolation problem is even more serious if the equipment is installed in a pod, as is often the case on air platforms.

The isolation problem is solved by separating the reception

and transmission times. The signal is received until the delay line is filled, when reception is interrupted, and the signal at the output of the delay line is then transmitted, suitably amplified and modulated. The on-off modulation of the transmitted signal causes a power loss, as shown in Figure 5.18. The delay line is generally a few tenths of a microsecond long, so that the on-off frequency is a few megahertz.

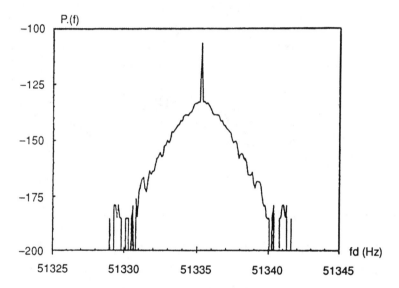

Figure 5.18 Spectrum of CW skin echo received by missile.

The signal received by the semi-active missile is a signal coherent with the aircraft skin return. In fact it is even cleaner, since the various phase components of noise caused by the vector summation over the single elementary scatterers of the aircraft surface are absent.

Figure 5.19 shows the signal received by a semi-active missile while tracking the aircraft to be protected. The missile receiver positions a Doppler gate, at a frequency on the order of kilohertz, on the target only, to avoid distraction by other signals such as clutter, which is almost always much stronger than the echo of interest (Fig. 5.11). By processing the signals at this Doppler frequency, the missile extracts all the angular information needed for guidance toward the target.

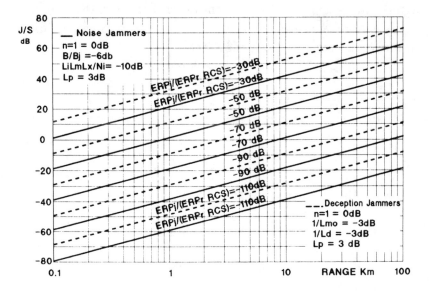

Figure 5.19 J/S provided by noise and deception jammers.

The objective of the deception jammer is to drive the Doppler gate (velocity gate) of the missile off the target, and possibly onto the clutter, so that the missile heads for the clutter instead of the target. In order to achieve this, the output signal of the deception jammer delay line is modulated, after preamplification, so as to cause a frequency shift. To obtain this frequency shift, the voltage of the TWT helix can be suitably modulated by serrodyning. One phase modulates the signal periodically by exploiting the fact that the transit time in the TWT depends on the voltage applied to the helix.

A more recent method is to use a phase modulator to produce the required frequency shift. At first, the deception signal is emitted without frequency modulation. Then, as in RGPO (section 5.3.7), it is frequency shifted with a suitable acceleration. Once the velocity gate has been pulled off, the jammer can be turned off, or made to start the cycle again. The effect will be to cause break-lock. Here too amplitude modulation can be superimposed, if it is thought that the missiles are equipped with COSRO/LORO seekers.

Obviously, this type of deception jammer is capable of dealing with all CW radar, not only with semiactive missiles, and is also effective against systems using interrupted CW.

Typical values of the parameters of CW deception jammers are the following:

- angular coverage $\pm 50°$ Az $\pm 20°$ El (airborne application).
- frequency coverage sub-bands between 5 and 18 GHz.
- sensitivity -40 to -60 dB m.
- Δf 0 to 50 kHz.
- power 50 to 200 W.
- antenna gain 0 to 5 dB.

When the transmitting antenna gain is on the order of 30 dB, and the beam is therefore very. narrow (about 3° by 3°), the deception jammer must use a passive tracking system to keep it pointed at the victim radar. For this purpose, a monopulse system capable of giving angular information during the look-through period can be used. In the same period, noncoherent CW deception jammers update the AFC circuit.

5.2.3.3 The deception jammer equation

For a deception jammer, in which the signal is a replica of that used by the radar and will therefore undergo an identical radar processing, the J/S is given by

$$\frac{J}{S} = \frac{(ERP)_j}{n(ERP)_R \sigma} \frac{4\pi R^2}{L_p L_{mo} L_d} \frac{F_p^2}{F_{pt}^4}$$

where n is the number of code elements (equal to unity if the radar is uncoded or, when it is coded, if the jammer is capable of replicating the code). L_{mo} are the losses due to moding. And L_d are the possible losses due to Doppler processing by

Figure 5.20 ECM system in Pod configuration.

the radar, in the case of a pulsed-Doppler radar to which the jammer does not respond coherently.

5.2.4 The pod

Airborne ED equipment can be either internal or pod mounted. If space is not provided inside the platform at the design stage, it must be found later, but this can be very difficult. In the latter case, a new certificate of airworthiness is also required.

Pod installation requires that the ED equipment be fitted in a streamlined container attached to the aircraft by means of pylons, normally intended to hold ordnance (Fig. 5.20). This type of installation is very useful especially when the ED equipment is retrofitted, and no space or supply of power is available inside the aircraft.

By the use of pods, an aircraft can avoid being encumbered by a weight of ED equipment not needed for a particular mission, but, on the other hand, during combat missions the aircraft will be forced to give up one of the few available pylons to the pod. However, the use of pods makes it possible to minimize the number of systems to be purchased; pods are fitted exclusively to platforms on the flight line, when the aircrafts are about to be engaged in missions requiring that particular ED.

Pods can be slung under the fuselage or under the wings.

When under the fuselage, they are generally heavy and bulky; however the equipment is not excessively stressed by the environmental conditions. When under the wings, pods are generally smaller and more manageable, but the electronics inside them must be capable of enduring a severe vibration regime. Therefore the quality of both pod design and pod manufacture must be extremely high to meet *mean time between failures* (MTBF) requirements.

An aircraft with a pod has a reduced flight capability; the maximum admissible speed as a function of altitude is lower than that of an aircraft without a pod. The reduction in airworthiness introduced by pods is so important that a special pilot-controlled system is required to jettison the pod, should the pilot need to regain full capability. Conformal pods are the least damaging to airworthiness. Once installed, they can match the geometry of the aircraft so well that they look as if they are part of it.

5.3 ECM Techniques

The basic equipment described above allows realization of many fundamental ECM techniques, of both noise and deception jamming types [9].

These techniques will now be described in greater detail, with emphasis on operational objectives rather than methods of realization. They are:

(1) spot noise.
(2) barrage noise.
(3) swept noise/CW.
(4) gated noise.
(5) amplitude modulated noise/CW.
(6) multiple false target generation.
(7) range gate pull-off.
(8) velocity gate pull-off.
(9) dual mode.

(10) inverse gain.

(11) count down.

(12) cooperative jamming.

(13) cross polarization.

(14) cross eye.

(15) terrain bounce.

(16) illuminated chaff.

5.3.1 Spot noise

This jamming technique can be used against search radars to mask either the range or the presence of targets, and against tracking radars to mask the range. It can be used for self-screening, that is, protection of the platform on which the equipment is installed; for mutual support missions, that is, protection of a formation flying in the same jamming radar beam; and for standoff missions, that is, to mask the presence of other targets and to create a corridor through the enemy air defense network by reducing the effective radar range, as shown in Fig. 5.21.

Spot noise can be used effectively only against fixed-frequency radar, in general old fashioned search radar equipped with MTI. Radar ECCM techniques that render this type of jamming less effective are:

- frequency agility.
- frequency diversity.
- track on jam.
- jammer strobe.
- sidelobe blanking.
- sidelobe canceling.

5.3.2 Barrage noise

When the victim radar is frequency agile or uses techniques of the spread-spectrum type, noise jamming must cover a wider band. In this case there is an additional loss due to the fact

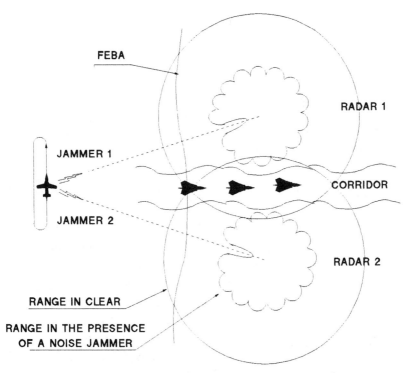

Figure 5.21 Creation of corridors in an air defense network by stand off spot noise jamming.

that the radar keeps its IF band, of width B_{IF}, matched to the duration of the pulse, while the jammer is forced to spread its power over a band broader than, or at least as broad as, the band used by the radar's frequency agility:

$$B_j \leq B_{RF}$$

In general the loss due to frequency agility can be considered equal to the ratio of the RF bandwidth occupied by the radar to its IF bandwidth

$$L_{ag} = \frac{B_{RF}}{B_{IF}}$$

This ratio can take very large values, between 100 and 1000, and compromise jamming effectiveness.

Radar ECCMs valid against this type of jamming were mentioned in the previous section.

5.3.3 Swept noise/CW

Swept noiseis often used to generate confusion in a search radar by creating such a large number of false targets on the *plan position indicator* (PPI) that the automatic detection system is overwhelmed. A CFAR receiver is not always able to maintain its performance in the presence of short bursts of jamming; the same can happen in the presence of swept CW. Moreover, in the latter case, when the swept frequency is very high, strong spikes can be generated in the radar receiver, which may induce oscillations in its first stages (Fig. 5.22).

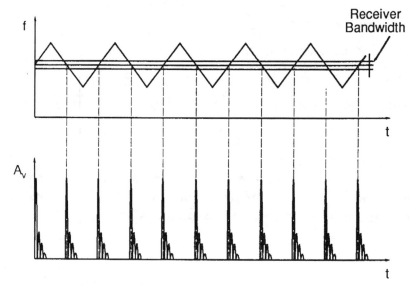

Figure 5.22 Swept CW jamming.

Swept jamming can be used against both search and tracking radars to mask the range of the platform being protected. Frequency agility is obviously not effective against this type of jamming swept over a very wide band. An effective ECCM device is the Dicke-fix receiver, described in section 6.2.2.1.

5.3.4 Gated noise

Gated noise is used to mask the range of a target; the noise is generated near the target (Fig. 5.23). It is a more complex jamming method than continuous noise, since there must be a circuit to predict the time of emission of the noise, which must be synchronized to the radar PRF, whether fixed or jittered. However, it is a more effective method. In fact it can elude ECCM devices capable of detecting the presence of jammers which normally cause a tracking radar to shift to its "track-on-jam" mode, because in general this shift occurs against continuous noise jammers only. Another advantage of gated noise, is that it makes possible the jamming of multiple threats simultaneously, in time-sharing mode [12].

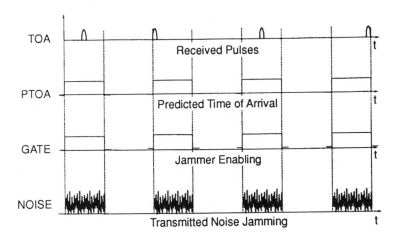

Figure 5.23 Gated noise jamming.

The gated noise may also be generated before the true target echo, which renders this technique very effective against a leading-edge tracker (section 6.3.2.2).

In this case it should be remembered that if the radar is frequency agile, the jammer noise power will have to be distributed over the full agility band. As already mentioned, this severely reduces the effectiveness of this technique.

5.3.5 Amplitude modulated noise/CW

The jamming techniques just described, and also the CW signals tuned to the frequency of the victim radar, can be amplitude modulated. *Amplitude modulation* (AM) is generally applied to jam a tracking radar, rather than a search radar.

AM can jam radar sensors of the conical-scan, lobe-switching, COSRO, and LORO types, if it is done:

(1) at a frequency near the radar sensor's scanning frequency to deceive the tracking radar coherent detector.

(2) at a very low frequency, to jam the servo loop. In this case, some weakness of radar design must be exploited: For the jamming to be successful, the AGC loop should be unable to compensate for low frequency amplitude variations, typically between 0.2 and 2 Hz, and at the same time the servo should overshoot at these frequencies.

Against victim radars of the conical scan, lobe switching, COSRO, and LORO types, CW or noise should be used, depending on the design of the radar receiver. When the radar receiver is ac-coupled to the processing circuits, the post-detection CW signal is eliminated (Fig. 5.24). However, tracking radar receivers are often gated, and in this case the jamming is again effective.

In any case, with both CW and noise jamming, the objective is to interrupt the radar tracking, by interfering with either the angular control loop or the range control loop. When subjected to continuous jamming, the tracking radar often switches to its track-on-jam mode, attempting to track the jammer in angle. To determine the range, if it has no back-up system, it uses devices that memorize target velocity with a relatively high time constant. The extrapolated distance will be given by the last range measured before the shift to track-on-jam, R_0, plus the memorized radial velocity V_0 multiplied by the elapsed time:

$$R = R_0 + V_0 t$$

In these conditions, angular jamming by AM can be extremely

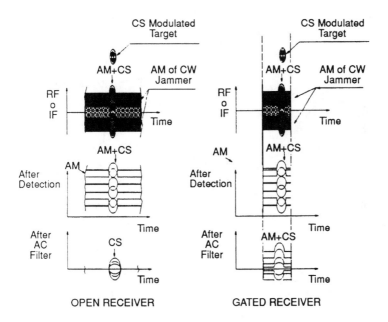

Figure 5.24 Effect of CW jamming on open and gated a.c.-coupled receivers.

effective.

When the radar scanning frequency is known, either because it has been measured instantaneously by the ED system, or because it is known *a priori,* a short with an AM frequency sweep around the scanning frequency is enough to produce deleterious effects (Fig. 5.25):

$$f_{am} = f_0 + \Delta f \left(1 - \frac{2t}{T}\right)$$

When the scanning frequency can be measured, inverse-gain modulation is also possible (section 5.3.10).

When the scanning frequency of the victim radar is not known, the following techniques can be employed:

• Wide sweep. In this case, significant angular jamming will be caused only when the modulation frequency passes through the radar scanning frequency. The problem is the dwell time,

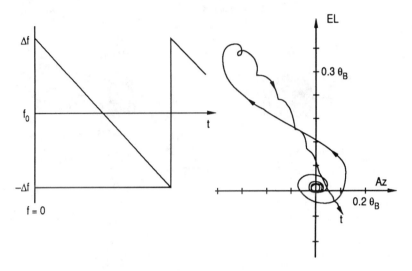

Figure 5.25 Angular errors introduced by swept amplitude modulation.

that is, the proportion of time during which the jammer dwells on the effective modulation frequency. Since the jammer must make a wide frequency sweep in a limited time, scanning must be rather quick, and the dwell time in the zone where jamming is effective could be too short.

- Wide sweep with a stop in the vicinity of the radar scanning frequency (Fig. 5.26). In this case the ECM system needs a sensor capable of determining where the jamming is effective (jog detector), possibly by measuring the amplitude variations of the radar signal that can be correlated with the AM frequency sweep.

If the radar antenna has an angular servo with a rather high band, as happens with missile seekers, then the first passage of the swept signal through the radar scanning frequency will induce angular errors that cause strong oscillations in the pointing of the antenna. Noticeable fluctuations will be detected in the radar signal received by the ED system. Such fluctuations indicate that the jamming is very near to the radar scanning frequency. The width of the sweep will therefore be reduced

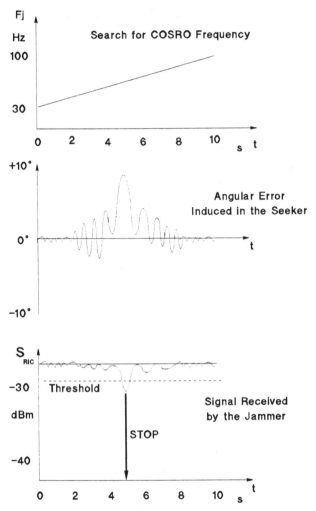

Figure 5.26 Search of the scanning frequency of COSRO or LORO radars by jog detection.

around this frequency value. This method is valid only when the radar has a fixed scanning frequency.

When the radar scanning frequency is not known, on-off AM, also called harmonic AM, may be used. For example, by on-off modulating at a low frequency, it is possible to generate a series of lines in the spectrum of the transmitted signal which interfere with the radar AM. This is harmonic jamming (Fig. 5.27). Often a jitter is inserted in the on-off frequency to be sure to "cover" the radar scanning frequency. In this process there is a strong conversion loss, so that this method should be used only when the J/S ratio is thought to be sufficiently high.

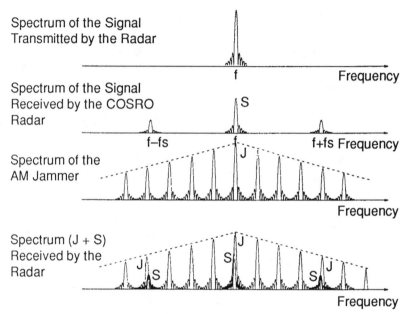

Figure 5.27 Harmonic jamming.

Another AM method that can be fruitfully employed when the radar scanning frequency is not known is the count down technique, which will be discussed in section 5.3.11.

5.3.6 Multiple false target generation

This technique is effective against search radars and the acquisition phase of tracking radars. It can be used both for self-protection and for standoff jamming.

When the ECM system is able to tune itself to the radar frequency and to synchronize with the radar PRF, it is possible to create a series of false targets on the search radar PPI (Fig. 5.28).

If the radar is frequency agile or is using a jittered PRF, false targets can be created only at a range greater than the range of the platform on which the jammer is installed. To create credible targets the jammer must be able to receive the radar signal from the sidelobes as well and to synchronize on the main lobe. With this synchronization one can give a false target credible motion, both radial and angular. By synchronization only on the PRF, one can create only radial motion of the false targets, delaying or anticipating the transmission in order to generate confusion.

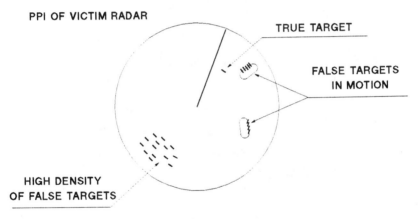

PPI OF VICTIM RADAR

TRUE TARGET

FALSE TARGETS IN MOTION

HIGH DENSITY OF FALSE TARGETS

Figure 5.28 Generation of multiple false targets.

This device for generating multiple targets could be used for RGPI deception jamming against fixed-frequency and fixed-PRF tracking radars, even if this type of radar is not much used nowadays. To achieve RGPI the deception generator is tuned to the radar frequency. By means of an amplitude modulator, a deception signal synchronized with the last received radar pulse, and at first coincident with the next pulse, is generated. By reducing the delay, the transmission of the deception signal is advanced, simulating the presence of a new target much faster than the true target, and therefore more threatening.

The radar is thus tempted to switch to tracking the false target (Fig. 5.29).

Figure 5.29 A very powerful and sophisticated naval ECM system able to generate many types of noise and deception jamming.

In the case of TWS radar, generally airborne, amplitude modulation of the retransmitted signal is performed in synchronization with the radar scanning frequency, to create echoes of higher intensity in directions different from the target direction (Fig. 5.30).

5.3.7 Range gate pull-off

Range gate pull-off (RGPO) is very effective against tracking radars. It is also known as range gate walk-off and range gate

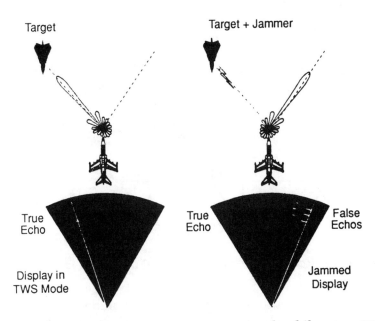

Figure 5.30 Deception jamming against *track-while-scan* (TWS) radar.

stealing (Fig. 5.31) [2]. Up to now, it has been the principal technique employed in self-defense. It is generally very effective especially against automatic tracking systems.

In the absence of an operator to spot the deception and lead the range gate back onto the true target, it is enough to lead the radar range gate away from the real echo and then to deactivate the transmission of the deception signal itself. Only the thermal noise of the receiver will then remain in the range gate, which implies break-lock. Once break-lock has been achieved, the radar must start the search and acquisition phases afresh before it can continue tracking, thus losing precious seconds.

Since this technique to generate the deception signal relies on an RF memory, which is tuned pulse-by-pulse is also effective against frequency-agile radars. Moreover, it is effective even if the radar uses random PRF.

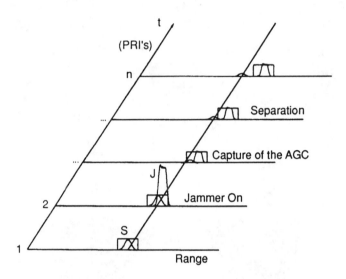

Figure 5.31 *Range gate pull-off* (RGPO).

Sometimes an amplitude modulation is added to the RGPO. This modulation simultaneously generates an angular error in a scanning tracking radar, but it is scarcely effective against a monopulse radar.

RGPO may be countered effectively by a radar with an ECCM technique called *anti-range gate stealing* (ARGS), or leading-edge tracking. This technique is discussed in section 6.3.2.2.

Since the classic deception jammer operates at a low duty cycle and with a relatively high peak power, it is not very effective against radars using very long, coded, or high duty cycle, pulses. Moreover, being that the RF memory loop is necessarily of limited duration, in the case of long pulse it is necessary to employ CW devices with long memories.

5.3.8 Velocity gate pull-off

Another very effective deception device against tracking systems exploiting CW signals and the doppler effect, such as semi-active homing missiles, is *velocity gate pull-off* (VGPO) [3]. It is a deception technique that operates in CW, therefore with a relatively low peak power.

In analogy with RGPO, VGPO uses a CW signal whose power is much higher than the power of the skin echo produced in the illumination of the target. Initially it will appear to the missile receiver at the same soppler frequency and will capture the AGC. Later the frequency of the deception signal is altered, dragging with it the velocity gate. The ideal is to lead the velocity gate to where there are already doppler lines generated by clutter (therefore at a lower Doppler frequency), so that the missile is "hooked" and heads for the clutter.

In any case, once the velocity gate has been lured away, switching off of the deception signal will cause break-lock, and therefore force the missile to recommence the search and acquisition phase in doppler. When the missile is already in flight, this re-acquisition phase is extremely difficult.

The effectiveness of this technique may be reduced by the use of *guard gates* (section 6.3.2.3) as ECCM in the missile receiver. In this case the simultaneous presence of a signal in the main gate (deception) and in the guard gate (skin return) switches the system to memory; that is, the missile proceeds on the basis of the last velocity values. If the deception persists, the missile will shift to the track-on-jam mode, trying to reach the target.

Here, too, amplitude modulation, which is useful against angle-tracking systems of the scanning type, can be superimposed.

5.3.9 Dual mode

Quite often, RGPO cannot be successful because of ECCM devices such as the leading-edge tracker, and gated noise is not very effective because of circuited frequency agility. A combination of the two techniques can however be successful. Gated noise substantially reduces the capability of the leading-edge tracker, while RGPO is insensitive to frequency agility.

By using the two techniques simultaneously, a break-lock situation can be achieved where before it was impossible using each technique on its own (Fig. 5.32). The dual mode of operation, pulse and CW, can be achieved either by using two TWTs, one pulsed and the other CW, perhaps with a single modulator, or by using a TWT capable of operating simultaneously in pulsed and continuous regimes [2].

Since this technique requires generation of the deception while the jamming transmission is under way, it is necessary, during installation, to ensure the correct isolation between transmitter and receiver. Otherwise the technique can be used only against fixed-frequency and fixed-PRF radars.

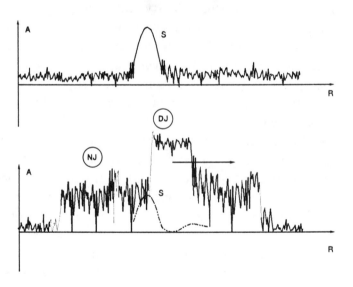

Figure 5.32 Dual mode. Both *noise jamming* (NJ) and *deception jamming* (DJ) are generated.

5.3.10 Inverse gain

This technique provides deception or noise jamming with amplitude modulation in phase opposition to that generated by the target (Fig. 5.33).

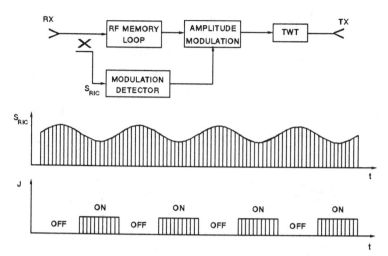

Figure 5.33 Inverse gain. The modulation of the deception signal is opposite to that of the radar tracking signal.

A *phase-lock loop* (PLL), or adaptive threshold, circuit determines the modulation induced by the radar, and coherently generates a modulation that in the simplest case is of the on-off type. This type of modulation is capable of making the victim radar move in a sense opposite to that needed for correct tracking, thus achieving break-lock.

5.3.11 Count down

The *count-down* technique is applied to angular jamming of a tracking radar which uses AGC [2, 3]. A noise or deceptive jamming signal of the on-off type is transmitted, with frequency and duty cycles such that the AGC is practically never at the right level. To make it more comprehensive, often the duty cycle is changed periodically. The technique is so named because originally a counter was used to perform a count down to determine the period of variation of the duty cycle.

To understand the effectiveness and the limits of this technique, it is necessary to recall briefly the mode of operation of an AGC circuit.

In general the gain control of a radar receiver operates at intermediate frequencies (Fig. 5.34). The detected signal of interest, on which the range gate is positioned, passes through a sample and hold circuit and is converted into a continuous voltage (see section 2.2.6.3). This voltage is compared with a reference value; if it is higher, an error signal is generated to reduce the IF gain; if lower, to increase the gain.

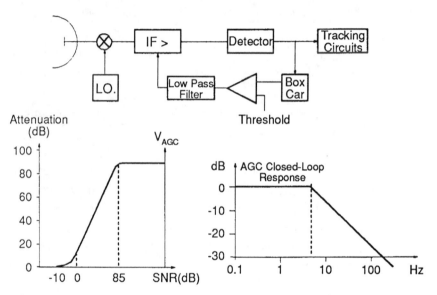

Figure 5.34 Block diagram of radar with *automatic gain control* (AGC) and AGC characteristic.

The AGC circuit tends to compensate amplitude variations of the echo signal arising from scintillation, range effect, target size, all fluctuations taking place at very low frequency.

If the radar is a conical-scan, or lobe-switching radar, care must be taken to limit the AGC bandwidth to at least one decade below the scanning frequency, so as not to cancel amplitude modulations arising from the conical scan or lobe switching (see section 2.2.6.1). In practice, for this type of radar the band is kept to a few hertz, with a maximum of 8 to 10 Hz when the scanning frequency is above 100 Hz. The characteristic curve of the AGC voltage of a radar with wide dynamic range is shown in Figure 5.34.

The output dynamics of a receiver of this type is on the order of 15 dB, around an operating level fixed by the AGC threshold. Beyond these values, there will be saturation or a null signal at the receiver output.

Consider the radar signals in the different stages shown in Figure 5.35. Since the AGC operates at low frequency, around the mean value of the signal, a few tens of milliseconds are needed before the AGC brings the receiver within its dynamic range after each discontinuity of the input signal.

If a jamming signal, either pulsed or CW, is transmitted in the on-off mode with a certain duty cycle and at high frequency, the AGC will position itself so as to receive correctly the mean value of this signal.

If the duty cycle is suitably chosen (Fig. 5.35), the radar will not be able to extract the modulation necessary for tracking either from the jamming, (preventing effective track-on-jam), or even from the real signal, which is too small.

To determine the validity of the count-down technique against a radar of the monopulse type, one must keep in mind that the angular information is contained within a single pulse; if the receiver gain is incorrect, because of disturbances in the AGC, the sign of the correction to be introduced in the angular loop will be correct, but the amplitude of the correction will be either too large (receiver saturated) or too small (receiver gain too low).

Taking into account the fact that in monopulse radar the AGC can easily have a bandwidth of a few tens of hertz, jamming of the count-down type must operate at hundreds of hertz. But at these frequencies, the angular loop surely will not respond but will average the errors provided by the radar sensors. As pointed out above, these errors are of the correct sign, although at times too strong and at times too weak; they are therefore on average at a correct level. Only if the monopulse radar has a very narrow AGC band, will count down have some

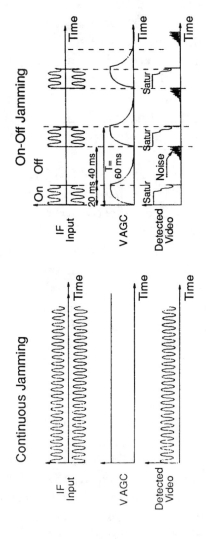

Figure 5.35 IF radar signals and AGC voltage with jammer always on or jammer on and off according to a certain duty cycle (count down). In this case the AGC is never implemented and the radar does not succeed in extracting angular information from the detected video signal.

real effect in degrading the capability of angular tracking.

5.3.12 Cooperative jamming

Cooperative jamming is an ECM technique that requires the cooperation of two platforms, each possessing either a deception or a noise jammer (Fig. 5.36). It is aimed at a tracking radar and is of a general nontarget specific type; that is, it does not depend on the tracking system used by the radar [3]. In particular it is valid against monopulse radars that, as already mentioned, are very resistant to angle countermeasures. The technique is to adopt a "blinking" or "buddy" mode, which induces the radar to point now at one platform and now at the other. The aiming point will therefore wander from one target to the other, and it is extremely probable that the projectiles or missiles will end up between the two platforms, without hitting either. Cooperative jamming requires a radio link between the two platforms to assure that the jamming is synchronized.

If the victim radar is the radar of an AAA battery, then the two platforms must remain constantly within its resolution cell, both in range, within a few tens of meters, and in angle:

$$\Delta R_{\max} = \frac{c\tau}{2}$$

$$\Delta L_{\max} = R\vartheta_B$$

If the victim radar is the seeker of a missile, the velocity of the two aircraft must remain within its velocity gate; ΔL must be such that both aircraft are able to receive the signal of the radar illuminator, and both aircraft must be within the missile's beam.

5.3.13 Cross polarization

This ECM technique is valid against all tracking systems [3]. It is also effective against some radar ECCM, such as the sidelobe blanking technique which will be discussed in section 6.2.2.4.

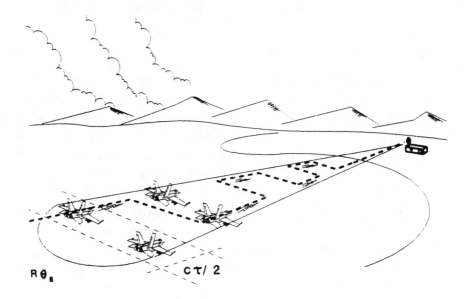

Figure 5.36 Cooperative jamming. When jamming emissions alternate between aircraft, the radar is forced to wander back and forth from one aircraft to the other, causing an oscillation in the line of sight that can annul weapon systems effectiveness.

Cross polarization exploits the fact that in the polarization orthogonal to its design polarization, every antenna has an antenna pattern in which there is a null instead of a maximum in the main lobe (Fig. 5.37).

If the antenna is of the monopulse type, the patterns of the Σ and Δ beams are as depicted in the figure, where it is clearly noticeable that in the orthogonal polarization it is as if Σ were exchanged with Δ. Because of this, in practice the target equilibrium tracking point is shifted by about a beam width (ϑ_B).

The first requirement for an operational cross-polarization ECM system is the ability to measure the polarization of the victim radar, and the second is an ability to retransmit a signal with an accurate orthogonal polarization (Fig. 5.38).

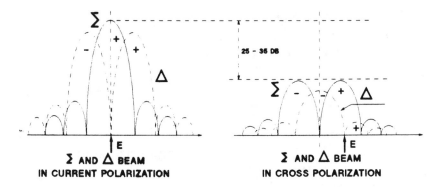

Figure 5.37 Σ and Δ antenna patterns in normal and orthogonal polarization. In the latter case the tracking radar equilibrium point is far off boresight.

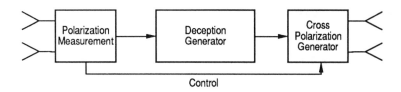

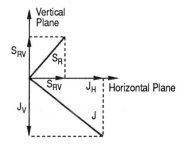

Figure 5.38 Block diagram of cross-polarization deception jammer.

The following considerations are needed to estimate the cost-effectiveness of the system:

- the antenna gain in cross polarization is about 25 to 30 dB below the gain of an antenna with the same polarization as the radar.

- if no RGPO has been previously performed or track-on-jam mode is not in operation, the cross-polarization signal must be stronger than the skin return by at least 25 to 30 dB.
- if the signal generated in cross polarization is not exactly at 90°, it will have a component in the polarization of the radar, which will contribute to successful tracking.

Assume, for example, that jamming of a vertically polarized radar is required (Fig. 5.39). A strong horizontally polarized jamming signal, P_j, must be generated. If there is a 2° error in generating the orthogonally polarized signal, a signal V_{jh} will be generated, given by

$$V_{jv} = V_{jh} \sin 2° = 0.035 V_{jh}$$

corresponding to a power

$$P_{jv} = 1.2 \times 10^{-3} P_{jh}$$

so that

$$\frac{P_{jv}}{P_{jh}} \simeq -29 \, \text{dB}$$

The vertical polarization component of the jamming signal, V_{jh} will aid the radar in its target tracking with a signal about 30 dB below the transmitted jamming power. Since an antenna generally attenuates the cross-polarization component by at least 30 dB, and the latter must be much higher than the real signal, the ECM system must be capable of measuring the victim radar's polarization with a precision higher than one degree. It must also be capable of transmitting it with the same precision whilst accommodating the normal relative changes in aspect between the victim radar and the target experienced during conflict maneuvers.

The jamming signal in cross polarization can be further attenuated by about 20 to 25 dB through a relatively minor modification of the radar sensors, namely the addition of a polarizing filter in the antenna system.

The error introduced by cross polarization must be used to reach break-lock. That is, once a sufficient displacement has been achieved, jamming should be suspended; otherwise, after a transient the missile can continue on its approach course.

In order to carry out a correct cost-effectiveness evaluation of this technique, its general validity, at least against old weapon systems, should be considered.

5.3.14 Cross eye

Another angle-jamming system of a general type, effective against all RF tracking systems, is the cross-eye technique [3, 13, 14]. A cross-eye system exploits the principle whereby it is possible to generate a phase distortion of the wavefront by employing two coherent sources, separated by a distance L, radiating signals that arrive at the victim radar matched in amplitude but opposite in phase. Since all tracking systems point in a direction orthogonal to the received wavefront, the victim radar will incur an angular error. The cross-eye technique is difficult to implement since it requires very advanced technologies.

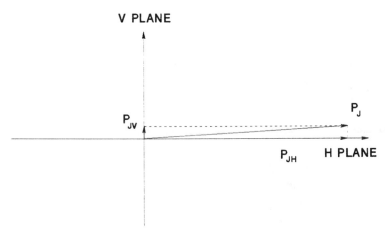

Figure 5.39 Precision of orthogonality required for cross polarization. A small misalignment of jammer polarization can assist the radar to pursue correct tracking.

Its effectiveness against tracking radars is demonstrated by a natural phenomenon which takes place when a target flying at low altitude above the surface of the sea is being tracked (section 2.2.6.4). In this case, the radiation reflected by the sea combines with the direct radiation (multipath) and causes oscillations in the elevation plane of the antenna (nodding). This occurs because whenever the two radiations are opposite in phase, the antenna tends to point much higher than the target's true position. Figure 5.40 shows the arrangement of an ECM system capable of implementing the cross-eye technique. In practice, what is required is the generation of two signals, of suitable amplitude, which are seen as opposite in phase by the victim radar. In the diagram, the signal received by one antenna is transmitted by the other and vice versa. This pattern is also called "retro-reflective."

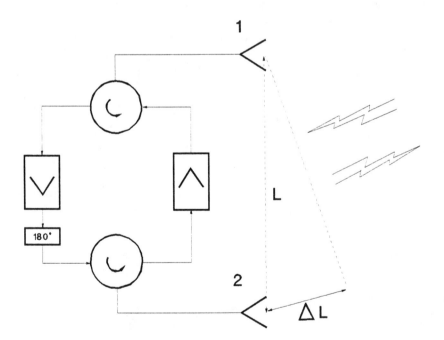

Figure 5.40 Diagram of a cross-eye deception jammer. The victim radar sees two deception signals opposite in phase which cause a distortion of the wavefront.

Starting from the wavefront, the signal received by antenna 2 will have a phase $\varphi = 2\pi\Delta L/\lambda$ will undergo a phase shift φ_0 in the system, to which a further phase shift of 180° will be added, and will emerge from antenna 1 suitably amplified.

The signal received by antenna 1 will also undergo a phase shift φ_0 (a phase match is assumed even where the paths are different) and in a distance ΔL will be phase-shifted by $\varphi = 2\pi\Delta L/\lambda$. The differential phase shift between the two transmitted jamming signals will be therefore

$$\Delta\varphi = \left(\frac{2\pi}{\lambda}\Delta L + \varphi_0 + 180°\right) - \left(\varphi_0 + \frac{2\pi}{\lambda}\Delta L\right) = 180°$$

However, as will now be shown, the two transmitted signals are not of equal power. It is known (see also chapter 2) that two coherent isotropic sources, separated by a distance L, produce interference patterns such that the received power at range R is given by

$$P_r = \left[\sqrt{P_1}\sin\left(\frac{2\pi}{\lambda}R_1\right) + \sqrt{P_2}\sin\left(\frac{2\pi}{\lambda}R_2\right)\right]^2 +$$

$$+ \left[\sqrt{P_1}\cos\left(\frac{2\pi}{\lambda}R_1\right) + \sqrt{P_2}\cos\left(\frac{2\pi}{\lambda}R_2\right)\right]^2$$

This is due to the fact that the phase shift between the two sources, as seen by the radar, varies. When the sources are in phase, there will be a power maximum; when they are opposite in phase, there will be a power minimum.

Diagrams of the phase and power received at range R are shown in Figure 5.41, where it is important to note that, when the two signals are received in opposite phase, the apparent radar center is shifted to the position I.

The angular error α produced by the cross-eye technique because of the distortion of the wavefront can be expressed in the following way, if it is assumed that the two sources radiate

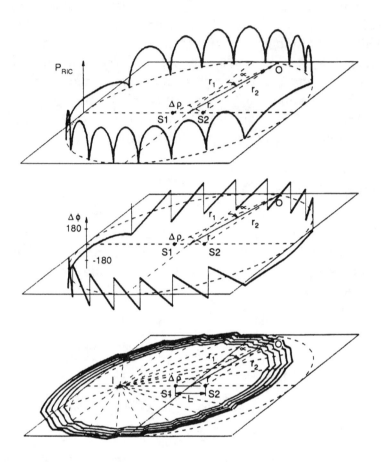

Figure 5.41 Power, phase and apparent wavefront of a system consisting of two coherent isotropic radiators radiating two in-phase signals, seen at range R.

isotropically and that the antenna beam of the tracking radar is very wide:

$$\delta = \frac{L \cos \alpha}{2} \frac{1 - k^2}{1 + 2k \cos \varphi + k^2}$$

where α is the angle between the normal to the system formed by the two sources and the line joining the midpoint of the line between the sources to the victim radar; k^2 is ratio of signal powers; φ is the phase shift between the signals emitted by the two sources, as seen by the radar.

The ratio $2\delta/L\cos\alpha = G_{CE}$ is called cross-eye gain, and in practice indicates the magnitude of the shift of the radar center as seen by the victim radar, in terms of apparent $L/2$, with respect to the true center of mass.

The maximum distortion of the wavefront occurs where the shift approaches 180°. By plotting G_{CE} as a function of φ, the well-known diagram depicted in Figure 5.42 is obtained.

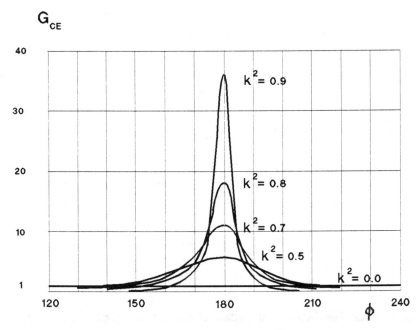

Figure 5.42 Error due to cross-eye jamming (G_{CE}), measured in terms of apparent half-baseline, as a function of the phase shift and power ratio between the two deceptive signals received by the radar.

The objective of the cross-eye technique is in all circumstances to keep the two jamming signals, as seen by the victim radar, in a power ratio equal to k^2 and with a phase shift between the two signals equal to 180°, so that the center of mass of the two sources, as measured by the radar, is shifted by G_{CE} times the half-distance $L/2$ between the sources.

The phenomenon is a position, not an angular, shift; so that at long ranges the error introduced in angle is scarcely significant. What counts is that it exists, and that if a projectile or a missile tries to approach the target, it will in fact undergo the mentioned shift, and finish in the wrong place.

Assuming $\varphi = 180°$ and $k^2 = 0.8$, one obtains

$$G_{CE} = \frac{1 - 0.8}{1 - 1.789 + 0.8} = 18$$

It should be noted that for the cross-eye technique to work properly, there must be an observable baseline distance between the two sources. When the deception is seen from a direction along the line joining the sources, the cross-eye effect is nil. A problem inherent in the technique is that the two signals produced, being opposite in phase and of nearly equal amplitude, tend to cancel each other. Because of this, the J/S required for a reliable cross-eye effect is of the order of 20 dB for each source.

Amongst the implementation problems is the need to maintain the correct phase and power relations between the transmitted signals in all operational conditions, including platform shifts, vibrations, aspect angles and so forth.

The transmitted waveform is either a replica of the received waveform or a noise waveform, generated with a suitable phase modulation so as to maintain coherence between the sources. The former can be used against all radar, including frequency agile ones, the latter only against a fixed-frequency radar wishing to exploit track-on-jam ECCM.

The cross-eye technique can be applied also against semi-active missile systems, but in this case, in order to establish the wavefront distortion suitable against the possive seeker, sophisticated measures are required.

Angular deception of the cross-eye type generally requires that the received signals traverse the baseline L, so that this type of deception has an inherent time delay.

A possible countermeasure against this type of ECM is therefore the use by the radar of extremely short pulses with frequency agility.

5.3.15 Terrain bounce

This technique can be used by low-flying aircraft as a defense against attack from above by AAMs equipped with active pulsed-Doppler homing or semiactive CW homing. It entails directing jamming signals toward the ground in such a way that the jamming signal scattered toward the victim seeker covers the true echo of the jamming aircraft (Fig. 5.43) [14, 3].

For the terrain bounce technique to be successful, the signal backscattered by the ground toward the missile must be stronger than the (unwanted) signal radiated by the antenna sidelobes of the jammer and received directly by the victim radar. The ideal antenna should thus have maximum emission in the direction of the ground and a null in the direction of the missile.

The technique must be activated while the missile is still sufficiently far away that both the echo signal and the jamming signal scattered by the ground are contained in the same resolution cell of the radar.

Initially the radar antenna will tend to track the centroid of the two received signals. With decreasing aircraft-missile range, a crucial point will be reached where the angular separation between the two signals is near the beamwidth of the radar antenna. Below this range, the radar will choose the signal that appears stronger. If the signal backscattered by the ground is the stronger, the missile will be lured away from the true target. This technique works also when the victim radar is a monopulse radar.

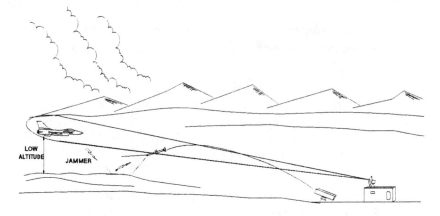

Figure 5.43 Terrain-bounce jamming exploits ground reflectivity to deceive the tracking radar or the seeker of a missile. It can be used in low-altitude flight.

5.3.16 Illuminated chaff

The *jammer-illuminated chaff* technique, also known in the literature as JAFF (jammer + chaff) of CHILL (chaffilluminated), can generally be employed for self-screening or as an aircraft support technique (for chaff see section 5.5.1.2). It requires illumination of previously ejected chaff clouds with a noise or deception signal, in order to present the victim radar with alternative, false targets, thus causing break-lock in angle (Fig. 5.44).

The principal attraction of this technique lies in the possibility of imposing on the echo reflected by the chaff toward the victim radar a suitable doppler frequency, for example, a doppler equal to that of the aircraft, or a suitable band of masking doppler frequencies. This helps to solve the main problem of the use of chaff against coherent radars which exploit doppler information to filter and discard static targets.

Fundamentally, the aim of this technique is to produce relatively easily cheap, off-board decoys which can be used against coherent radars, pulsed-doppler, CW, or with MTI, and against monopulse radars.

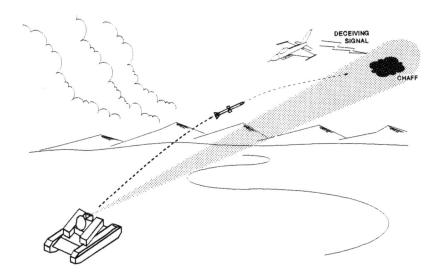

Figure 5.44 Chaff is strongly attenuated by an MTI filter. However, it becomes a very good target, endowed with the correct doppler frequency, if illuminated with a coherent deceptive signal.

5.4 Infrared countermeasures (IRCM)

5.4.1 Modulated sources

At present the prime threat to aircraft is the IR-guided missile. The majority of deployed missiles use a relatively simple IR seeker, able to extract angular information from the amplitude modulation of the received IR signal.

The reticle can be of the type shown in Figure 5.45 and the signal at the output of the sensor, after suitable amplification, is as shown. By correlating modulation with the position of the reticle it is possible to extract the information needed for missile guidance.

Angular jamming can be introduced into the missile guidance loop by high intensity sources, radiating in the same part of the IR spectrum as the engines and exhaust gases of the aircraft. These generate an IR signal whose intensity is higher than the intensity of the signal emitted by the aircraft, but

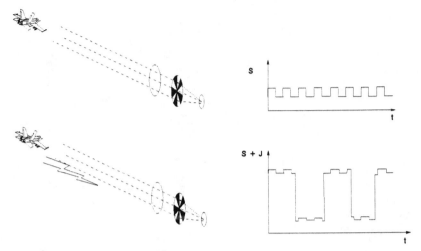

Figure 5.45 Angular jamming generated by amplitude-modulated IR radiators against an IR seeker is similar to that caused, in the analogous situation, a conical-scan radar.

differently modulated. This is analogous to the AM jamming of a COSRO system. At present these sources require high feed power and have relatively short lifetimes. From the manufacturing point of view, it is necessary to choose a source capable of emitting the spectrum needed for defense of the platform. ECCM against this type of jamming source employ focal plane-array sensors, able to extract angular information instantaneously in a monopulse manner, without resorting to mechanical scanning.

5.4.2 Laser IRCM

For on-board defense of platforms, systems exploiting laser technology are under study, with the aim of degrading the performance of the IR seeker. These systems are based almost exclusively on the destructive effect of the high energy that can be transmitted by an extremely narrow laser beam (directed energy weapons).

5.5 Off-board ECM systems

5.5.1 Passive systems

5.5.1.1 Passive decoys The basic concept underlying the passive decoy technique is a simple expendable device capable of reproducing a radar signature at least as attractive to enemy weapon systems as the signature of the platform to be protected.

In the naval context, corner reflectors should be mentioned; they may be inflatable, and are either buoy-based, towed or free. The RCS of a corner reflector (Fig. 5.46) can be easily obtained by thinking of it as a surface whose area is equal to its base area, always able to emit with maximum gain. This follows from its geometry. In fact, being a 90° trihedral, a corner reflector reflects all radiation falling on its surfaces over very wide angles of arrival.

Suppose that a corner reflector has edges of length l. The area subtended by a triangle of side l is

$$A = \sqrt{\frac{3}{4}} \, l^2$$

The effective area of this aperture is

$$A_{\text{eff}} = \sqrt{\frac{3}{4}} \, l^2 \times \eta$$

Assuming $\eta = 0.7$, the effective area is well approximated by

$$A_{\text{eff}} = 0.3 \, l^2$$

Recalling the RCS formula

$$RCS = \frac{A_{\text{eff}}^2 (4\pi)}{\lambda^2}$$

Figure 5.46 Corner reflectors can generate a very large RCS in a large angular sector.

one obtains for the RCS of the corner reflector

$$RCS = \frac{4\pi \, (0.3 \, l^2)^2}{\lambda^2}$$

In the airborne context there are decoys that can be dispensed behind an aircraft, or dispensed forward by means of rockets (forward-launched decoys). In order to create a sufficiently high RCS, they will contain either small corner reflectors or Luneburg lenses.

5.5.1.2 Chaff

One of the first radar countermeasures is the dispensing of chaff [15, 16]. *Chaff* is made of clouds of minute dipoles that are dispersed into the atmosphere to create a high reflectivity zone to mask the presence of true targets. Figure 5.47 shows a classical example of the use of chaff to create a corridor or a "window" through which aircraft pass in concealment.

Chaff is used against search radars, and also against tracking radars to cause break-lock. In order to be effective, as soon as it is launched, chaff must develop a high reflectivity of great persistence. To obtain this, the dipoles must be tuned to the

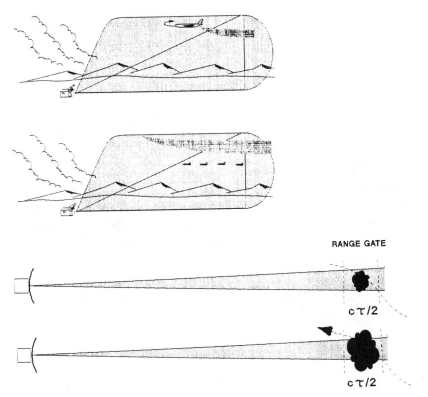

Figure 5.47 Chaff can be used to seed corridors for strike aircraft or for self protection.

radar frequency. Generally they are tuned to $\lambda/2$ on the center frequencies of the radar bands. They must be extremely light and dispensed in huge numbers.

Today dipoles consisting of extremely thin aluminium-covered fiberglass filaments are used. The density achieved in the cartridge to be dispensed is of a few millions of dipoles per cm^3, and the persistence in the atmosphere is of several tens of minutes.

The RCS of a dipole averaged over all possible orientations is given by

$$4\sigma_1 = 0.15\lambda^2$$

over a band of about $\pm15\%$ around the resonance frequency.

This is so if a length-to-width (l/d) ratio of about 1000 is assumed for the dipoles.

The RCS of N dipoles spaced more than twice the wavelength apart will be

$$\sigma_n = N0.15\lambda^2 = N\sigma_1$$

When a cloud of chaff is generated, initially the average distance between dipoles will be much less than 2λ, as if each dipole were slightly screening another. The result is that the RCS will not be equal to $N\sigma_1$, but smaller. This can be approximated by saying that initially the RCS will be given by the RCS intercepted by the radar as if the chaff cloud were a solid, and will reach a maximum where the whole of the radar cell is filled with dipoles spaced 2λ apart. In this case, there will be present in the cell the limiting number of dipoles, $N_{\max}$, for the maximum RCS:

$$RCS_{\max} = N_{\max}\sigma_1$$

The maximum number of dipoles spaced 2λ apart in the radar cell volume $L_1 L_2/L_3$ will be given by

$$N_{\max} = \frac{L_1 L_2 L_3}{2\lambda 2\lambda 2\lambda}$$

However, $N_{\max}$ will be a very large number. For a tracking radar with a 2.3° by 2.3° beam and a pulse of $0.5\,\mu s = 75\,\text{m}$, at 10 km, the volume is $12 \times 10^6\,\text{m}^3$, and $N_{\max}$ is about 50 billion. This is much higher than the normal number of dipoles in a cartridge and is matched to a given bandwidth, so that the maximum RCS developed in one band is effectively

$$RCS_{\max} = \sigma = N\sigma_1$$

The RCS presented by a chaff dispersed in the radar cell can be expressed by the same formula as rain clutter:

$$\sigma = \eta \frac{R\vartheta_{az}}{\sqrt{2}} \frac{R\vartheta_{el}}{\sqrt{2}} \frac{c\tau}{2}$$

where η, which has dimensions m^2/m^3, is now the reflectivity of the deployed chaff $r\vartheta$ is the dimension in azimuth and elevation of the radar cell $c\tau/2$ is the radial dimension of the radar cell.

The determination of the parameter η is rather complicated. Within the scope of this book it will suffice to recall that approximated values for this parameter vary between 10^{-7} and 10^{-10}, m^2/m^3.

As to the volume and weight of a chaff package, the example which follows deals with a case in which chaff is to be used against a radar in the X band ($\lambda = 0.03\,m$). The length of the dipoles will be 0.015 m. Since the diameter must be about 1/1000 of the length, the diameter of the aluminium-coated fiberglass must be 15 m. Therefore in a small volume

$$V = 13.5 \times 10^{-6}\,m^3$$

such as is depicted in Figure 5.48, there can be 4×10^6 dipoles, with a packing density of 0.8, or 2.5×10^6 dipoles with a packing density of 0.5. In this case, considering that the aluminium-coated glass density is 2.5, the total weight of the package will be only 16.8 g.

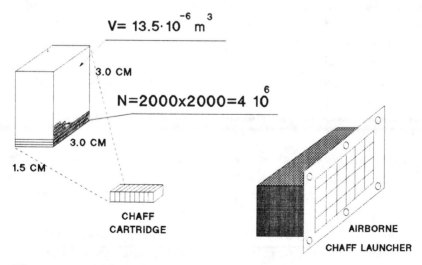

Figure 5.48 Millions of small dipoles are packed into a chaff cartirdge.

To determine chaff lifetime in the atmosphere, fall rates of between 0.3 and 1 m/s should be considered, depending on the altitude and the material.

For airborne use, chaff cartridges are dispensed by more or less standardized chaff launchers. Ejection can be either mechanical (by springs), or pyrotechnical (by squibs), towards the aircraft trail to obtain a fast chaff blooming.

Chaff may be tuned to a single band, or to more than one band. Sometimes it is possible to tune it in flight to a suitable band, by means of a special device in the launcher (chaff cutter).

In the naval case, the cartridges are much larger and heavier, since the chaff must protect a target whose RCS is several thousands of square meters. In order to obtain fast blooming, the noval chaff cartridge may contain small explosive charges.

Chaff can be employed either to seed corridors or for self-defense of the platform that launches it. When chaff is employed for self-protection the most widely used techniques are distraction and dilution.

Distraction tries to create false targets at locations different from that of the protected platform (Fig. 5.49). The main objective is to confuse the acquisition system of a tracking radar. In general chaff is launched in a coordinated way after break-lock has been achieved by a jammer, or before acquisition by the seeker of the missile.

Dilution involves dispensation of the chaff in the radar cell where the target is (Fig. 5.50). It may be used when the RCS developed by chaff is surely higher than the RCS of the platform.

The positive aspect of chaff is its simplicity and ease of use. A single, simple *radar warning receiver* (RWR) on board the platform to be protected suffices to give warning and to initiate the launching of chaff at the right moment.

A negative aspect for aircraft protection is that, on ejection, chaff will almost immediately decelerate and come to rest because of the low mass of the individual dipoles. This means

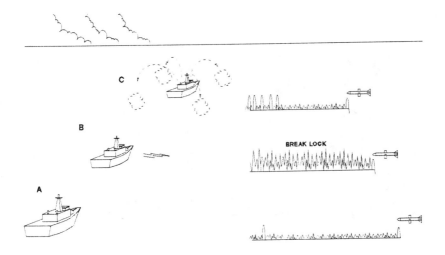

Figure 5.49 Distraction chaff. After break-lock has been achieved by a noise jammer, the ship distracts the seeker of the missile by generating false targets using chaff.

that if the radar has an MTI, or is a pulse-doppler radar, the echo produced by chaff will be reduced, with respect to that of the aircraft, by about 30 to 40 dB, depending on the quality of the radar doppler filter. This drawback may be partly obviated if, before launching the chaff, the pilot maneuvers so that the aircraft presents a very low doppler to the enemy radar. The chaff is then launched, and finally the aircraft resumes its course (Fig. 5.51).

In the naval case, the negative aspect is the high RCS that has to be achieved to protect the large platforms involved.

5.5.2 Active systems

The main objective of these systems is to lure the approaching threat toward a decoy target ejected from the threatened platform, either by retransmitting an echo very similar to, but more attractive than, the echo of the true target, or by produc-

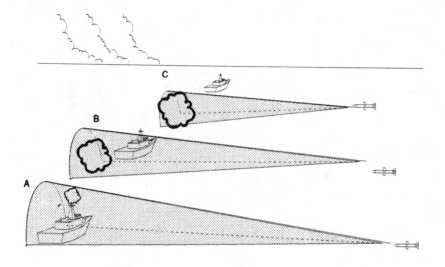

Figure 5.50 Dilution chaff. Cartridges of chaff with suitable RCS are exploded in the radar cell where the ship is. Subsequently the ship maneuvers to separate itself from chaff.

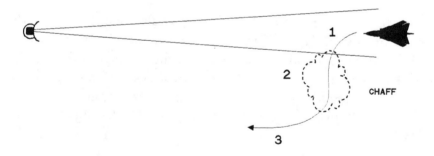

Figure 5.51 Chaff for self-protection. The aircraft maneuvers to generate a low doppler echo and ejects chaff whose large RCS lures the radar range gate.

ing such a strong noise as to activate the radar's track-on-jam ECCM.

These decoys can either be towed by the platform (towed decoys) or launched (expendable decoys). The protection of a naval platform and of an airborne platform must be distinguished and will be discussed separately.

5.5.2.1 Decoys for naval platforms

In this case, the main problem that needs to be solved is the production of a signal with an adequate ERP which can mask the ship's high RCS.

The required ERP can be obtained either with an antenna gain, or with RF power generated by a TWT. However, in the former case, it is highly doubtful whether a decoy of limited size could carry a very directive antenna and an on-board pointing system. In the latter case, on the other hand, the decoy must carry a rather heavy payload, consisting of electronic circuitry, a high voltage power supply and a TWT with an adequate cooling system.

If the decoy is towed on a buoy, the problem of weight does not arise, but there remains a problem of credibility, since anti-ship missiles have a remarkable range discrimination.

Corner reflectors may conveniently be used to decoy trackers away from the true target, since they can normally produce a very strong echo much more cheaply than an active system.

If the threat consists of two missiles, one has to assume that the decoy may be destroyed by the first missile; the timely availability of a second towed decoy in a suitable position will therefore be of crucial importance.

For the expendable naval decoy, besides the problem of the limited volume and weight available for the payload, the problems of missile-target-decoy kinematics, and of decoy credibility, have to be solved. The decoy will have to be launched at the right moment, to capture the attention of the anti-ship missile, and to divert the missile onto a course which does not threaten the ship (Fig. 5.52). In this case also, the additional problem created by the coordinated launching of two anti-ship missiles has to be considered.

5.5.2.2 Decoys for airborne platforms

Defense of an aircraft by means of active off-board ECM

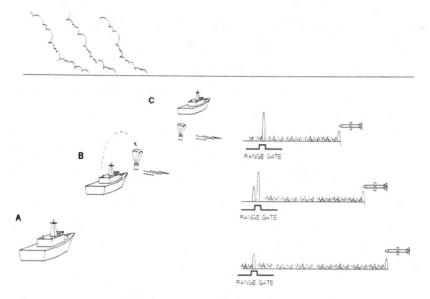

Figure 5.52 An expendable naval decoy can generate strong deception signals that attract the range gate of a seeker.

systems must first of all overcome the problem of emulating the doppler frequency of the aircraft. Since almost all anti-aircraft systems exploit powerful doppler filters, an ECM system not capable of entering the filter is ineffective.

In the following discussion, towed and expendable decoys will be distinguished [17].

Towed Decoys

A very simple method of solving the doppler problem is to use a towed decoy. Being tied to the platform, it will have essentially the same speed and the same doppler. It therefore suffices to generate the wanted deception or noise signal with a suitable ERP in order to capture the tracking gates of the threat.

However, a towed decoy always faces the problem of coverage in angle. If the tracking system of an anti-aircraft missile has been lured toward a towed decoy, the missile will home onto that, but in certain angular sectors to the front and rear, defense of the platform cannot be assured. To the front, the fuzing system of the missile may in any case cause the warhead to damage the aircraft (Fig. 5.53). To the rear, it can happen that the missile homing onto a towed decoy does not explode, because of the small size of the latter, notwithstanding the short miss distance, and continues to head for the aircraft.

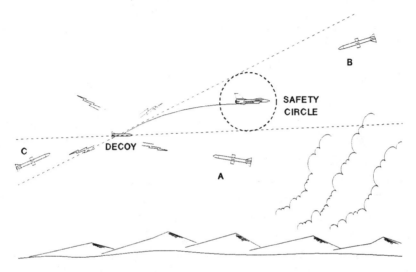

Figure 5.53 The towed decoy is long lasting and produces an echo at the same Doppler as the true target. In the front and rear sectors (missiles B and C), however, it is not capable of assuring a valid defense.

However, if the missile fuze is activated and the towed decoy is destroyed, the launching system must immediately deploy a second towed decoy to deal with the possibility that there is another missile.

The towed decoy is very effective against a semiactive missile system, since a very modest ERP suffices to provide adequate defense against the majority of the systems now deployed. Towed decoys are likely to be less effective against command missiles, whose operator can easily discriminate between the aircraft and the decoy. In general it is difficult to provide a towed decoy with an ERP large enough to make the radar guiding a command missile shift to its track-on-jam mode. Another problem of the towed decoy is its potential unacceptability to pilots, who are forced to take the aerodynamic penalties of these appendages and to take them into account when maneuvering.

Expendable Decoys

These ECM systems are based on the ejection of mini-missile-shaped objects whose payloads generate deceptive signals to lure the tracking gates of the threat.

The two main problems to be solved here are those of doppler frequency and period of effectiveness. The first problem has already been mentioned; possible solutions to it will be discussed next. The second problem is that, once the expendable decoy has been launched, it is difficult to maintain its effectiveness for long, since it will be abruptly separated from the aircraft. Therefore, before launching an expendable, one should be sure that the missile is already approaching and is at the right range. In the absence of this information, it is necessary to launch expendable decoys at regular intervals, starting from the moment at which CW emission, probably the illuminator of a semiactive missile, is detected. To this end, the aircraft must be equipped with a large number of expendables; new microwave technologies (MMIC) may make this feasible.

To increase the effectiveness of these devices, systems based on forward-fired active decoys, which are launched toward the approaching missile in order to increase the period of effectiveness, have been developed.

The doppler problem can be solved either by transmitting to the decoy in a suitable way the computed frequency that has to be radiated to lure the velocity gates of the anti-aircraft missile, or by providing an open loop correction on board of the expendable. In case the problems of isolation of received and emitted signals in such a small object must be solved.

5.5.2.3 Flares

A very good countermeasure against IR-guided weapons is the launching of *flares* (Fig. 5.54). Flares are contained in special cartridges, normally stored in the same launchers as are used to dispense chaff. Once ejected, they can, for a few seconds, generate a radiant intensity higher than that of the protected platform. The missile is lured by the flare and homes onto it, forgetting the true target.

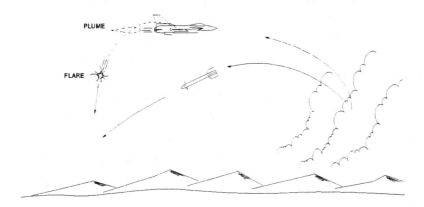

Figure 5.54 The launching of flares (expendable IR sources) at the right moment can ensure a valid defence against missiles with IR guidance.

The characteristics that distinguish flares are:

- intensity (W/sr) in the seeker bandwidth.
- activation time.
- persistence.
- weight.

The problem presented by countermeasures of this type is that a platform cannot carry an infinite number of cartridges and, since it is not known whether a missile is approaching or not, does not know when to launch the flares.

To solve this problem, *missile launch warning* (MLW) or *missile approach warning* (MAW) systems may be used. The former operate passively, exploiting IR radiation, and are able to give warning of the launch of a missile. The latter use radar techniques to give warning of the approach of a missile. In this way the launching of flares can be postponed until it is really useful.

5.6 Communications countermeasures (COM-ECM)

The main objectives of *communications countermeasures* (COM-ECM) equipment, which can be installed on terrestrial, naval or air platforms, are the following:

(1) to deceive the enemy C^3 system by introducing false information into the enemy communications network.

(2) to analyze and evaluate reactions to jamming.

(3) to degrade the enemy C^3 system by jamming communications as far as possible without being observed.

(4) to designate the most important nodes of enemy networks to ARMs so that radars there emplaced can be destroyed.

A good example of a COM deception jammer would be a system capable of intercepting and recording enemy communications in order to cause confusion by retransmitting them after a delay.

Figure 5.55 explains the functions of COM-ECM systems. It shows a block diagram of a system for jamming the guidance link of a fighter aircraft. The aim of this equipment, generally installed on board an aircraft flying at a safe distance, is to render unintelligible to an enemy fighter pilot the information transmitted from a control tower for guidance against attacking aircraft.

The factors influencing the intelligibility of a radio conversation subjected to jamming are:

(1) human factors.

(2) jamming-to-signal ratio at the demodulator of the victim receiver.

(3) multi-frequency jamming capability, including number of contemporarely jamming emissions, channel dwell time, duration of look through.

(4) frequency shift between victim signal and jamming signal.

The first of these emphasizes that intelligibility depends on the conditions of stress in the operational situation as well as on the training of operators.

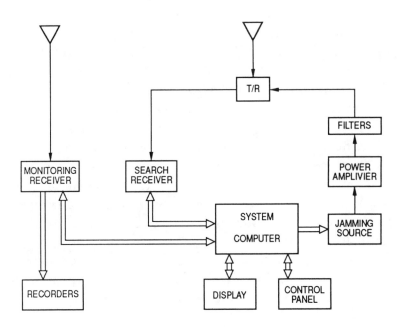

Figure 5.55 Block diagram of a communications noise jammer.

The second factor emphasizes the statistical nature of the jamming problem, as the JSR depends on quantities that cannot be calculated with extreme accuracy, such as transmission losses. Normally, an effective JSR should be near 0 dB [18].

The third factor recalls that, if multi-frequency jamming uses a single time-sharing jammer, the necessary power increases with the number of the emissions to be jammed. Jamming effectiveness depends also on the duration of look through, which must be as short as possible and adaptable to the operational situation, on the dwell time in each channel, on the mean value of the jamming period and on the tuning velocity of the jamming source.

The fourth parameter, frequency shift between victim signal and jamming signal, in practice implies that the jammer receiver must be able to measure the frequency of the victim emission during the look through period with an accuracy of a few hundred hertz, and that the jammer source be tunable with the same accuracy in a few tens of microseconds. In AM, if the frequency shift is higher than 1 kHz, effectiveness is reduced by approximately 1 dB for each additional 100 Hz shift.

A high-power jammer, such as the one illustrated in the preceding example, generally designated by a COM-ESM system, may have an ERP which, taking into account antenna gain, can be higher than 2 kW. Besides the high ERP jammers just mentioned, there are also low-power camouflaged jammers that are positioned, near the emitters to be jammed and therefore cannot be retrieved, so that they are called *expendable noise jammers* [19]. They are positioned either manually during withdrawal operations, by means of special artillery projectiles, or by launching from aircraft.

The jamming is activated either by preprogramming a suitable clock, or by sending a command signal, or by priming with a mini ESM system contained in the jammer itself.

REFERENCES

[1] J. Boyd, *Electronic Countermeasures*, Los Altos, CA: Peninsula Publishing, 1978.

[2] L. Van Brunt, "Applied ECM," Vol. 1, EW Engineering Inc., Dunnloring, VA, 1978.

[3] D.C. Schleher, *Introduction to Electronic Warfare*, Norwood, MA: Artech House, 1986.

[4] R.N. Johnson, "Radar Absorbing Material: A Passive Role in an Active Scenario," *International Countermeasures Handbook*, 1986, pp. 375-381.

[5] H.F. Harmuth, "On the Effect of Absorbing Materials on Electromagnetic Waves with Large Relative Bandwidth," *IEEE Transactions on Electromagnetic Compatibility*, Vol. EMC 25, No. 1, February 1983.

[6] F.M. Turner, "Noise Quality Optimizes Jammer Performance," *Electronic Warfare/Defense Electronics*, November-December 1977.

[7] J.B. Knorr, "Simulation Optimizes Noise Jammer Design," *Microwave Journal*, May 1985.

[8] E. Arcoumancas, "Effectiveness of a Ground Jammer,"*IEE Proceedings*, Vol. 129, Pt. F, No. 3, June 1982.

[9] S.A. Hovanessian, "Noise Jammers as Electronic Countermeasures," *Microwave Journal*, September 1985.

[10] C.W. Deisenroth, "Analog Countermeasures Memories: Key to Delay and Replication," *International Countermeasures Handbook*, 1986.

[11] H.F. Eustace, "What Jams What?" *International Countermeasures Handbook*, 1978.

[12] R.E. Marinaccio, "Self Protection CM: Present and Future," *Microwave Journal*, February 1987.

[13] V.A. Vakin and L.N. Shustov, "Principles of Countermeasures and Reconnaissance," *Osnovy radioprotivodeysviya i radiotekhnichescoy razvedky*, Moskow. (Translated by: Foreign Technology Division, Wright-Patterson Air Force Base, OH, 1969.)

[14] G.R. Johnson, "Passive Lobing Radars," *EW*, March-April 1977.

[15] D.R. Armand, "Chaff Primer," *Microwaves*, December 1970.

[16] *IEE Proceedings*, Special Issue on Electronic Warfare, Vol. 129, Pt.F, No. 3, June 1982.

[17] M. Liebman, "Expendable Decoys Counter Missiles with New Technology," *Defense Electronics*, October 1986.

[18] L.E. Follis and R.D. Rood, "Jamming Calculations for FM Voice Communications," *EW*, November-December 1976.

[19] A.G. Self, "Expendable Jammers Prove Indispensable," *Microwaves & RF*, September 1984, pp. 143-149.

Chapter 6

Electronic Counter-Countermeasures Systems

6.1 Introduction

As soon as the effectiveness of ECM systems became apparent, weapon systems had to be protected by means of additional electronic devices that could counter the countermeasures. These devices were therefore called *electronic counter-countermeasures* (ECCM) systems [1, 2, 3].

This chapter will discuss the ECCM that weapon and communication systems can employ to operate in a hostile environment. The systems in question are:

- search radar.
- tracking radar.

- electro-optical systems.
- communications systems.

6.2 Search radar counter-countermeasures

As was shown in Chapter 3, a search radar can be used either as an element of an air defense network covering a broad area, such as the *Nato Air Defense Ground Equipment* (NADGE) system, to provide a wide panoramic view, or, on a smaller scale, to organize the defense of a small area, for example a ship or a convoy. In both cases, attack aircraft will attempt to avoid detection by adopting suitable flight paths and profiles, and, above all, ECM techniques [4].

For an attack by a single aircraft, the ECM that can be used against a search radar are few and of limited scope. They cannot do much more than attempt to cancel the platform echo, while simultaneously trying to conceal the direction of arrival of the jamming.

For a coordinated attack supported by either an escort jammer, whose mission is electronic defense of other platforms in a formation, or a standoff jammer, which protects other attacking platforms while remaining at a safe distance, the variety of usable jamming methods is much greater. It is primarily against this type of ECM that warning radars are equipped with ECCM devices.

Some ECCM are simple spinoffs from techniques used to improve radar performance, while others are introduced specifically to counter jammers, although their adoption requires an increase in weight or a slight degradation of some radar function. In what follows, the most widely used ECCM techniques will be examined. Very often many other techniques are denoted by exotic names but can be reduced to these basic types.

6.2.1 Induced counter-countermeasures

6.2.1.1 Sensitivity time control

The *sensitivity time control* (STC) is used to realize a radar receiver whose sensitivity, or gain, varies with range (Fig. 6.1). In the first place, it is used to limit the dynamic range of the radar signals sent to the processor. Recall that the power of a received signal varies as $1/R^4$; the echoes of large targets, or of clutter at short range, can quickly saturate a radar receiver; large white spots will appear on the *plan position indicator* (PPI) of a simple radar, where no targets will be discernible.

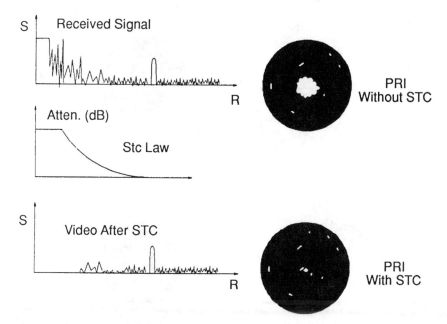

Figure 6.1 The *sensitivity time constant* (STC) eliminates saturation of the signal at short range.

In most cases, an STC is used to keep not only the target signal, but also the signal consisting of target plus clutter, within the dynamic range, so that the *moving target indicator* (MTI) that follows can adequately increase target visibility with respect to clutter.

In practice, the STC, implemented in a radar for other purposes, is a no-cost ECCM device capable of preventing saturation in the presence of a jammer at short range.

6.2.1.2 Fast time constant

The *fast time constant* (FTC) is a differentiating circuit matched to the length τ of the radar pulse. It chops all signals whose duration is longer than τ (Fig. 6.2); when there are big banks of clutter, or intentional, long-duration jamming signals, only the front will pass.

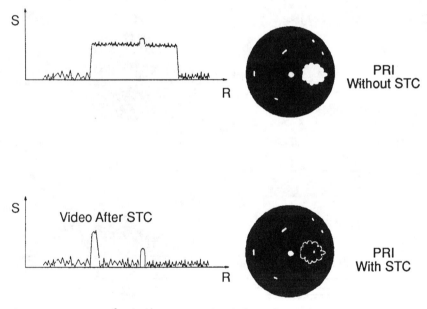

Figure 6.2 The *fast time constant* (FTC) eliminates the negative effects of signals of long duration.

Although the signal is chopped in order not to disturb the PPI, in the presence of a strong interference signal the sensitivity of the receiver is reduced almost to the level of the interfering signal. This means that a CW signal will not interfere with the PPI, but only signals higher or a few decibels lower than the CW signal will be discernible. A radar device used to cancel CW jamming without strong sensitivity degradation is the back-bias receiver, which will be discussed in section 6.2.2.2.

6.2.1.3 Staggered/random pulse repetition frequency

In all commercial radars, *pulse repetition frequency* (PRF) is either fixed, or can be changed by the operator when a change in radar range is required.

In more sophisticated radars, where MTIs are used to cancel clutter, a single PRF is inadequate; many PRFs are required in order to give the MTI filter the desired characteristic.

A radar can resort to *staggered* PRF, that is, several PRFs in a definite sequence. This provides it with an ECCM capability against those jammers which attempt to create false targets at ranges shorter than the range of the jammer itself (Fig. 6.3). The jammer lacks the ability to create range-correlated pulses, unless it can detect and exploit the levels and sequences of the stagger. To prevent the jammer from achieving this, where MTI canceling of second- and third-time-around clutter echoes is not required, the most sophisticated MTI radars will use a completely random PRF. In simpler radars, with no MTI, random PRF is used merely to decorrelate the second-time-around echoes.

In military radar, the PRF can always be varied by at least 10 to 20%; random PRF should be expected as well, since jammers have great difficulty in predicting it, and it is easily realized.

6.2.1.4 Moving target indicator

The *moving target indicator* (MTI) did not start its career as an ECCM. Its main task is to attenuate echoes from ground, sea, and rain (section 2.2.4.1).

It can however become a powerful ECCM circuit when the search radar is jammed by chaff [5, 6]. Chaff launched in space will float in the air (section 5.5.1.2); it will therefore resemble rain driven by the wind and will be canceled to an extent determined by the characteristics of the MTI itself. For example,

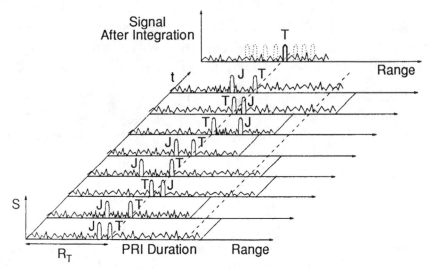

Figure 6.3 The use of jittered or staggered PRF decorrelates the deceptive signals created in advance of the time target.

an MTI double canceler can attenuate the signal from chaff by 20 to 30 dB.

6.2.1.5 ECCM capabilities of pulse compression

The capabilities of a coded pulse compression radar were discussed in Section 2.2.4.4. Pulse compression gives a search radar a high clutter cancellation capability and high range discrimination that allow the radar detector to output reliable, computer-manageable data. Computerized data management ensures that the detection system has a short reaction time, which is indispensable against present day threats flying at very high speed and at extremely low altitude.

In the following, the spinoff ECCM capabilities of pulse-compression will be discussed. There are essentially two of them:

- reduced interceptability of coded waveforms, and therefore *low probability of intercept* (LPI) characteristics [7].
- reduced susceptibility of coded waveforms to deception jamming [8].

Consider two radars, one with pulse compression and one

without, both having the same range performance, subjected to jamming by a standoff aircraft, which attempts to create a corridor for attack aircraft.

The pulse-compression radar will have a peak power n times lower than that of the uncoded radar, n being the number of elements in the code, which may compromise the *range advance factor* (RAF) of the RWR-ESM system on board the aircraft generating the standoff jamming.

The standoff jammer is normally positioned at a safe distance, outside the range of the weapon system, so that, notwithstanding its significant RCS, it can correctly detect the signals radiated by the victim radar without being detected in its turn.

Assuming that the jammer's RWR/ESM system has a sensitivity s_0 and that the uncoded radar transmits a power P_T, one may infer that the range from the radar at which the aircraft must position itself to intercept the signals correctly is

$$R = \sqrt{\frac{P_T G_T G \lambda^2}{(4\pi)^2 s_0}}$$

In the case of a coded radar, the peak power transmitted is n times lower, so that the range at which the standoff jammer must position itself is $\sqrt{n}$ lower, while the radar range remains unchanged.

For example, if the radar uses a 13-element code (with P_T reduced by 11 dB), the standoff range will be roughly one fourth of that afforded by an uncoded radar, and quite probably will be such that the aircraft is detected by the radar before its RWR-ESM system is capable of receiving the radar's signals.

To avoid detection, the standoff aircraft must improve the sensitivity of its ESM, or rather move to a superior ESM, since an increase in sensitivity causes an enormous increase in input traffic for the ESM computer.

Thus, a coded radar can have good LPI characteristics. It has only to use a code with a very large number of elements, that is, with a high value of the product τ_B.

Since normally it is impossible to receive during transmission, unless a receiver with a separate and well-isolated antenna is used, the duration of the transmission should be kept within acceptable limits, which are a few tens of microseconds. This means that if a large number of code elements is required, each element must be extremely short.

Because of this, the spectrum of the transmitted pulse becomes very broad and the latter comes to resemble spread-spectrum transmission [9]. Detection of such a waveform requires high sensitivity ESM receivers.

If the pulse-compression product $\tau \times B$ is increased, in order to obtain the same radar range, being B unaltered, the required peak power must decrease the transmission time increases. In the limit, the transmission can be nearly continuous and the code may have an almost random pattern and an extremely high number of elements [10]. In this case, to ensure the necessary isolation between transmission and reception, separate antenna systems must be used. The transmitted waveform becomes noise-like. Detection of signals of this kind by ESM systems is extremely difficult.

The other ECCM characteristic provided by pulse compression derives from the fact that deception jammers are normally unable to transmit pulses as long as the coded pulses or to retransmit the code (Fig. 6.4). Because of this, the traditional deception jammer will not be able:

(a) to deceive the radar, since only a few elements of code will be jammed, and therefore the effect will only be a slight increase of the sidelobes and a slight fluctuation of the maximum intensity of the true signal.

(b) to achieve *range gate pull-off* (RGPO) by generating false targets, or to create targets to confuse the radar PPI.

To recover effectiveness, the jammer must be upgraded and advanced technologies such as *digital radio-frequency memory* (DRFM) must be employed.

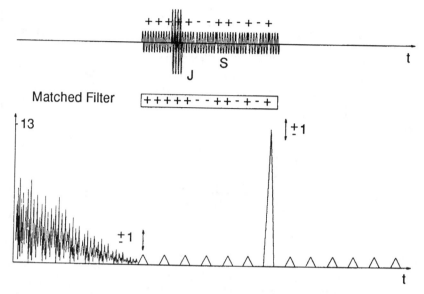

Figure 6.4 ECCM capabilities of a coded radar. Jamming of one single element of the code causes fluctuation of just one unit on the peak and the sidelobes of the compressed signal.

6.2.1.6 CFAR receivers

In order to react quickly enough against a low-flying, high-speed attack aircraft, a search radar must be able to extract plots that can be used by a computer. The main problem for the computer is saturation by an excess of false alarms.

Consider for example a radar with a 200 km scale, a 100 m range bin, a 500-Hz PRF, and a 4 s antenna scan period. In one scan over 360 degrees in azimuth, this radar will receive

$$N = 4 \times 500 \times \frac{200\,000}{100} = 4 \times 10^6$$

samples. This means that if the P_{fa} is 1×10^{-6}, then at each scan the computer will have to process four false returns due to thermal noise, as well as the genuine targets. Considering ground, sea, and rain clutter residues after MTI filtering, the number of false targets detected could be higher by several factors of ten.

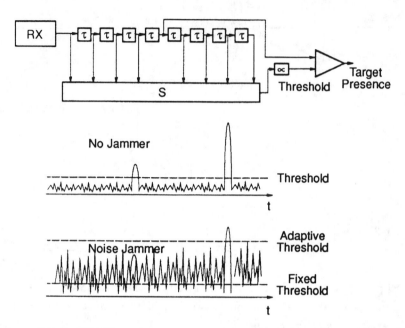

Figure 6.5 A CFAR receiver ensures correct detection of target signals stronger than the jammer's signals without producing too many false alarms.

If a jammer is present, this number could become extremely high and could severely saturate the computer dedicated to *threat evaluation and weapons assignment* (TEWA), which is responsible for extraction of radar tracks, evaluation of their threat level, and selection of a reaction.

To obviate this risk, a search radar, an air defense radar, or a point defense radar must have a receiver equipped with a device that keeps the probability of false alarm at a steady rate, for example, not higher than 1×10^{-6}. A device of this kind is said to have constant *false-alarm rate* (CFAR) characteristics (Fig. 6.5) [11]. As described in Section 2.2.4.2, a widely used CFAR device is the autogate, which generates an adaptive threshold by averaging the signals around each range bin, in order to fix the probability of false alarm at the desired value by the adjustment of a single parameter (α).

Other devices can be used, again based on an adaptive threshold, but realized by a loop process. The detection threshold is increased up to the point at which the number of the signals to be processed reaches the maximum manageable by the computer.

Some other special devices are also considered to be CFAR devices, when they tend to reduce false alarms arising in specific ways. The Dicke-fix receiver is an example; as explained in Section 6.2.2.1, it tends to reduce the effects of very powerful, very short pulse jamming. Also, the back-bias receiver (section 6.2.2.2), which eliminates false alarms caused by CW signals, can be considered as a CFAR device. Since these devices are dedicated exclusively to countering the effects of jamming, they will be discussed later in the chapter as pure ECCM.

Hard-limited receivers can also be included among those with CFAR characteristics. For example, in the presence of noise created by a jammer at a much higher level than the radar thermal noise, the effect of a hard limiter is to guarantee an unchanged probability of false alarm at the output (Fig. 6.6).

6.2.1.7 Frequency agility

Like the others already mentioned, this technique is used, in the absence of ECM, to improve the normal performance of radar and, in the presence of ECM, as a powerful ECCM.

Frequency agility, that is, the shifting of carrier frequency on a pulse-to-pulse or a burst-to-burst basis, offers enormous advantages over fixed frequency (section 2.2.4.3). These advantages are:

(a) increase in range, other parameters being equal, up to 35%.
(b) clutter reduction in radar not equipped with MTI.
(c) angular glint reduction in tracking radars.
(d) lobing reduction or elimination in search radars.
(e) nodding reduction in tracking radars.
(f) reduction of the effectiveness of jammers [12, 13].

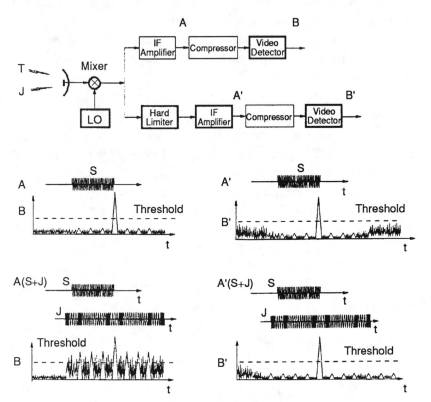

Figure 6.6 CFAR characteristic of hard-limited receivers. After a hard limiter a jammer will not be able to induce mistaken crossings of a threshold.

Frequency agility endows a radar with great resistance to countermeasures of the noise type and sometimes even to those of the deception type. It is effective, for example, against *range gate pull-in* (RGPI) and against multiple false target generation.

Because the frequency used by the radar cannot be predicted, if the agility bandwidth is B_a, the jammer will need a bandwith $B_j \leq B_a$ in order to be effective. The power of the jammer must therefore be distributed over the full bandwidth B_a, much wider than the band B instantaneously occupied by the radar pulse, which is about $1.2/\tau$ Since the agility band can be 500 to 1000 times wider than the radar band B, the noise jamming power injected into the radar band is substantially reduced.

The noise power density p_j of the jammer is defined by

$$p_j = \frac{P_j}{B_j}$$

and the jamming power J received by the radar is

$$J = \frac{P_j(B/B_j)G_jG_R\lambda^2}{(4\pi)^2 R^2 L_p}$$

When the radar is at a fixed frequency, B_j is equal to about $4B$, and the loss arising from the bandwidth ratio is about 6 dB.

When the radar is frequency agile, B_j, which must be higher than or equal to B_a, can be 500 to 1000 times B, with a loss from the bandwidth ratio of about 27 to 30 dB (Fig. 6.7).

The ECCM effect of frequency agility can be clearly seen in the diagram of Fig. 5.19, where, considering the appropriate value for B/B_j, one deduces that the self-screening range $(J/S = 0)$ increases enormously. In such cases a noise jammer can lose all its effectiveness.

Frequency agility prevents a deception jammer from creating false targets at a range below the range of the jammer, because the frequency in the next PRI will not be the same and will not be known until the jammer receives the next pulse from the radar.

6.2.1.8 Frequency agility compatible MTI

To perform good clutter cancellation the transmit-receive equipment must have high phase stability and clutter must not fluctuate excessively from pulse to pulse. When frequency agility is used, clutter is decorrelated, which prevents its cancellation by the MTI. On the other hand, frequency agility

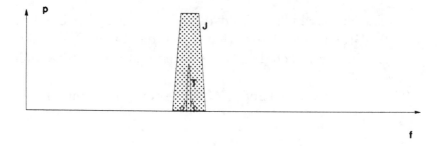

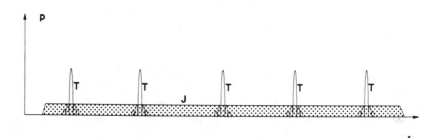

Figure 6.7 Frequency agility is a very powerful ECCM technique because it forces the jammer to distribute its power over a very wide band. By giving the ratio B/B_j the appropriate value, it is possible to evaluate the loss of effectiveness of the jammer with the help of the diagram of Figure 5.19.

yields enormous advantages in dealing with targets and jammers.

The following technique renders frequency agility and MTI compatible [14]: Once a pulse has been transmitted, the bipolar video signals are digitized and memorized, over the whole range axis. In the next PRI, the frequency is shifted, and again all returns are memorized; this is repeated for many PRIs, care being taken to maintain coherence for each frequency used. Then MTI filtering, simple, double, and so on, is performed on all returns at the same frequency. With this method, although the MTI performance can undergo noticeable degradation, frequency agility can be used and all the advantages mentioned will accrue.

6.2.2 Dedicated counter-countermeasures

6.2.2.1 The Dicke-fix receiver

The function of the Dicke-fix receiver is to act as an ECCM against high intensity wideband jamming [15]. This type of jamming can be achieved by a CW rapidly swept over a very wide band, for example, 2 GHz in 50 νs, to jam a radar receiver with a 2 MHz band. When the CW sweeps over the radar band, high intensity spikes, of 50 ns duration, are produced at the receiver input. Often these pulses are able to initiate an oscillation in the first stages of the amplifier, thus blinding the radar receiver (Fig. 6.8).

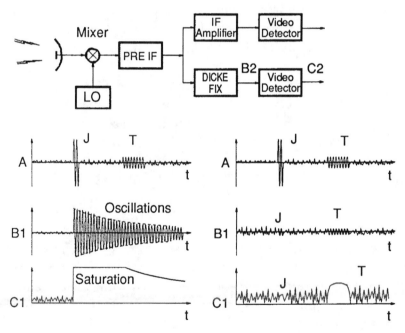

Figure 6.8 Effectiveness of the Dicke-fix receiver against CW jamming.

The Dicke-fix receiver consists of two parts. In the first, a filter of broad bandwidth B_W is followed by a hard limiter. The second part is a matched filter of narrow bandwidth B_s. The first section is designed to amplify all wideband signals without starting to oscillate, and to limit them within a preset value (normally a few decibels below the noise). In the second part, amplification is completed after narrowing the bandwidth to a value matched to the pulse. The gain in the signal-to-jamming ratio S/J is equal to the ratio of the two receiver bandwidths B_l/B_s.

6.2.2.2 The back-bias receiver

The back-bias receiver is an ECCM device against narrow-band CW or spot noise jamming [16]. This receiver (Fig. 6.9) has a circuit for detection of CW signals or signals whose peak-to-peak amplitude is quasi-constant for a given period of time, and is able to suppress these signals, leaving only pulse signals. This device is extremely effective against CW-like signals appearing within its band, such as spot noise generated by frequency-modulated CW with a relatively narrow bandwidth.

Jamming suppression of between 50 and 60 dB can be attained. For this reason, noise jamming produced by interference between two equal signals, phase-modulated in a decorrelated way, is sometimes preferred to that produced by frequency-modulated CW. Insertion of a back-bias receiver will reduce the radar sensitivity on the order of 3 dB below its maximum level.

6.2.2.3 The jammer strobe

The jammer strobe is a device used to give search radar the direction from which spot jamming is arriving. There are many methods for its implementation. The most widely used requires an auxiliary omnidirectional antenna (Fig. 6.10) to make the comparison between signals received by the main channel and those received by the auxiliary channel. The direction to the jammer is the direction for which the jamming signal outputs from the main channel (A) is greater than that of the auxiliary channel (B). This type of circuit can be used to realize another ECCM device, called sidelobe blanking, described in the next section.

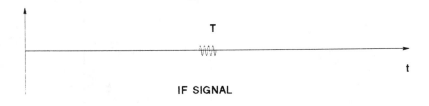

IF SIGNAL

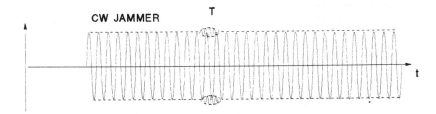

IF SIGNAL IN THE PRESENCE OF JAMMING

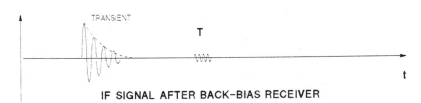

IF SIGNAL AFTER BACK-BIAS RECEIVER

Figure 6.9 Effectiveness of a back-bias receiver against CW jamming.

6.2.2.4 Sidelobe blanking

Whenever the signal in B (Fig. 6.10) becomes higher than the signal in A, the output is blanked, because it must come from a sidelobe and not from the main lobe. Sidelobe blanking can be used for pulse signals, and therefore against deception jamming, and also against noise jammers. Since the sensitivity is reduced by the jamming-to-noise ratio J/N, it is necessary in the latter case to be sure that the predicted jamming will not be such as to cause too great a loss of sensitivity [17, 18].

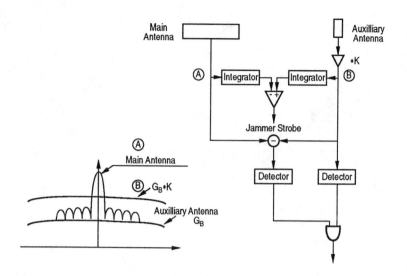

Figure 6.10 The jammer strobe is a device identifiying the direction of arrival of the jammer. Sidelobe blanking, on the other hand, inhibits the receiver output whenever the signal from the sidelobes is higher than the signal from the main antenna.

6.2.2.5 The sidelobe canceler (null steering)

This ECCM technique places a null in the sidelobe structure in the direction of the jammer. In this way the effectiveness of standoff jamming, which creates corridors in the radar air defense network, is reduced.

The signal sample to be canceled is extracted at the point A of the radar receiver (Fig. 6.11). In the case of noise jamming it will be a quasi-CW signal. By means of the gain G, the amplitude and the phase of the signal arriving from the omini-directional auxiliary antenna are adjusted so as to minimize the CW signal. This is done with a suitable time constant, so that the null is maintained in the direction of the jammer while the radar antenna keeps scanning. Jamming attenuation of up to 20 dB is achieved [19, 20, 21].

6.2.2.6 Automatic frequency selection

The most sophisticated radars, capable of changing transmission frequency relatively quickly, can be equipped with a

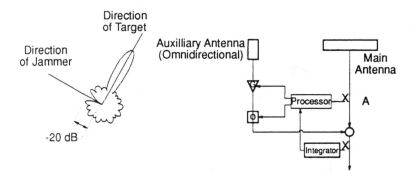

Figure 6.11 Sidelobe canceler or null steering is a technique allowing attenuation of a signal coming from the sidelobes by means of interference with the signal from the main antenna.

special device, called *automatic frequency selection* (AFS), that allows automatic selection of a frequency not being jammed [13]. There are two fundamental types of this powerful ECCM. The first is based upon the analysis of the spectrum which a radar can use, performed in a special PRI during which the radar pulse is not transmitted. Upon completion of the analysis, it is possible to decide whether to maintain the same transmission frequency or to change to another where jamming is absent. The second method is based upon the analysis of both the signal-to jamming ratio, which must be suitable for the functions that the radar has to perform, and the jamming-to-noise ratio, which must be minimum. This last analysis is performed during dead time, that is, in the interval between the maximum range time and the transmission time of the next pulse. Because of band limitations of the jammer, the shifting of the sidelobe nulls, the limitations of the tuning speed of the jammer, and the effect of multipath, it is generally possible to find a frequency that improves the signal-to-jamming ratio.

6.2.2.7 Multibeam antenna
The possibility of using more than one beam to cover the elevation plane is a powerful ECCM for a search radar. When

there is only one beam in elevation, the presence of a jammer reduces radar capability at all elevations (Fig. 6.12). However, if the radar has more than one beam, its capability is reduced only for the beam that is jammed, while the performance at all other elevations is unaffected. Multibeam capability can be achieved with *frequency sensitive antennas* (FRESCAN) [22] by the transmission of a long pulse, or a train of pulses, divided among a number of frequencies, or, with the greater flexibility of phased-array antennas, simply by the formation of multiple beams in the elevation plane. Often radars use scanning or multiple beams only on reception, while on transmission the beam is not divided [23, 24, 25].

6.3 Tracking radar counter-countermeasures

Before starting to operate in tracking mode, a tracking radar must be able to see and acquire the target. Therefore, many of the ECCM already examined for search radars are employed also in tracking radars. The most important and most widely used are:

- fast time constant.
- sensitivity time control.
- random PRF.
- frequency agility (with frequency laws of various types).
- moving target indicator.
- CFAR receiver.
- back-bias receiver.
- Dicke-fix receiver.
- automatic frequency selection.
- sidelobe blanking.
- pulse compression (code).
- jammer strobe.

In the following, those ECCM capabilities of tracking radars which result from the careful design for the radar function will be discussed first; then devices dedicated specifically to ECCM,

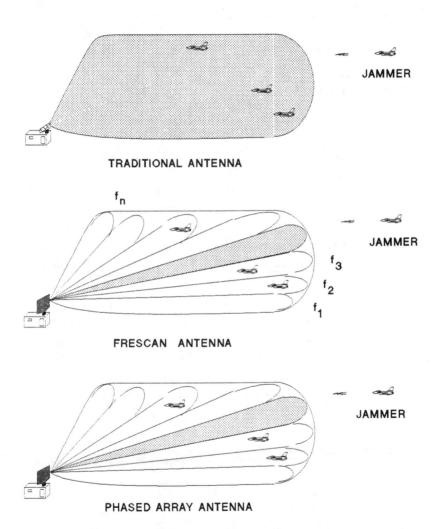

Figure 6.12 ECCM capabilities of a multiple-beam radar. A jammer can jam only one of the beams, leaving unaltered the radar capability of all the others.

and required to maintain tracking in range and velocity, and for angular tracking, will be reviewed.

6.3.1 Induced counter-countermeasures

6.3.1.1. Preselective filters

Since thermal noise is present at all frequencies, it can enter the *intermediate-frequency* (IF) band from either the signal band or the image band. If the image band is eliminated, the noise will be lowered by 3 dB, thus improving the SNR (Fig. 6.13). It is therefore good practice to build radars without an image band. This improves the resistance of the radar to countermeasures, even in the presence of a wideband noise jammer. A monopulse tracking radar is extremely sensitive to jamming in the image band, because suitable matching of amplitude and phase outside the normal operating band is not assured, so that the jamming might even invert the sign of the angular gradient.

To eliminate the image band one can use preselective filters. If the radar is frequency agile, one single preselective filter is not enough because in general the IF band is much narrower than the band exploited by the agility. In this case the radar may use two or more intermediate frequencies. A simpler method, which however attenuates the image band response only by about 20 dB, is to use an image-rejection mixer.

6.3.1.2 Low sidelobes

A good method for minimizing the effects of jamming on a tracking radar is to use antennas with low sidelobes [26, 27]. Noise or deception jamming arriving through the sidelobes could cause a lock-on situation, with deleterious consequences for the aiming of the associated weapon system. An example of an antenna with small sidelobes is the polarization-twisting Cassegrain antenna, where the first sidelobes are about −30 dB down from the main lobe.

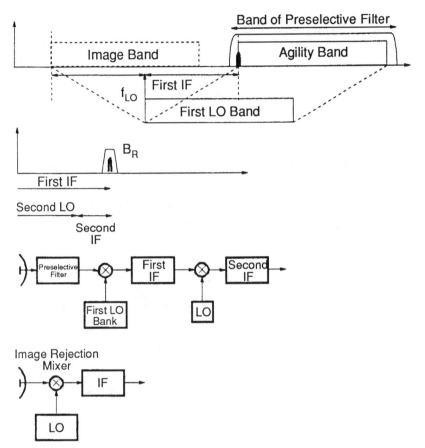

Figure 6.13 Image band rejection.

6.3.1.3 Intermediate-frequency filters

Another technique that reduces the effects of a jammer is to give a high slope to the matched IF band filter. This is especially important in monopulse radars, where, to avoid skirt jamming, the slopes are designed to more than 60 dB per decade.

Skirt jamming is deception jamming transmitted with very high power and at a frequency shifted by about half the IF bandwidth. The radar will receive pulses on the edges of the IF filter band, where the matching in amplitude and phase is difficult to check, so that tracking precision cannot be ensured.

6.3.1.4 Fast automatic gain control

Fast automatic gain control (AGC) is very effective against jammers exploiting amplitude modulation. It is easily implemented by monopulse radars, but not by conical-scan radars.

To achieve fast AGC performance with a conical-scan radar, a special configuration of the kind shown in Figure 6.14 is needed. In Figure 6.14a, the cutoff frequency is higher than the conical-scan frequency. The amplitude modulation due to the scanning is derived directly from the AGC voltage by suitable filtering. In Figure 6.14b, on the other hand, AGC voltage can compensate all fluctuations of the signal, with the exception of those at the conical-scan frequency, by means of a notch filter placed right on this frequency.

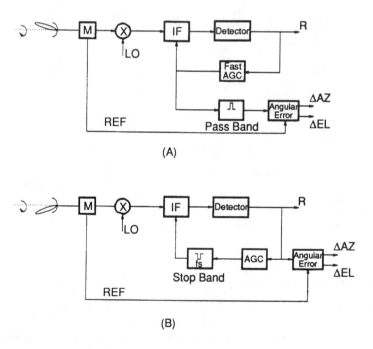

Figure 6.14 Fast AGC for conical-scan radar. a) Without notch filter; b) With notch filter.

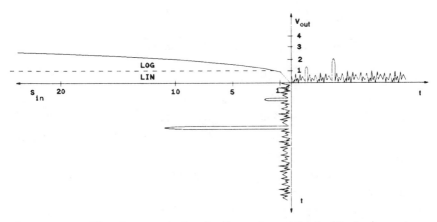

Figure 6.15 The characteristic of a linear-logarithmic (lin-log) receiver allows all output signals with low dynamics.

In any case, a normal design for a tracking radar provides for a cutoff AGC frequency much higher than the servo loop cutoff frequency.

If a conical-scan radar uses a frequency of about 30 Hz, the cutoff frequency of the AGC is limited to between 1 and 3 Hz, and the servo bandwidth is less than 0.1 Hz, which results in rather slow tracking. This is why modern conical-scan radars use special devices to increase their scanning frequency.

6.3.1.5 Logarithmic receivers

To avoid the damaging effects of modulated jamming, receivers with logarithmic characteristics are often employed (Fig. 6.15). Although in this case the sensitivity and the linearity of the angular gradient will be degraded, all the shortcomings resulting from waiting for the implementation of the AGC, saturation, capture of the AGC by deception jammers, and so forth, will be avoided. Where the signals are very small, the receiver often behaves linearly; it is then known as a "lin-log receiver."

The effect of compression on amplitude modulation is generally too strong for a scanning radar, so that logarithmic -

receivers are normally employed in monopulse radars.

6.3.2 Dedicated counter-countermeasures

6.3.2.1 The jamming detector

Tracking radars are frequently equipped with devices able to detect the presence of a noise jammer. Such devices generally allow activation of track-on-jam ECCM, and warn the operator of the threatening situation.

Jamming detectors can use gates during the radar dead times, or use gates open around the target. The "jammer present" signal is exploited by updating the range tracking system from the velocity memory, and by deriving data for angular tracking from the jamming signal itself. The most sophisticated radars are able to detect the presence of a look-through period. In this case the radar waits for the period in which the jammer is off to update its tracking parameters correctly. This device is sometimes also called "look-through ECCM."

6.3.2.2 Anti-range gate pull-off

The *anti-range gate pull-off* (ARGPO) ECCM is also known as *anti-range gate stealing* (ARGS).

Two fundamental configurations achieve this ECCM performance. The first is based on the leading-edge tracking of the true echo; the second, on the loss of balance of the range tracking loop, with early (E) and late (L) gates weighted differently.

In the first configuration (Fig. 6.16), the first receiver has low output dynamics to avoid giving the deception signal more weight than the true signal. For example, the capture of the AGC is avoided. Limited receivers, such as the Dicke-fix receiver, and logarithmic receivers are of this type. The receiver output is sent to a differentiating circuit that practically abolishes all signal tails beyond a certain prearranged value.

Assume that the radar has either a random PRF or is frequency agile to avoid range gate pull-in. Then, as long as the

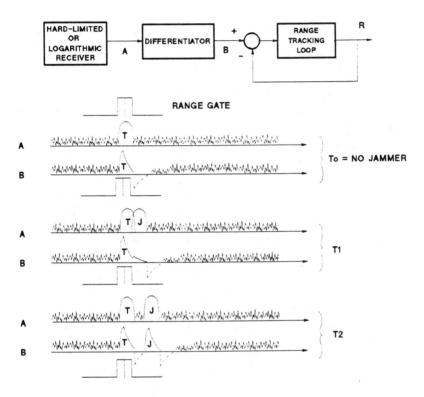

Figure 6.16 *Anti-range gate stealing* (ARGS) based on differentiator.

true echo and the deception are not separated, when the deception jammer attempts to delay the echo, practically only the first part of the true echo will appear at the output of the differentiating circuit (Fig. 6.16). When true echo and deception are separated, the range gate will not detect the second echo, since it is still hooked to the first.

In the second configuration (Fig. 6.17), again after a limited or a log receiver, the value of the early gate is weighted more than that of the late gate, which results in a forward shift of the range: The measured range is shorter than the actual range, but by a predetermined and therefore unimportant amount. Figure 6.17 illustrates the operating principle. Since there is a forward shift of the gates, the range-tracking system is not affected by the presence of the deception echo.

In both configurations, the time constant of the tracking loop acts like a small ECCM, in the sense that it tends to ensure that the gates do not shift velocity.

Against these devices, which are inexpensive, RGPO has no hope of success. The only path open to the jammer is to exploit a noise cover pulse.

Sometimes, in the presence of noise jamming, the radar can purposefully narrow the band of the tracking loop, thus succeeding in tracking in range with signal-to-jamming ratios below $-10\,\mathrm{dB}$. If the radar combines frequency agility with ARGPO, the only thing the jammer can do is to "attack" not in range but in the angular loop.

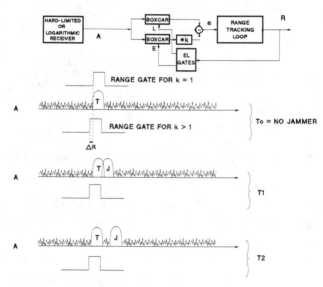

Figure 6.17 ARGS based on differential weighting of the E/L gates.

6.3.2.3 Guard gates

The *guard gates* technique is an ECCM which entails presetting sensors around the gate in which tracking is performed so that as soon as the presence of an additional echo is detected, the tracking system switches to memory for a short time, and

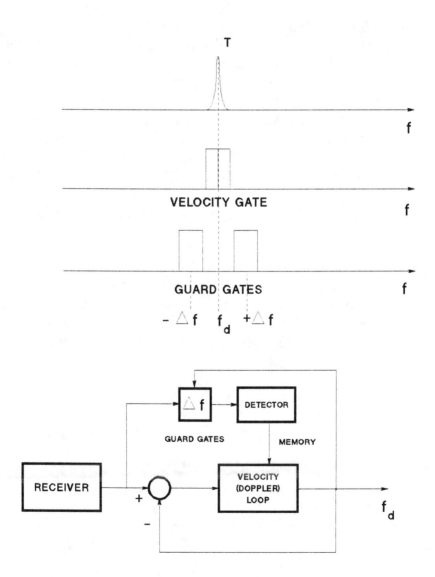

Figure 6.18 Guard-gate ECCM.

then reacquires the old target (Fig. 6.18). Accordingly, when a deception jammer tries to lure the tracking gate to a false target, as soon as the true echo and the deceptive echo separate, the true echo will enter the guard gate, thus blocking the tracking gate. When the sensors indicate that the deceptive echo has gone, the gates will again position themselves correctly. If the deceptive signal stays beyond the time-out, the system switches to the track-on-jam mode (Section 6.3.2.5).

This technique is frequently used in semiactive missile systems, in which tracking is performed by extracting angular information and velocity from the doppler signal generated by the target toward which the missile is homing. If the victim aircraft attempts *velocity gate pull-off* (VGPO) (Section 5.3.8), the guard gates will prevent it from working. For this reason, the guard gate technique is also called *anti-velocity gate pull-off* (AVGPO).

6.3.2.4 Double tracking

Often in airborne radar the *fast Fourier transform* (FFT) is used to process the signal on both the range and the velocity axes. In this way the target produces an echo which, being characterized in both range and velocity (doppler) allows double tracking (Fig. 6.19).

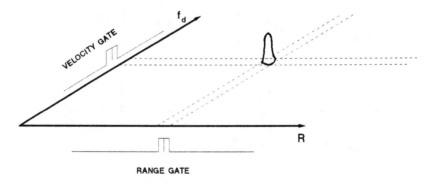

Figure 6.19 Sophisticated radars can exploit double tracking, in range and velocity.

If a jammer attempts to open a range gate not coherent with the doppler, it is ignored by the logic of the pulsed-doppler radar. In this sense the FFT can be regarded as a powerful ECCM.

6.3.2.5 Track on jam

It is often desirable not to lose angular tracking of the target, even when a noise jammer denies range information. In many cases, for example in the case of anti-ship missiles, precise range information is not of paramount importance.

Therefore, when noise jamming is very strong, the sensor detecting the presence of a jammer can update the range-tracking loop from the velocity memory, while for angular tracking the radar receiver can extract the required information from the jamming signal itself. This is "track on jam."

6.3.2.6 Random conical scan

This ECCM is exploited by conical-scan radars to counter amplitude modulation in jamming signals which otherwise could cause large angular errors and even lead to break-lock. The random conical-scan technique also effectively counters ECM systems of the inverse-gain type, when they are based on conical-scan frequency measurement. However, it is not effective against instantaneous inverse-gain jammers.

This simple ECCM is achieved by varying the velocity of the motor responsible for conical scan in a pseudorandom way within a given domain. Scanning-frequency shifts on the order of 10 Hz per second can be achieved.

6.3.2.7 COSRO-LORO

Conical-scan on receive only (COSRO) and *lobe-switching on receive only* (LORO) radars have been realized entirely for ECCM motives.

During transmission the beam of a COSRO radar antenna is generated by a feed centered on the boresight, while during reception an offset and rotating feed is used, capable of modulating the signal appropriately for angular tracking.

A LORO radar, on the other hand, usually transmits the sum beam, but scans the separate beams on reception [28].

The antenna system of a COSRO radar generates a slight modulation during transmission [28] and much care must be taken to ensure that, however slight, it cannot be detected by a jammer. The angle ECM normally used against radars of these types has already been mentioned. It entails generation of amplitude-modulated jamming signals at variable frequency, while attempting to detect the radar reaction frequency by measuring the amplitude variations during the look-through period. As soon as this frequency has been identified, the sweep is stopped and modulation of the jammer continues at this frequency. Taking into account the servo bandwidths, this process can go on for tens of seconds.

6.3.2.8 Monopulse

The most effective ECCM against angular deception and jamming is monopulse. If there are no faults in the radar design, it is practically impossible to generate errors in angular tracking merely by amplitude modulation.

Up to the present, a frequency agile radar with a logarithmic receiver, ARGPO, and monopulse is practically immune to jamming. Frequency agility prevents the generation of enough noise power inside the receiver to reduce its detection and acquisition capabilities to any significant extent. The logarithmic receiver ensures that all the output dynamics are compressed and can be managed by the range and angular loops servos, without any fear of the AGC being captured. The ARGPO circuit ensures correct range tracking even in the presence of a powerful deception jammer. Finally, monopulse angular tracking is resistant to any type of modulation.

The only ECM techniques that can be successful against the monopulse technique are:

- chaff and platform maneuvers, if the radar is not equipped with an MTI.
- cooperative jamming.

- terrain bounce.
- cross polarization.
- cross eye.
- decoys.
- illuminated chaff. These are described in Chapter 5.

6.4 Infrared counter-countermeasures

IR sensors leave little scope for countermeasures. Up to the present, infrared ECM consist essentially of modulated sources and flares (sections 5.4.1 and 5.5.2.3).

Against these techniques, which are characterized by spectra different from those of the protected platforms, ECCM systems can resort to spectral analysis of the received signals to distinguish them from signals of interest.

Against modulated sources, monopulse IR systems can be employed. Although much more costly, these systems do not require sequential scanning in order to determine angular errors.

Against flares, one may use seekers with focal plane arrays of thousands of elementary sensors, which are capable of distinguishing by image processing techniques between targets and flares.

6.5 Communications counter-countermeasures (COM-ECCM)

In order to avoid the problems caused by jamming, or to avoid interception and exploitation of one's own signals by the enemy, modern communication systems resort to the following ECCM or masking methods:

(1) frequency hopping (spread spectrum).
(2) burst transmission.
(3) null steering .

(4) direct sequence.
(5) uniformity of message format.
(6) encryption.

6.5.1 Frequency hopping

Frequency hopping has many channels available for message transmission, and in a pseudorandom way selects one, which it occupies for an extremely short time before hopping to another (Fig. 6.20) [29, 30].

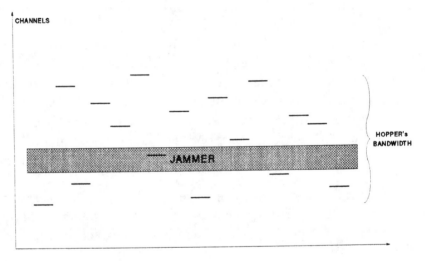

Figure 6.20 Frequency hopping is one of the most powerful ECCM techniques in the telecommunications domain.

To achieve this, both transmitter and receiver know which frequencies will be used for transmission by the use of pseudorandom prearranged codes, and by rephasing in real time when the transmit-receive equipment is turned on.

Frequency hopping can be:

• slow, with on the order of 50 hops per second.
• fast, with on the order of 500 hops per second.

The advantages of this technique are several:

(a) Communications systems can be intercepted and located only with great difficulty. It is not easy for COM-ESM equipment to perform *direction-finding* (DF) measurements in a few milliseconds, when the frequency used is not known *a priori*.

(b) The enemy is prevented from intercepting the transmitted message, because it is impossible to track or monitor an emission which makes random use of a large number of channels.

(c) Communication systems are practically immune to jamming, since the jammer is forced to distribute its power over too wide a band. For example, with frequency hopping at 100 Hz, the link channel will be occupied for less than 10 ms. In order to introduce a significant error rate, the jammer should be able to tune itself in a few milliseconds, which is possible only if the channel to which it should be tuned is known. It is extremely difficult to track the frequency hopping of a given emitter in a dense environment such as a battlefield, in which hundreds of emitters can be present.

6.5.2 Burst transmission

The burst transmission technique requires the transmission of information in an extremely short time (100 ms) to avoid location, interception, or jamming by an enemy.

The operator types the message and the address of the recipient on a keyboard, checking them on a display that, in the case of a portable radio, will be of the liquid-crystal type. The computer associated with the radio codes the message suitably and transmits it to the recipient in a very short time.

This type of transmission is used, for example, to guide aircraft from the ground during attack-on-the-ground actions, or for anti-aircraft weapons coordination during a defensive action.

6.5.3 Null steering

With techniques similar to those described for radars it is possible, by means of an auxiliary antenna, to generate an interference signal which severely desensitizes the system in the direction of the jammer, thus allowing correct reception of a useful signal arriving from a different direction [31].

6.5.4 Direct sequence

The direct sequence technique transmits the message embedded in noise. This type of transmission cannot be detected even by ordinary channelized search receivers.

Techniques for detection of these signals might be based on the correlation of the signal received by one or more receivers. However, direct sequence transmission is not widely used, at least in the communications domain, essentially because of the "near-far" effect, so that adequate detection techniques have not yet been developed. The near-far effect is the unintentional jamming of friendly nearby equipment because the radiated signal is so strong that the processing gain of the system does not succeed in canceling it. Friendly equipment is thus prevented from receiving messages correctly from faraway stations to which they are linked.

6.5.5 Uniformity of format

In order not to permit an enemy to perceive differences between the various operational phases, for example in case of an imminent attack, an attempt should be made to present the enemy with transmissions all organized in the same way. For example: no distinction being made between transmissions of slight relevance and those of great importance. With the same objective, significant variations in the volume of traffic should also be avoided.

6.5.6 Encryption

A message liable to be intercepted by an enemy can be made unintelligible to those to whom it is not addressed. In order to do this, a device capable of coding the signals according to an apparently chaotic code (cryptographic code) is introduced (encryption). This device is added to an ordinary radio before the modulator.

The intended recipient of a message, that is, a station intentionally linked to the transmitter, is able to decrypt the message by means of a suitable decoder at the receiver output.

Encryption codes, whose number is large, can be changed according to a preestablished strategy, or according to code keys distributed to authorized users of the network.

REFERENCES

[1] S.L. Johnston, "WW-II Radar ECCM History," IEEE Radar-85 Conf. Rec., May 1985, pp. S-2, S-7.

[2] S.L. Johnston (ed.), *Radar Electronic Counter Countermeasures*, Norwood MA: Artech House, 1979.

[3] "Methodology for Specifying Jammer System Parameters," *International Countermeasures Handbook*, Palo Alto: EW Communication 4th Ed., 1978, pp. 406-410.

[4] D.C. Schleher, *Introduction to Electronic Warfare*, Norwood, MA: Artech House, 1986, pp. 109-183.

[5] M. Mahaffey, "Electrical Fundamentals of Countermeasure Chaff," *International Countermeasures Handbook*, 2nd Ed., 1976, pp. 512-517.

[6] S.L. Johnston, "Radar Electronic Counter Countermeasures Against Chaff," International Radar Conference, Paris, May 1984, pp. 517-522.

[7] E.J. Carson, "Low Probability of Intercept (LPI) Techniques and Implementations for Radar Systems," *IEEE Proceedings National Radar Conference*, 1988, pp. 56-60.

[8] R.M. Raines and S.A. Blankenship, "The Impact of Advanced Modulation Techniques on EW," *Defense Electronics*, October 1986, pp. 81-97.

[9] R.A. Dillard and G.H. Dillard, *Detectability of Spread Spectrum Signals*, Norwood, MA: Artech House, 1989.

[10] D.F. Albanese, "Pseudo-Random Code Waveform Design

Trade-Off for CW Radar Application," Radar-77, London, 1977, IEE Conf. Pub., No. 155, pp. 513-514.

[11] P.S. Tong and P.E. Steichen, "Performance of CFAR Devices in ECM Environment," DDRE Radar Symposium, 1976.

[12] B. Bergkvist, "Jamming Frequency Agile Radars," *Defense Electronics*, January 1980, pp. 75-83.

[13] S. Strappaveccia, "Spatial Jammer Suppression by Means of Automatic Frequency Selection System," Radar-87, IEE Conf. Pub., No. 281, 1987, pp. 582-587.

[14] G. Petrocchi et al., "Anti-Clutter and ECCM Design Criteria for Low Coverage Radar," *Proceedings International Radar Conference*, Paris, December 1987, pp. 194-200.

[15] G. Picardi, *Elaborazione del segnale radar*, Section 4.12, Franco Angeli Editore, Rome.

[16] G. Picardi, *Elaborazione del segnale radar*, Section 4.13, Franco Angeli Editore, Rome.

[17] P.O. Aranciba, "A Sidelobe Blanking System Design and Demonstration," *Microwave Journal*, Vol. 21, No. 3, March 1978, pp. 69-73.

[18] L. Maisel, "Performance of Sidelobe Blanking Systems," *IEEE Transactions on Aerospace and Electronics Systems*, Vol. AES-4, No. 2, March 1968, pp. 174-180.

[19] D.J. Chapman, "Adaptive Array and Sidelobe Cancellers: A Perspective," *Microwave Journal*, Vol. 20, No. 8, August 1977, pp. 63-64.

[20] T. Bucciarelli, M. Esposito, A. Farina, and G. Losquadro, "The Gram Schmidt Sidelobe Canceller," *IEE Radar-82*.

[21] M.H. Er, "Techniques for Antenna Array Pattern Synthesis with Controlled Broad Nulls," *IEE Proceedings*, Vol. 135, Pt. H, No. 6, December 1989.

[22] I.W. Hammer, "Frequency-scanned Arrays," Chapter 13 in M.I. Skolnik (ed.) *Radar Handbook*, McGraw-Hill, New York, 1970.

[23] P. Barton, "Digital Beam Forming for Radar," *IEE Proceedings*, Vol. 127, Pt. F, No. 4, August 1984.

[24] R.A. Mucci, "A Comparison of Efficient Beam Forming

Algorithms," *IEEE Transactions on Acoustic, Speech and Signal Processing*, Vol. ASSP-32, No. 3, June 1984.

[25] P.A. Valentino, "Digital Beam Forming: New Technology for Tomorrow's Radar," *Defense Electronics*, December 1984.

[26] G.E. Evans and H.E. Schrank, "Low Sidelobe Radar Antennas," *Microwave Journal*, Vol. 26, No. 7, July 1983, pp. 109-117.

[27] N.F. Powell, "System for Obscuring Antenna Sidelobe Levels," U.S. Patent No. 4,435,710, March 1984.

[28] J.H. Dunn, D.D. Howard, and K.B. Pendleton, "Tracking Radar," chapter 21 in M.I. Skolnik (ed.) *Radar Handbook*, McGraw-Hill, 1970, pp. 21-31.

[29] P.E. Van, "New Concepts in Battelfield Communications. Part 1," *International Defense Review*, 3/1987.

[30] G.S. Sundaram, "New Concepts in Battlefield Communications. Part 2," *International Defense Review*, 5/1987.

[31] R.A. Hodges, "Interference Cancellation Equipment, an alternative ECCM," *International Defense Review*, Suppl. Vol. 17-5, 1984.

[32] D. Edward, "The Requirements of Cryptography," *International Defense Review*, Suppl. Vol 17-5, 1984.

Chapter 7

New Electronic Defense Techniques and Technologies

7.1 Introduction

The *electronic defense* (ED) techniques and technologies currently in use have been amply treated in this book. In this chapter the discussion will center on the way in which traditional techniques evolve whenever technology provides designers with new, more powerful and increasingly small devices.

This applies both to the analog-digital domain, with its silicon (Si) and gallium arsenide (GaAs) hybrid and semi-custom circuits, and to the microwave domain, where the advent of *monolithic microwave integrated circuits* (MMIC), that is, monolithic GaAs circuits, makes possible the design of complex networks in minute volumes, totally unthinkable only a few years ago (Fig. 7.1) [1]. It applies also to software, where the languages have been developed which permit the exploitation of *artificial intelligence* (AI) techniques for more reliable management and interpretation of data.

Possible countermeasures against the *anti-radiation missile* (ARM) will be examined. Because of its hard-kill capabilities, this weapon will be considered as a threat rather than as an ECM.

Figure 7.1 Comparison between an MIC and an MMIC in GaAs.

Finally, new systems designed to counter the recent threat embodied in stealth aircraft will be discussed.

7.2 New electronic defense techniques and technologies

Several years are needed for the development of a new ED system, so that a successful modern system must balance the risk inherent in its use of the most advanced technology against the risk of being out-of-date by the time it is developed. The block diagram of Figure 7.2 shows the essence of an ED system. In this context the techniques and technologies available now, or expected to be available in the near future will be discussed.

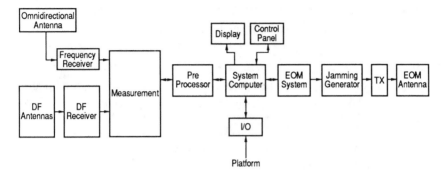

Figure 7.2 Block diagram of an integrated ED system.

7.2.1 ESM antennas

A wide choice of antennas capable of meeting all requirements is available at present. Physical limitations and the maturity of the technology are such that no particular breakthroughs are expected in antenna systems for either frequency or *direction-finding* (DF) receivers.

7.2.2 Complex receivers: cued receivers

Easily classifiable, and at times completely autonomous, architectures are generally adopted for traditional receivers. On the other hand, the most advanced front ends resort to hybrid integrated structures which allow the optimization of system performance to deal with the expected enormous density of traffic and the required precision of measurement. A quantum jump has been achieved in microwave processing performance, thanks to MMIC technology, with size, weight, and sometimes cost being greatly reduced. Automation of manufacturing permits realization of complex microwave components on a single chip, and an impressive improvement of performance repeatability. Cost reduction is not automatic, but it is possible, provided that structures utilizing a large number of identical chips are manufactured.

The exploitation of advanced technologies allows realization of complex receivers called "cued receivers" [2]. In these structures, the frequency receiver and the DF receiver are strongly integrated. The former is able to examine the spectrum, or part of it, in a panoramic way, in order to decide rapidly, with the help of cuing logic, which signal is of interest and at what frequency. Once this has been decided, the panoramic receiver and the DF receiver are tuned to make measurements on the signal in protected mode (Fig. 7.3).

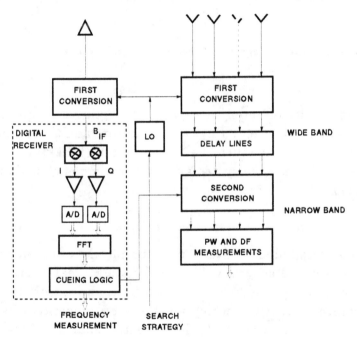

Figure 7.3 Advanced cued ESM receiver. The panoramic receiver used to tune the DF receiver instantaneously is of digital type.

The instantaneous tuning of the receiver can be achieved by means of ultrafast frequency synthesizers or by means of microwave filters employing GaAs devices (MMIC), which are tunable electronically in negligible times. Thanks to the technological evolution of GaAs digital components and ultrafast silicon components, a panoramic frequency receiver exploiting digital techniques (digital receiver) can be realized. A digital receiver can perform simultaneous spectral analyses of the signals present in a very wide band (on the order of gigahertz), in real time.

This can be achieved with chips able to perform digital conversion of the in-phase (I) and quadrature (Q) signals (Fig. 7.3) with long words and at very high speed (for example, 6 bits at 1 GHz), and chips able to perform the *fast Fourier transform* (FFT) almost in real time (a few tens of nanoseconds). Among the various panoramic receivers, the digital receiver offers the best prospects in terms of performance, weight, bulk, and cost.

7.2.3 Signal processing

The front-end technology presently available shifts the problem of ESM toward signal processing. With a traffic of 5 to 10 million pulses per second assumed at the input of the ESM in critical situations, it is easy to calculate the through-put necessary for signal processing.

To define a pulse completely, the following word lengths are typically required:

- PW 12 bits.
- frequency 12 bits.
- amplitude 6 bits.
- D 10 bits.
- code 5 bits.
- TOA 32 bits.
- spare.

Altogether, leaving adequate margin, at least seven 12-bit words will have to be generated; taking into account other potential requirements, this necessitates the capability to generate a message of at least 100 bits. If an input traffic of one million pulses per second is assumed, the number of bits produced in one second is 1×10^9. That is, the processor which follows must be able to accept and process a billion bits per second.

This means that it will be necessary to resort either to prescalers, to compact information so that it can be managed by easily available computers, or to very-high-speed computers. Some very fast computers are today able to perform large numbers of operations: throughputs of 100 *million instructions per second* (MIPS) with 32-bit words are possible. In the past, *transistor-transistor logic* (TTL) integrated circuit technology has been exploited; today, attention has shifted to *N-channel metal oxide semiconductor* (NMOS) and above all to *complementary metal oxide semiconductor* (CMOS) technologies. NMOS allows integration of up to 35,000 logic gates in one chip; CMOS, while also allowing a large number of gates (about 30,000 per chip), has the additional advantage of a much lower heat dissipation (about 0.02 mW/gate against 0.77 with N-MOS and 0.88 with TTL).

When speed is required, there is no doubt that *emitter-coupled-logic* (ECL) is the preferred technology. ECL allows a gate density of about 25,000 per chip, with a delay per gate of only 0.2 to 0.3 ns, as against the 10 to 35 ns of CMOS. The problem with ECL is power consumption, which may be as much as 2.4 mW/gate [3].

GaAs technology will shortly allow manufacture of moderately complex integrated circuits with a delay per gate on the order of 0.05 to 0.1 ns and a power dissipation on the order of 0.05 to 0.1 mW/gate. It is known that silicon technology is being improved; complex bipolar Si circuits with 20 GHz clock rates have been tested.

The TTL, CMOS, and ECL technologies already permit the manufacture of semi-custom chips, that is, industrial mass-production chips that answer the requirements of individual customers. Complex logic circuits that in the past required many stock chips interconnected by printed circuits can now be realized on a single chip. Moreover, many manufacturers now offer a hybrid custom service, with realization on single chips of complex analog circuits designed by the client. These circuits are realized by *computer-aided design* (CAD) and *computer-aided manufacturing* (CAM).

7.2.4 Artificial intelligence and expert systems

Often in ED operations it is necessary to take decisions in a highly complex situation, for example:

- to decide which of the tracks provided by an ESM system are real and which are merely reflections caused by the environment.
- to decide which tracks are part of the same complex emission.
- to identify an enemy platform on the basis of its ESM tracks.
- to choose the best jamming program to be performed by ECM equipment, in view of the operational scenario.

At present, to avoid an excessive number of errors, the task of disentangling ambiguous and complex situations devolves exclusively onto the operators. Relying on their experience, they integrate and correlate available information from all sources, formulate hypotheses and test them, perhaps even modifying some of the system parameters. However, all this may take a long time.

It is a facile prediction that in the future the complexity of the electromagnetic scenario will increase and that, at the same time, it will be necessary to react very quickly. An operator will not be able to perform manually the operations just described, and there will be a need to resort to fast machines capable of emulating human reasoning.

Human beings can perform two types of reasoning: One is of a deterministic, or rational, type ($2 + 2 = 4$, cause and effect), the other of an empirical, intuitive or heuristic type, based on impressions and memories of analogous situations, that, by nature, is neither rigorous nor certain.

Traditional computers operate deterministically, in essence applying iterative processes to numerical data, and are incapable of integrating this with heuristic reasoning.

An integrated machine capable of using both types of reasoning is a machine endowed with *artificial intelligence* (AI) characteristics [4].

AI is the branch of computer science that studies the fundamental methodologies and techniques required for design and manufacture of systems capable of performance similar to that of human intelligence. Fields of application of AI technology are:

- Expert systems, that is, systems whose rules are drawn from experience.
- Robotics.
- Natural language processing.
- Automatic programming.

Expert systems find an effective application in ED systems such as ESM and ECM. In fact, an expert system is a program which can be implemented on suitable computers and can solve problems related to a given situation, by exploiting rules drawn from expert knowledge and experience. Figure 7.4 shows the architecture of an expert system. The blocks related to the knowledge base, the inference engine, and the user interface are the system shell, that is, its supporting frame.

The knowledge base consists of suitable representations (formal logic and language) of the characteristic elements (data, links, rules drawn from experience, and expertise, deductions, implications, correlations, analogies, etc.) of the situation (for example, platform identification) on which the system is called to operate. The user interface transforms the current data into symbols that can be managed by the system.

The inference engine "guides" the comparison process between facts concerning a specific case and the knowledge base, allowing correlations, the testing of hypotheses, and so forth, and finally arriving at possible decisions or solutions.

When data concerning an actual situation arrive at the system input, they are suitably translated by the user interface so that they can react with the rules, reasoning, and situations memorized in the knowledge base, with the guidance of the

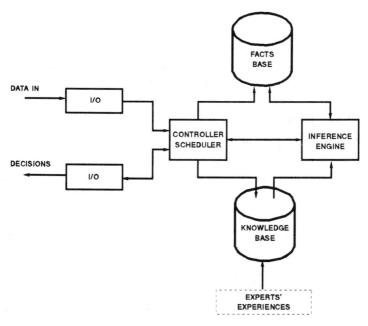

Figure 7.4 Architecture of an expert system.

inference engine. Hypotheses are formulated and tested until, all processes being exhausted, the solutions are presented, together with the reasons that led to them.

The use of expert systems is at present very limited because of:

- incapacity to learn automatically.
- difficulty of managing a very large knowledge base.
- difficulty in the management of uncertainties.
- excessively low speed of present day computer technology.

This last limitation arises from the slowness of the components used in computers and from the fact that normal computers operate sequentially, performing one operation after another. However, it may be predicted that in the future the utilization of new components and new parallel computer architectures, for example neural networks, will improve the speed of expert systems to the point that they can usefully execute AI programs.

7.2.5 Computers

The spread of complex software requiring large computing resources, as for example AI applications, has driven the development of hardware with continually improving performance.

Processor technology is evolving at high speed, with computing capability typically doubling each year. In recent years, microcomputers based on 32-bit *central processing units* (CPU) have spread widely. Besides high performance in data processing, they offer solutions to the problems of signal processing.

Particularly promising from the point of view of increased performance is the *reduced instruction set computer* (RISC) architecture. These microcomputers are based on a simple and effective architecture. Studies of a large number of software applications show that a remarkable performance increase can be achieved by speeding up the execution of simple and frequently used instructions, even at the cost of slowing down more complex ones. Following this, RISC designers have chosen to realize only one set of simple instructions and to execute most of them in the CPU, rather than use subroutines. Implementation of more complex instructions is left to the compiler.

In this way CPUs with very high average speed can be realized. The speed of these computers is measured in MIPS. Already machines with a throughput of 100 to 200 MIPS, achieved by integrating RISC and ECL technologies, are available.

Another means of obtaining a very high computing performance is the use of parallel processing. This tendency is supported not only because there are physical limitations to improvement of the performance of one single computing unit, but also because complex processes are often intrinsically parallel and lend themselves to execution in parallel architectures.

In this context, much interest has been devoted to transputers (transistor computers), so called because a CPU is integrated into the structure as if it were a single component. These computers have been developed with the precise objective of solving simultaneously both the parallel processing and the more serious programming problem [5, 6].

In the field of parallel processing, neural networks are of particular interest. These are still under study; up to now there exist simulated neural networks, hard-wired and of average size, but with only a few tens of neurons. Neural networks can be seen as computer systems with fine grain parallel structure pushed to the limit and highly interconnected nodes. These networks are developed from the old idea of trying to emulate the structure of the human brain, which, by comparison with a computer, is totally inefficient for numerical computation but unparalleled in solving problems such as image recognition.

Neural networks appear to be very promising in AI applications dealing with problems of pattern recognition and for the future realization of expert systems.

If from a certain point of view the future of processing entails the use of highly parallel hardware, as in neural networks, recent technological progress will offer a workable alternative to traditional electronic computing. This is the optical computer which seems to offer interesting solutions to many problems such as image recognition, data transmission, and vector analysis.

In the near future, the use of neural networks with optically interconnected nodes could offer extremely high performance and provide adequate processing support for AI applications worthy of the name.

7.2.6 Display

The display, together with the console, or keyboard, is the human/machine interface of the integrated ESM-ECM system.

The information that must be displayed is plentiful and varied, and requires various kinds of presentation: tabular, polar, and so forth. Among the mass of data, friendly, hostile, and unknown tracks have to be identified. In such a context, it is understandable that the ideal display for an ESM-ECM system must be in color. For this reason, currently raster TV displays best meet these requirements.

A color *cathode-ray tube* (CRT) is currently used, but the in future, liquid-crystal displays will be available. There have been huge investments in the realization of sufficiently large liquid-crystal displays with a very large number of pixels. When this technology has been developed further, the cost of displays should compete favorably with the cost of CRTs, but with the advantage that high voltages are not required.

At present, plasma displays still have the defects of being monochromatic, relatively slow, and of low luminous intensity.

7.2.7 Generation of jamming programs: DRFM

There is another technique which may now be used to generate the jamming signals discussed in Chapter 5. In this area, researchers working in ED have set up a revolutionary device, a classical breakthrough, called *digital radio-frequency memory* (DRFM) (Fig. 7.5) [7, 8, 9, 10]. Its operating principle is very simple. All the difficulties concern its technological realization.

The input signal is down-converted into a utility band, for example around a carrier frequency of about 1 GHz, by means of an appropriate *local oscillator* (LO). It is then converted into baseband by a 1 GHz LO, generating two beat signals I (in-phase) and Q (quadrature). The two signals are sampled, converted very quickly into digital format, (e.g., at a clock rate of a few nanoseconds), and stored in a digital *random access memory* (RAM). In the example in the figure, the use of 2 bits, one for I and one for Q, has been assumed.

If the oscillators used in the conversions are sufficiently stable, for example crystal-controlled oscillators, the memorized signal preserves all the phase information of the input signal.

To reproduce the signal, it suffices to read out the memory at the same clock rate: two signals, I and Q, replicas of the stored ones, will be generated; at the converter output there will be a signal identical to the incoming one, with some spurious components due to sampling quantization.

To keep the spurious components low, the number of bits must be carefully chosen. Since an increase the number of bits

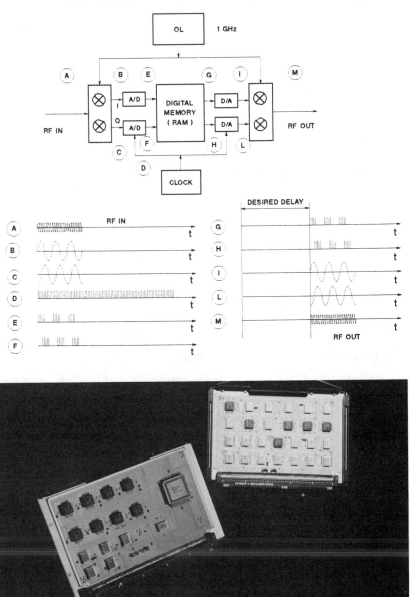

Figure 7.5 Diagram of a DRFM and its principal signals. The simplicity of the diagram hides the high technology required for its realization. The photograph shows the digital part, wich operates with a clock rate of more than 500 MHz.

increases the bulk, and especially, the amount of heat to be dissipated, it is necessary to find a compromise. The bandwidth of the memory depends on the clock; the maximum frequency that can be sampled is a function of the sampling frequency:

$$f_{\max} = \frac{f_{\text{clock}}}{2}$$

Since there are two channels, I and Q, the bandwidth of the digital memory is

$$B = f_{\text{clock}}$$

DRFM allows:

- storage and coherent reproduction of received signals.
- delay of the output signal for as long as is desired.
- generation of noise around the frequency of the received signal by altering the stability of the read-out clock.
- synthesis of any frequency within the band.
- generation of coherent jamming.

To sum up, DRFM is a device directly controlled by a computer, capable of carrying out the jamming discussed in Chapter 5 such as noise jamming, RGPO, VGPO, cross eye, and so forth.

7.2.8 Transmitters

The noise and deception signals generated in the way described in the previous section need to be amplified and made powerful enough to produce, together with the antenna gain, the desired *effective radiated power* (ERP).

The power amplifier most used in ECM equipment is the traveling-wave tube (TWT) (Section 5.2.2.1), whose performance is remarkable in terms of:

- bandwidth (for example, 6.5 to 18 GHz).
- both continuous power (200 to 400 W) and high-duty pulse power (2000 W peak power with 7 to 10% duty cycle).
- efficiency (15 to 30%).

To generate higher powers, TWTs may be combined, for example, by means of a Butler matrix. As is shown in Figure 7.6, the power of each transmitter can go to the four correspondent output branches, or all the power may be concentrated in only one output, according to the commands sent.

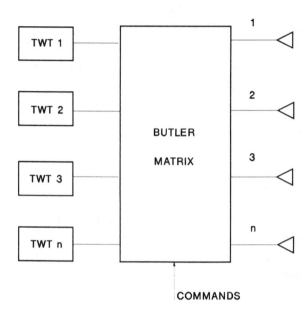

Figure 7.6 Combination of the power of several transmitters by means of a Butler matrix.

The shortcomings so far recognized in the use of TWTs are:

- the need for high voltage (7000 to 25,000 V, depending on the peak power), with danger of discharges in the presence of humidity, or at high altitude.
- the intrinsically short, although nominally long (1000 to 2000 hours), lifetime.

For immediate availability, the cathode must be kept heated, with a number of precautions to avoid damage such as "cathode poisoning." The equipment must be kept on stand-by, even if no transmission is required. This means that the nominal life of the TWT is quickly exhausted.

Spare tubes in storage must be switched on briefly every few months, to prevent the loss of vacuum which tends to occur because of the permeability of the TWTs metal parts.

However, from the technological point of view, solid-state power generation is also now making its mark. In fact, *field-effect transistors* (FET) able to generate a few watts even at the highest frequencies are now available. By putting many solid-state amplifiers in parallel, with suitable care in phase matching, it is nowadays possible to obtain powers on the order of 100 W. This means that in the future it will be possible to avoid those transmitter problems related to lifetime, high voltage use, and logistics that today arise from the use of TWTs.

7.2.9 ECM antennas

The antennas used for ECM applications are very varied. The simple horn is used when a very wide beam is required in both azimuth and elevation. When a high ERP is required, high-gain antennas are used. In this case, since the antenna beam is very narrow (pencil beam) a very precise DF measurement system is needed to point it accurately. The angular information may be taken from the ESM subsystem, or from the ECM itself, by equipping it with an automatic-tracking device. During the look-through period, this device must be able to detect the presence of the emission to be jammed and to measure angle-pointing errors, in order to direct the antenna correctly by means of a servo (Fig. 7.7).

Another type of antenna is the multibeam antenna, capable of generating a series of narrow and slightly overlapping beams to cover the sector of interest (Fig. 7.8). It can be realized, for example, by means of a Rotman lens that is today normally produced by photoengraving on a dielectric substrate. It is equipped with radiating and receiving elements and can insert a different delay between the elements in such a way that at the output a phase front pointing in the desired direction is

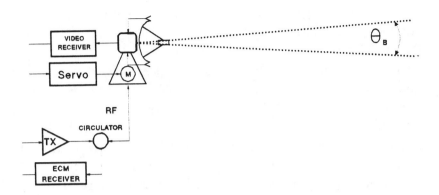

Figure 7.7 It is possible to obtain a high ERP by using a high gain antenna. However, precise angular tracking of the victim radar is required.

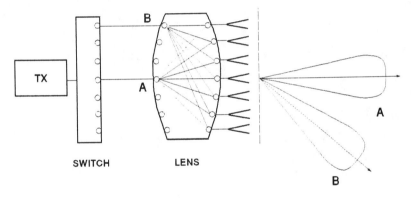

Figure 7.8 Multibeam ECM antenna. The direction of pointing of the beam depends on the input chosen.

generated. The beam position does not depend on frequency, which is a great advantage for ultra wideband systems.

7.2.10 Phased-array antennas

Phased-array antennas are used in ECM equipment to achieve high scanning rates, in order to operate simultaneously against several threats [11]. In contrast to radar, where the need for very low sidelobes, narrow beam, and absence of grating lobes implies that the number of radiating elements must be on the order of several thousands, in an ECM system the number of the array elements can be much lower: typically being 10 to 250, according to the desired performance.

In angular jamming, it will not be too disadvantageous to emit radiation in another direction with reduced power, for example, by a factor of 20 (sidelobe level the order of 13 dB). On the other hand, a radar with dynamic range of 90 dB, receiving signals from the sidelobes attenuated with respect to the main lobe by only 26 dB (13 dB on transmission plus 13 dB on reception), would detect the presence of other threats at completely erroneous angles, which would be unacceptable. Figure 7.9 shows the diagram of a phased-array antenna.

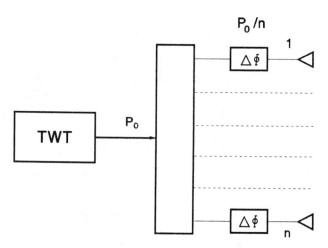

Figure 7.9 Phased-array antennas for ECM applications do not necessarily require a very high number of radiators.

Phase shifters are very important elements of a phased-array antenna. They can be either ferrite phase shifters, if the power passing through them is of many tens of watts, or pin-diode phase shifters, if the power is only a few watts.

Ferrite phase shifters have the advantage of a low insertion loss (0.5 to 1 dB), but require relatively high currents for phase control and are relatively slow in executing commands

($\simeq 10\,\text{ms}$). Pin-diode phase shifters are faster, but their insertion losses are much higher (3 to 6 dB).

To direct the beam in angle, unless the phase shift is achieved by means of a true time-delay device, the frequency too must be considered. To point the beam in the α direction, there will have to be a phase shift

$$\varphi_n = \frac{2\pi}{\lambda} nd \sin \alpha$$

in the nth element, where d is the distance between the elements (Fig. 7.10).

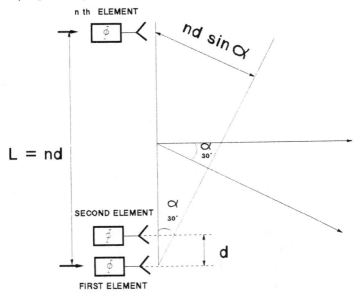

n th ELEMENT

$L = nd$

$nd \sin \alpha$

α
30°

SECOND ELEMENT

α
30°

d

FIRST ELEMENT

Figure 7.10 If the phase shift is achieved by a path difference, the direction of the wavefront is independent of frequency.

If the frequency is changed, that is, if λ is changed to $\lambda_1 1$, then in order to keep the α phase front constant, d being constant, the element n will have to be phase-shifted by

$$\varphi'_n = \frac{2\pi}{\lambda} nd \sin \alpha$$

7.2.11 Active phased-array/Shared-aperture antennas

In phased-array antennas the power is distributed over a great many elements. Instead of generating all the power in a single transmitter and then dividing it, one may generate power locally for direct feed to one or more elements by using a battery of mini-TWTs. In this case the antenna and the transmitter are integrated, and one may speak of *active phased-array antennas*. However, this type of configuration has the disadvantage that it does not readily lend itself to reciprocal antenna structures.

The ideal solution for an active phased-array antenna is shown in Figure 7.11, where instead of mini-TWTs, solid-state transceiver modules have been used [12]. In this case, solid-state RF power generation is achieved locally in the antenna, thus avoiding many transmission losses.

A problem to be overcome is the efficiency of the transceiver module. At present, efficiency is very low ($\simeq 10\%$), which means that 90% of the power is absorbed and transformed into heat, thus creating dissipation problems.

The use of MIC and, above all, of MMIC technologies, will allow the transceiver module to be realized with wideband (for example, 6 to 18 GHz) and high power (up to a few watts), at a decidedly low projected cost. This in turn will allow manufacturing of more and more complex and perfectly phased array antennas, which will contribute to their wider use. Substantial investment in volume production of integrated transceiver modules will reduce the manufacturing cost, thus rendering cost-effective the realization of active phased arrays with thousands of elements.

Phased-array antennas with thousands of elements will be applicable to both radar and ED functions, thus realizing shared-aperture antennas that can be used for integrated systems.

In an integrated system, there can be a single power generation, transmission and signal reception system, used by both radars and other on-board systems (Fig. 7.12). With an -

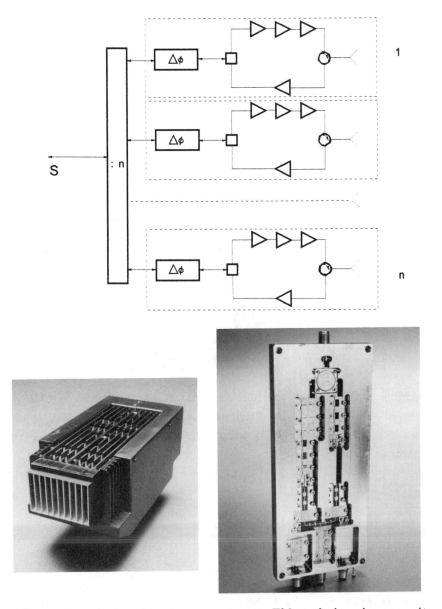

Figure 7.11 Active phased-array antenna. This technique is very suitable for ECM objectives. The photographs show the array and the corresponding solid-state ultra wideband transceiver module.

antenna of this type, the problems of high voltage and limited lifetime of the TWTs are eliminated.

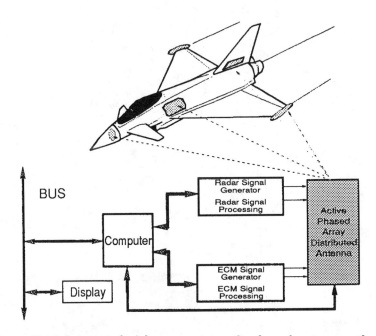

Figure 7.12 Integrated airborne system. Conformal antennas of active phased-array type can be distributed and used in time sharing by the various radar subsystems, ESM, ECM, and so forth, to realize a powerful integrated system.

The *mean time between failures* (MTBF) achieved with this type of active antenna is at least one order of magnitude higher than that obtained with TWT systems. Moreover, in this case graceful degradation will be achieved, that is, a fault-tolerant system capable of high performance even after the breakdown of 10 to 20% of its elements.

This type of antenna-transmitter allows a remarkable reduction of power consumption, and therefore of the total weight of an ECM system. See for example Figure 7.13, where two

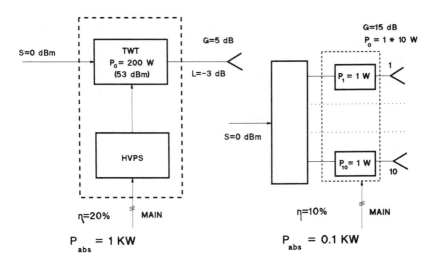

Figure 7.13 Comparison of ERP generation by TWTs and solid-state modules.

ways of generating a 55 dB m ERP are shown.

	TWT	Solid State
ERP	55 dBm	55 dBm
Angle cov.	120° Az	120° Az
	60° El	60° El
Power	53 dBm	40 dBm
Losses	3 dB	0 dB
Gain	5 dB	15 dB
Efficiency	20 %	10 %
Absorbed power	1 kW	100 W
Dissipated power	800 W	90 W
Weight	15 kg	5 kg

7.2.12 The ERP of an active phased-array antenna

The *effective radiated power* (ERP) is defined as the power supplied multiplied by the antenna gain:

$$ERP = P \times G$$

Consider a rectangular $m \times n$ array (Fig. 7.14) of radiators each of dimension d_x in the x direction and d_y in the y direction. Assume that the radiant efficiency of a single radiator is η and the power supplied to it is P_{el}. Since

$$G = \frac{4\pi}{\lambda^2} A_{\text{eff}}$$

we have

$$ERP = \frac{4\pi}{\lambda^2} d_x d_y \eta \, nm P_{el} \, nm = \frac{4\pi}{\lambda^2} d_x d_y \eta P_{el} (nm)^2$$

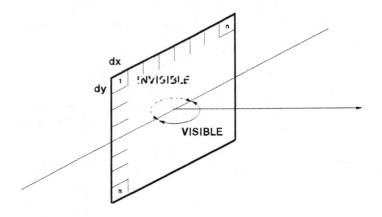

Figure 7.14 Planar array

If no grating lobes are to be visible during scanning to α in azimuth and ϵ in elevation, the maximum dimensions of the radiators will be

$$d_x = \frac{\lambda_{\min}}{1 + \sin \alpha}$$

$$d_y = \frac{\lambda_{\min}}{1 + \sin \epsilon}$$

where $\lambda_{\min}$ is the wavelength corresponding to the maximum frequency at which the array is to be used.

Defining the scanning factor F_s by

$$F_s = (1 + \sin\alpha)(1 + \sin\epsilon)$$

one may write the ERP of the active array at wavelength *lambda* in the form

$$ERP = \frac{4\pi}{F_s}\left(\frac{\lambda_{\min}}{\lambda}\right)^2 \eta P_{el}N^2$$

where $N = nm$ is the number of radiators.

7.3 Anti anti-radiation missile techniques

Countermeasures are available also against *anti-radiation missiles* (ARM).

From the radar point of view, the best ECM against an ARM are:

- extremely low peak power, so that the aircraft is forced to launch the ARM from short range, thus exposing itself to anti-aircraft defenses.
- well-randomized parameters, and therefore frequency agility, change of PW, change of PRF, and so forth.
- extremely low sidelobes.

At a radar site, ECCM can be organized, such as:

- highly reflective points to generate false targets credible to the ARM.
- decoys simulating radar emission to confuse the ARM.

7.4 Anti-stealth techniques

ECCM techniques have been developed also against stealth technology. They are:

- low-frequency radar.
- very-high frequency radar.
- carrier-free radar.
- bistatic radar.

Among these anti-stealth techniques the most promising is the use of low frequency, which exploits the reduced effectiveness of absorptive materials at low frequencies. Moreover, the engines and some areas of the airframe of an aircraft will, for various reasons, still be built mostly of metal; as a result, there will be areas of the aircraft of sizes on the order of one meter that will necessarily behave like traditional materials. If the wavelength of the radar carrier is also about one meter, resonance phenomena will occur, and the law of reflection appropriate to stealth aircraft will no longer apply. Under these conditions, it is unlikely that a stealth aircraft could present an RCS much lower than one square meter.

It will however be necessary to solve all the problems by which low frequency radar are afflicted, including:

- lobing.
- low discrimination.
- short range at low altitude.

This might be achieved by redesign of a low-frequency radar, exploiting all the new technologies currently available.

An anti-stealth technique that uses very high frequencies (millimeter wave radar) exploits the scattering produced by each small fissure or unevenness on the surface of the aircraft. The increase in RCS could be rather significant, but the reduction of range by atmospheric attenuation, especially during weather disturbances, is disadvantageous to this technique.

Carrier-free radars, transmitting very short video pulses, which therefore have a very wide spectrum, can be used as anti-stealth radars. The current problem for radars of this type is that the power associated with the transmitted pulse is very low, and therefore the radar range is limited. They will be more successful when carrier-free transmitters of adequate power become available.

Bistatic radars too can be used as anti-stealth radars (Fig. 7.16). In these radars, transmitter and receiver are not colocated, but are deployed a few kilometers apart. The principle underlying this anti-stealth technique is that stealth aircraft try to minimize the residual signals backscattered directly toward the radar, usually by increasing reflections in other directions. Therefore if the radar receiver is not colocated with the transmitter, it has a greater probability of detecting the aircraft. However, all the problems of bistatic radar, such as synchronization and coverage, come into play. The problem of coverage may be solved either by accepting long scanning times, or by using transmitters or receivers with wide beam antennas, or by limiting coverage for example only a warning belt.

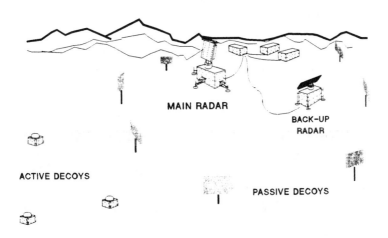

Figure 7.15 ECM against ARMs.

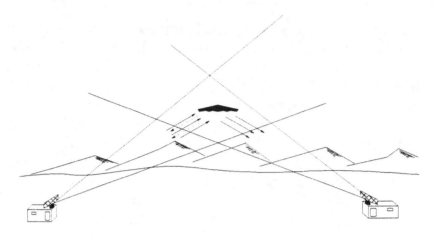

Figure 7.16 Bistatic radar, as well as low-frequency radar, can be used as an ECM against stealth aircraft.

7.5 State of the art and perspectives

Since the end of World War II, there has been a huge and impressive development of weapons based on electronic systems, and, with a certain delay, the development in turn of formidable electronic systems that have tried to render harmless such weapons. In the discussion which follows, only conventional weapon systems are taken into account. Nuclear weapons and intercontinental ballistic missiles are excluded.

To draw a balance, considering the conflicts that have occurred in this period and their various outcomes, one can say that:

- hard-kill weapon systems, for both offense and defense, have today become so effective that impressive new systems are not foreseen in the near term; continuing developments today concern *precision guidance munitions* (PGM) and improvements in IR seekers, ARMs, and missile speed.
- search radars have achieved optimum levels of performance. When the technology of solid-state transceiver modules is mature, search radars will all be of the active planar phased-array type.
- tracking radars, especially those used in missile seekers, have

have greatly enhanced their performance by the use of the monopulse technique, as the episodes of the Bekaa Valley, the South Atlantic, and various other conflicts have demonstrated. In tracking radars too, widespread use of the active phased array technique is foreseen.

- current short-range defense systems very often have not shown themselves to be easily manageable.
- current ESM systems appear to be able to interpret the electromagnetic scenario correctly, if it is not too dense and complex, and if they are adequately protected from interference caused by other friendly or hostile systems.
- current ECM systems are effective against the most traditional weapons, but do not seem to be effective against missile attacks, especially if the missile seeker resorts to such techniques as monopulse or frequency agility, or exploits leading-edge tracking.
- IR-guided missiles, in practice, cannot be stopped.

On the basis of what has been said in this chapter about the newly available ED technologies, it can be assumed that the following developments will soon take place:

- ESM systems will be much more protected (cued receivers) and much more sensitive (higher channelization thanks to digital receivers), for correct detection of coded radars. The extracted electromagnetic scenario will be much more reliable, thanks to the intervention of expert systems (AI). Expert systems will be integrated into lines of air defense and into IR detection systems.
- airborne threats will be of the stealth type and equipped with ARMs.
- air defense systems will have to envisage the use of anti-stealth radar and at the same time will have to be equipped with ECM against ARMs.
- ECM systems on-board platforms will be genuinely effective, because they will have almost unlimited resources (active phased-array antennas with instantaneous beam positioning), will be capable of generating extremely effective

noise and deception jamming signals (generated by DRFM) and angular jamming against monopulse, with or without frequency agility, without the need to resort to RGPO or VGPO techniques, both of which can be countered by suitable ECCM.

- miniaturized expendable active systems will be developed for backup.
- IR-guided missiles will be detected by *missile launch warning* (MLW) systems based on the IR signal emitted by the missile when it is launched, and by *missile approach warning* (MAW) systems, usually based on radar techniques. Both MLW and MAW will provide exact timing for the launching of flares, and in the future will probably designate the action of a laser to reduce the guidance capability of the IR seeker.
- finally, tracking radar, ECM, and other functions will be integrated, at least at the level of the transmitter antenna. In each case a computer-controlled active phased-array antenna will be used.

REFERENCES

[1] J.G. Tenedorio, "MMIC Reshape EW System Design," *MSN & CT*, November 1986.

[2] J.B.Y. Tsui, *Microwave Receivers with Electronic Warfare Application*, New York: John Wiley & Sons, 1986.

[3] [Ed.] "EW Design Engineers' Handbook," Supplement to the *Journal of Electronic Defense*, December 1987.

[4] D.A. Waterman, *A Guide to Expert Systems*, Reading, MA: Addison Wesley, 1986.

[5] D.E. Rumelhart and J.L. McLelland, *Parallel Distributed Processing, Explorations in the Microstructure of Cognition*, Vol. 1-2, Cambridge, MA: MIT Press, 1986.

[6] J. Aleksander and H. Morton, *Introduction to Neural Computing*, London: Chapman & Hall, 1990.

[7] D.C. Schleher, *Introduction to Electronic Warfare*, Norwood, MA: Artech House, 1986.

[8] S.C. Spector, "A Coherent Microwave Memory Using Digital Storage: The Loopless Memory Loop," *EW*, January-February 1975.

[9] W.J. Schneider, "Digital Countermeasures Memories: New Techniques Possible," *International Countermeasures Handbook*, 1986.

[10] G. Webber, "DRFM Requirements Demand Innovative Technology," *Microwave Journal*, February 1986.

[11] I. Bardash, "Phased Array for ECM Applications," *Microwave Journal*, September 1982.

[12] D. Boyle, "Phased Array Going Active. GaAs Expanding Fast," *International Defense Review*, 9/1989, pp. 1249-1251.

Chapter 8

Design and Evaluation Criteria

8.1 Introduction

In this chapter, criteria for the specification of *electronic defense* (ED) equipment are discussed. The objective is to try to remove the over-specification problem which generally arises from emotional considerations and which, neglecting compatibility, effectiveness and, most important, cost/effectiveness, requires an improvement of *all* parameters: more sensitivity, more frequency precision, more ERP, more coverage, less weight, less cost, and so on.

Design criteria, seen by a contractor who must design and manufacture a piece of equipment, starting from its specification, will first be analyzed. Then the point of view of the user, who must determine operational requirements for the equipment, will be discussed.

8.2 Design criteria

8.2.1 Generalities

The logical process leading to the specification of ED equipment should be as follows [1, 2]:

(1) The objective of the equipment should be defined, taking into account both the operational environment and the platform on which the equipment will be fitted.

(2) The analysis of the operational environment will determine the performance required of the equipment: Which type of signal must be intercepted? Will it be LPI or normal? Which type of radar must be jammed? Will it be monopulse or not? and so on.

(3) In order to achieve the required performance against an identified threat, a thorough knowledge of the threat is necessary. Here, one of two criteria may be followed, depending on whether the threat is known specifically in all details, or is known only in a generic way.

In the first case, *ad hoc* devices can be designed. In this way, however, one incurs the risk that if a few parameters of the threat are changed these devices will no longer be effective.

In the second case, the theory underlying the operation of the threat must be known in depth, so that general countermeasures may be found, which in their turn can be countered only with difficulty. This can be done only if all the possible threats and the available techniques and technologies are thoroughly known.

Once a technical solution which apparently achieves the objective has been identified, its validity must be tested. This can be done by a series of analyses and especially by accurate and exhaustive simulations to check the operation of the equipment in both typical and borderline situations. In addition, a cost-effectiveness analysis is necessary, to ensure that the solution found is not excessively costly.

But this is not all. The chosen solution might entail the taking of substantial technical or technological risks that could render it either infeasible, or realizable only in an unacceptable timescale. Such risks must be identified, evaluated, and if necessary mitigated by the search for alternatives.

A proper design sequence should comprise the following phases:

- definition of the system objective.
- analysis of the operational environment.
- definition of possible architectures.
- testing of achievable performance by analysis and simulation, and comparison with specifications.
- cost/effectiveness analysis.
- risk analysis.
- detailed design, including specification of back-ups for critical parts.

8.2.2 System objective

The system objective is generally indicated in the requirements defined by the user, namely, the armed forces.

The possible operational functions of ED equipment were discussed in Chapter 1. It may happen that new, unpredicted requirements arise. These must be analyzed and clearly understood case by case.

8.2.3 Analysis of the operational environment

The environment in which the equipment must operate is effectively determined by the platform and its missions.

The intended location of the equipment will determine its electrical characteristics, as well as the environmental norms to be considered and the limitations of weight, bulk, power consumption and power dissipation. It is of paramount importance to be aware of the installation problems. If the weight or bulk of an ED system prevents the platform from performing its intended mission, it will not be chosen, no matter how effective it may otherwise be. The problem of weight, bulk, and power consumption will be most severe in airborne applications, moderate for small ships, and only slight for large ships or ground installations.

The platform mission will determine the quality and quantity of the threats which have to be confronted. This will lead to an understanding of the complexity necessary in the system and establish characteristics of a typical mission, which are useful in evaluating system effectiveness.

In the following table some characteristics of operational environments are listed. Once the predicted threats have been identified, it is possible to proceed to the design of an ED system capable of meeting them.

Since designers of ED equipment often tend to attribute to radars capabilities higher than they actually possess, or to forget the limitations within which radars can operate, chapters 2 and 3 discuss the operating principles of radar and the problems of sensors and weapon systems.

8.2.4 Possible architectures

Once the operational objective, the platform, the environment, and the threats have been defined, it is possible to determine a system architecture for the performance of an assigned task.

This is a very crucial phase, which greatly depends on the ability of the design team. Too optimistic a team will put forward either excessively simplistic or excessively risky solutions that in the end will lead to unsatisfactory or nonfunctional equipment. Too pessimistic a team will design very complex equipment which offers solutions even in borderline cases, but which will be excessively heavy and too costly.

Plenty of common sense is needed to find a compromise based on the average real operational environment, and in borderline cases to accept some degradation, which must also be explained and made acceptable to the client.

EXAMPLES OF OPERATIONAL ENVIRONMENTS					
EQUIP-MENT	PLAT-FORM	OPERA-TIONAL ROLE	THREATS	EM CHARACT-ERISTIC	TRAFFIC
RWR	Aircraft	Interdiction	Sites : SAMs -command -semiactive -active AA Artillery: AIRCRAFT: -AAMs	RADARS and ILLUMINATORS in lock-on mode -Pulse -Pulse-doppler -Codes/LPI -CW/ICW -Agility -TWS	1×10^6 to 5×10^6 pps (From enemy radar)
SPJ	Aircraft	Interdiction	As above	As Above + -Conical Scan -COSRO-LORO -Monopulse	As above
ESM	Ship	Escort	AIRCRAFT: -Anti-Ship Missiles SHIPS -Anti-Ship Missiles Fire Control SUBMARINES	SEARCH and TRACKING RADARS -Pulse -Pulse-doppler -Codes/LPI -Parameter agility -TWS	0.1×10^6 to 0.3×10^6 (from enemy radar and friendly interferences)
ECM	Ship	Escort	As above	As above + -Conical Scan -COSRO-LORO	As above

System architecture must be designed so that the system will be able, first of all, to perform all the assigned missions, one at a time (what it does). Then it will be seen whether it can perform several missions simultaneously, and the potential degradation will be evaluated.

Finally, borderline conditions such as performance in the presence of unusually heavy traffic or interference, will have to be evaluated to check that they do not cause the equipment to fail completely (what it does not do).

For successful design of system architecture, all modern ED techniques and technologies must be known in detail. These techniques and technologies are described in chapters 4, 5, and

7. To be precise, chapters 4 and 5 discuss fundamental, more traditional, techniques. Chapter 7 deals with the most recent technological advances in both the digital domain and the microwave and applied software domains. Moreover, it is important to remember that the ECM effectiveness must be evaluated in the presence of ECCM techniques which might be in the hands of the enemy.

8.2.5 Verification by simulation

Before it is possible to assert that system architecture can satisfy the operating requirements, an extensive simulation of the architecture must be performed [3, 4, 5, 6].

With the powerful computational methods available today, it is preferable, before moving on to the hardware design phase, to run simulations of achievable performance. The simulations of radar, missiles, RCS, receivers, and so forth, available today correspond closely to the real systems, so that the answers given by simulation programs can be extremely realistic, accurate and reliable.

By simulation it is possible to verify the behavior of systems at various levels, from the operational level of a whole platform to that of the units and subsystems, to the most elementary hardware functions such as amplifiers, transistors, digital circuitry, and so forth.

It is thus possible to avoid manufacturing large numbers of test circuits. Circuits are designed with the support of *computer aided design* (CAD) from which the documentation needed for the manufacture of prototypes is directly obtained.

8.2.6 Cost/effectiveness analysis

Before deciding on the final architecture of an ED system, one must evaluate the cost/effectiveness ratio. From the point of view of design two cases should be distinguished:

• design aimed at satisfying specifications required by a client.

- design of a new product to meet a potential client need, not yet clearly expressed.

In the first case, the cost-effective solution is that which answers all significant requirements of the client and at the same time:

- minimizes the need for design of new techniques and technologies, by the utmost exploitation of systems components already in production, or of new technological elements already produced by the contractor which can be used as building blocks.
- minimizes recurrent production costs.
- offers possibilities for expansion or growth which will lead to the achievement of performances higher than those required by the client's specifications.

Clearly, for procurement of many identical systems, what counts is the sum of the recurrent and nonrecurrent costs. The way in which the latter are taken into account will depend on the innovations introduced by the new system, the delivery time, and so forth.

In the second case, when a new product has to be defined, work must begin with a cost/effectiveness analysis to determine specifications or requirements which the system must have. In practice, this approach coincides with the analysis which the client should carry out to determine the requirements of a new system (section 8.3).

The steps in a cost/effectiveness analysis from the point of view of design are the following:

- evaluate specifications: each aspect of required performance should be weighted according to its importance; a global value of 100 may conveniently be assigned to the whole system. This must be done by putting oneself in the position of the client, not the designer.
- evaluate possible architectures, examining how far they are able to respond to each performance requirement; these response capabilities are conveniently expressed as percentages.

- calculate a weighted average showing the extent to which each possible architecture responds to the specification.
- evaluate the effectiveness of the system as a sum of the weighted averages.
- evaluate the procurement cost of the different architectures; here the cost of the solution which permits total satisfaction of the specifications may conveniently be assumed to be 100. Should there be potential spinoffs to other products, the total cost should be reduced accordingly.
- calculate the cost/effectiveness ratio; the solution which minimizes this ratio is the one to be preferred, all other conditions being equal.

8.2.7 Risk analysis

Risk analysis, in terms of both intensity and probability, serves to exhibit the real feasibility of a particular program or piece of equipment. The presence of many high-intensity and high-probability risks is most certainly an indication of the infeasibility of a program.

Often, however, it is necessary to proceed with a program notwithstanding some risks, otherwise new equipment will be old at birth. In this case, it is better to foresee some alternatives so that if the risks develop unfavorably, an alternative program is ready.

In the choice of possible architectures, there are two extreme cases:

(a) Very complex architecture with distributed low risks.
- a complex architecture is always a problem.
- the system can be understood only with difficulty.
- the system can be maintained only with difficulty.
- the system has complex interconnections.
- the main *risk* is in the final integration of the system.

(b) Very simple architecture with risks centered on a limited area.
- the system can be understood easily.

- the system can be repaired easily.
- the system can be interconnected easily.
- the *risk* can be of a technological type.

In the first case, the risk can be mitigated by complex simulations, but detection of potential problems occurs in the final part of the program, when there is little time to react.

In the second case, the risk must be mitigated by readiness to accept, in the limit, some degradation of performance, but this allows efforts to be concentrated in the crucial area from the start of a program.

8.3 Evaluation criteria for the choice of a system

8.3.1 Generalities and objective

The cost/effectiveness ratio of equipment has been examined above from the viewpoint of a manufacturer. Here it will be examined from the viewpoint of the user, that is, the armed forces.

Before choosing among the various types of ED systems, a General Staff must first of all decide whether ED systems are really needed, since their employment entails a cost, and at times a limitation of platform performance. Once a positive decision has been made, it is necessary to determine the exact characteristics required of the ED systems.

In what follows, a method will be examined which makes possible an evaluation of the need for ED systems, a determination of system requirements, and finally an objective evaluation of the different systems offered in response to a call for tenders. Only an objective evaluation can determine whether an expense is or is not appropriate and avoid the acquisition of equipment of limited practical use.

One example will illustrate this. If from an analysis of the operational environment it had been determined that 40%

of the threats are monopulse-guided SAMs, 30% monopulse-guided AAMs, 20% monopulse-guided artillery systems, and only 10% systems with scanning guidance, the correct system will have principally to counter monopulse systems. A system capable only of countering threats with scanning guidance should cost, for given effectiveness, one tenth as much as a system able to counter all the threats.

To perform a cost/effectiveness analysis, the following steps should be taken:

- evaluation of the situation, or rather definition of a reference scenario in which a platform fitted with the ED device will on average have to operate.
- definition of the important events which will take place in the course of the mission and which will contribute to the kill probability of the platform.
- calculation of the attrition rate A_r, that is, the proportion of platforms lost during each mission; the attrition rate can be taken to be equal to the kill probability of enemy weapon systems met on the average in the course of the typical mission, neglecting platforms retrieved for repair.
- calculation of the number of platforms lost in the first days of hostilities as a function of the attrition rate.
- calculation of the new attrition rate on adoption of a particular ED system with respect to the same reference operational environment.
- calculation of the number of platforms saved thanks to the adoption of the ED system; the value of these platforms expresses the effectiveness of the ED system.
- evaluation of the procurement cost of the ED equipment.
- calculation of the cost/effectiveness ratio (saving); if it is less than unity, the equipment is said to be cost-effective, otherwise not.

A brief review of probability theory [7] will help one to understand the following discussion, which illustrates the method.

The probability p that a favorable event takes place is the ratio of the number of favorable cases to the number of all possible cases. The probability of drawing the ten of spades from a pack of 52 cards is

$$p = \frac{1}{52}$$

The joint probability is the probability that i independent events with probabilities $p_1, p_2, ...p_i$ will all occur:

$$p_T = p_1 \times p_2 \times ... \times p_i$$

The probability that event 2 takes place, together with the probability that event 1 does not take place, is

$$p_c = p_2(1 - p_1)$$

This formula is used to express the probability that a target is hit if several missiles are launched against it. Assume that two missiles are launched against an aircraft. The probability that the aircraft is hit is given by the sum of the probability that it is hit by the first missile plus the probability that, if it has not been hit by the first missile, it will be hit by the second missile:

$$p_{kt} = p_{k1} + p_{k2}(1 - p_{k1})$$

8.3.2 Definition of the reference operational environment

As stated briefly above, for evaluation of the effectiveness of ED equipment, a reference scenario, with respect to which it is thought that the missions will take place, must first be defined.

This is the most delicate step of all because, if it is too pessimistic, the resulting specifications for the ED system will be too onerous and the system will be heavy and expensive. If it is too optimistic, the system may be useless.

In any case, the reference scenario must be defined as a list of events which on average will occur in the course of a typical mission to which an ED contribution is required. In their turn, the events will be characterized by average situations.

Consider for example a platform of the attack aircraft type. The typical mission for such a platform is to penetrate into enemy territory in order to destroy some important objective (Fig. 8.1) [8, 9].

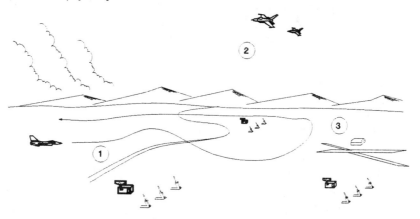

Figure 8.1 While approaching a target, an aircraft meets with several defense lines, each with a given kill probability. (1) Long-range SAMs; (2) fighter aircraft; (3) point defense systems.

The reference scenario might consist of the following events which would contribute to the hit:

- detection of the aircraft by the search radar of the enemy defense line.
- acquisition of the aircraft by the enemy tracking radar.
- precision guidance of enemy weapons.
- effect of the weapon warhead (projectile or missile).

If the mission is complex, the platform may meet more than one defense line in the course of its mission, so that the above reasoning must be iterated.

If the various defense lines are characterized by different weapon systems, it will be desirable, if possible, to consider hypothetical systems characterized by average performances.

8.3.2.1 Probability of survival Each of the events mentioned will occur with greater or less probability [2], according to the effectiveness of the enemy equipment. To evaluate the probability that an event occurs, an estimate of the quality of the average performance in a given situation is required (proportion of search radars with MTI, proportion of tracking radars of monopulse type, and so forth).

The parameters characterizing the effectiveness of the enemy systems, which are the following, may be expressed in terms of the probability that they achieve their mission:

- probability of detection of platform by enemy search radar, P_d.
- probability of acquisition of platform by tracking radar, P_a.
- probability of precision guidance of weapons, P_t.
- kill probability of the weapon, P_{kw}.

The kill probability of the enemy system will be

$$P_k = P_d P_a P_t P_{kw}$$

If during its mission the platform will have to confront n weapon systems, each with a kill probability P_k, the *total kill probability* of the n systems will be

$$P_{kt} = P_{k1} + P_{k2}(1 - P_{k1}) + P_{k3}[1 - (P_{k1} + P_{k2}(1 - P_{k1}))] + \cdots$$

Assume that no damaged aircraft has been retrieved for repair. Then the attrition rate will be equal to the total kill probability

$$A_r = P_{kt}$$

The probability of survival of a platform after one typical mission will be

$$Q_1 = 1 - P_{kt}$$

and after two typical missions

$$Q_2 = (1 - P_{kt})(1 - P_{kt})$$

After n typical missions it will be

$$Q_n = (1 - P_{kt})^n$$

8.3.2.2 Losses in the absence of electronic defense equipment

The usefulness of ED equipment becomes apparent only on the hypothesis of military conflict. Normally it is assumed that there will be a need to confront a sudden conflict situation in which platforms are called upon to perform a large number of missions within a few days, perhaps 10 to 12 for an aircraft, for example.

The graph in Figure 8.2 shows the probability of survival of a platform after n missions, with various attrition rates. Multiplying the probability by the number N of platforms initially available, one obtains the number N_s of platforms that will have survived after n missions, with an attrition rate A_r:

$$N_s = N \times Q_n = N \times (1 - P_{kt})^n$$

$N - s$ is used as a reference value in calculating the effectiveness of the ED system.

8.3.3 Effectiveness of electronic defense

Since ED equipment can influence the detection capabilities of search radar, and the acquisition and precision aiming capabilities of tracking radar, it will be capable of reducing the attrition rate and, therefore, the number of platforms killed in a given number n of missions. The economy thus achieved is a measure of the effectiveness of the ED equipment, which is to be compared with the procurement cost in order to evaluate its cost-effectiveness.

8.3.3.1 Parameters of electronic defense effectiveness

The effectiveness of ED equipment will therefore depend on its ability to reduce the P_d, P_a, and P_t of enemy weapon systems. Since P_{kw} can be influenced only by interfering with the

REMAINING PLATFORMS/INITIAL PLATFORMS Q_n
AFTER n MISSIONS

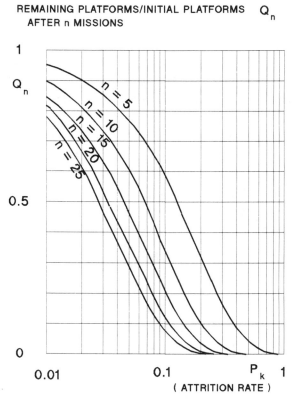

Figure 8.2 Percentage of remaining platforms after n missions as a function of attrition rate.

fuzes of enemy weapons, which have not been discussed in this book, it will be considered constant in what follows.

The parameters which express the *effectiveness* of an ED system are:

- effectiveness against search radar, that is, the probability of impeding or sufficiently delaying detection by enemy search radar, E_d.
- effectiveness against acquisition, that is, the probability of impeding or sufficiently delaying acquisition by the enemy tracking radar, E_a.
- effectiveness against tracking radar, that is the probability of preventing precision guidance of weapons, E_t, where $E_t = 1$

if the ED system causes the warhead to explode outside the lethality range of the weapon.

These three parameters are very useful because they can be easily calculated, simulated and tested in the laboratory and in test ranges and operational sites, for the different types of radar.

These parameters expressing the effectiveness of an ED system must be evaluated as weighted means with respect to the reference operational scenario. If, for example, in the operational scenario there are N_{tot} tracking radars, $N_1 + N_2 + N_i$ against which the ED system has effectiveness

$E_1 + E_2, ..., E_i$ respectively, the mean effectiveness of the system will be

$$E = \frac{E_1 N_1 + E_2 N_2 + ... + E_i N_i}{N_{tot}}$$

Denial factors may be defined for the victim radar in terms of these effectiveness parameters:

$$D_s = 1 - E_s$$

$$D_a = 1 - E_a$$

$$D_t = 1 - E_t$$

If the denial factors are high, the effectiveness of the ED system will be low, and vice versa. The probability of success of the victim radar's operations should be multiplied by these denial factors.

8.3.3.2 The force multiplier of electronic defense

In the presence of ED equipment, the attrition rate for a given weapon system becomes

$$A'_r = P'_k = S_s P_d S_a P_a S_t P_t \times P_{kw}$$

As before, it is possible to calculate the number N'_s of platforms, now protected by ED equipment, which have survived after n missions. The difference between N'_s and N_s gives the number of platforms saved by the use of ED systems, and therefore the savings achieved after n missions.

If

$$Q'_n = (1 - P'_k)^n$$

and

$$N'_s = NQ'_n$$

then the number of platforms saved is given by (Fig. 8.3)

$$S = N'_s - N_s = N(Q'_n - Q_n)$$

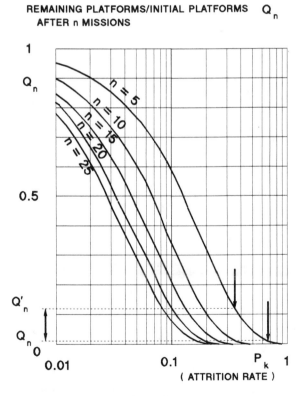

REMAINING PLATFORMS/INITIAL PLATFORMS Q_n AFTER n MISSIONS

Figure 8.3 The difference $(Q'_n - Q_n)$ multiplied by the initial number of platforms gives the number of platforms saved by the ED system.

In practice, the effect of the ED equipment can be considered as a force multiplier available at the beginning of hostilities. On the simplifying and limiting hypothesis that A_r does not depend on the number of platforms available, this multiplier M may be expressed

$$M = \frac{NQ'_n}{NQ_n} = \frac{Q'_n}{Q_n}$$

From this, one may infer that a modest fleet of platforms equipped with effective ED systems can have greater value than a numerically superior fleet lacking such systems or only lightly equipped with them.

8.3.4 Cost/effectiveness ratio

The ratio of the saving to the cost of procurement describes the extent to which the use of ED equipment is worthwhile.

The method discussed is particularly appropriate when a choice has to be made among various ED systems. The effectiveness parameter of each system is evaluated with respect to the reference scenario, and the attrition rate and the ratio of total procurement cost to saving are calculated. The best system will be the one for which, all other conditions being equal, the ratio is minimum.

It follows that the saving R will be given by the number of saved platforms S multiplied by the unit cost C_p

$$R = SC_p$$

If it is assumed that the life-cycle cost of each piece of ED equipment is equal to C_{ED}, the total procurement cost will be NC_{ED}. The cost/effectiveness ratio may be written

$$\frac{C}{E} = \frac{NC_{ED}}{N(Q'_n - Q_n)C_p} = \frac{C_{ED}}{(Q'_n - Q_n)C_p}$$

To sum up: Equipment is cost-effective if its cost is lower than the cost of the platform multiplied by the expected increase in the probability of survival after n missions.

The following examples are chosen to illustrate the theoretical approach rather than the rigorous quantitative analysis. The examples consider:

- attack aircraft defended by an ED system consisting of RWR + chaff.
- a ship under missile attack protected by an ED system comprising RWR + chaff.

The same methodology can be applied to the case of platform defense by means of self-protection equipment, expendable devices, and so forth.

The method illustrated is extremely effective not only in deciding whether to procure ED equipment or not, but also in choosing between two competing systems. It is very important in this case to define a reference scenario which reflects the real-life situation as accurately as possible, and to evaluate the effectiveness parameters of the competing systems and the savings achievable by each of them.

When the ratio of total procurement cost to total savings during the first days of a conflict is calculated, the system which has the optimized (that is, the smallest) cost/effectiveness ratio, all other things being equal, will be the one to be given further consideration.

For strategic ED equipment, the criterion of evaluation is the same. In this case it is more difficult to calculate in quantitative terms the negative consequences of the absence of ED equipment, which may be catastrophic. However, since the total number of such devices is in general limited, an evaluation must be made case by case and as a function of the strategic requirements which the equipment must satisfy.

8.3.4.1 Examples of evaluation of the cost/effectiveness ratio

Example No 1: Airborne RWR + Chaff System

Suppose one wants to evaluate an ED system comprising an RWR integrated with a chaff dispenser.

First of all, it is necessary to define the reference scenario. From the discussion in chapters 1 and 3, it follows that it will suffice to examine the case in which the aircraft must overcome defenses consisting of a search radar coordinating a SAM battery which protects the target (Fig. 8.4).

Figure 8.4 Avionic reference scenario.

For simplicity, suppose that the attrition rate is equal to P_k, that is, to the probability that the aircraft is hit. The kill probability is obtained by enumerating the events contributing to the kill of the aircraft. They are:

- detection of the aircraft by the search radar: $P_d = 0.95$.
- acquisition by the tracking radar: $P_a = 0.95$.
- precision targeting of weapons: $P_t = 0.95$.
- probability of missile kill: $P_{km} = 0.8$.
  The kill probability is given by

$$P_k = P_d P_a P_t P_{km} = 0.686$$

After n, say 5, missions, the number of surviving platforms is reduced, at an attrition rate $A_r = P_k$, to

$$N_s = NQ_n = N \times 0.003$$

where Q_n is the probability of survival after n missions (Fig. 8.3):

$$Q_n = (1 - P_k)^n = 0.003$$

If the missions are repeated with ED equipment consisting of RWR + chaff, the attrition rate is modified. Since appropriate launching of chaff can deceive the tracking radar range gate, the system will have an effectiveness E_t in this respect (Fig. 8.5). The effectiveness parameter E_t of this type of system will depend on the maneuverability of the aircraft, the effectiveness of the chaff (rate of development of RCS, magnitude of the developed RCS, lifetime) and on the effectiveness of the RWR.

In its turn, the effectiveness of the RWR is given by its ability to detect the threat correctly, notwithstanding the potentially heavy electromagnetic traffic, at the right moment. The right moment is measured either by range, which must be higher than the maximum weapon range, or, in the case of a delayed radar acquisition, by a reaction time lower than the weapon implementation time. When, in the approach to the target, the RWR detects the presence of a tracking radar in lock-on mode, the pilot will maneuver so as to give the chaff a high probability of success, and will generate the chaff launching command. With a probability equal to the effectiveness parameter E_t of the ED system, assumed here to be $E_t = 0.5$, the tracking radar will be forced to break lock, and the pilot will restore the aircraft to its normal course and continue its mission. The kill probability will now be

$$P'_k = P_d P_a P_t (1 - E_t) P_{km} = 0.343$$

The difference between the number of platforms outfitted with ED equipment remaining after n missions:

$$N'_s = NQ'_n = N \times 0.122$$

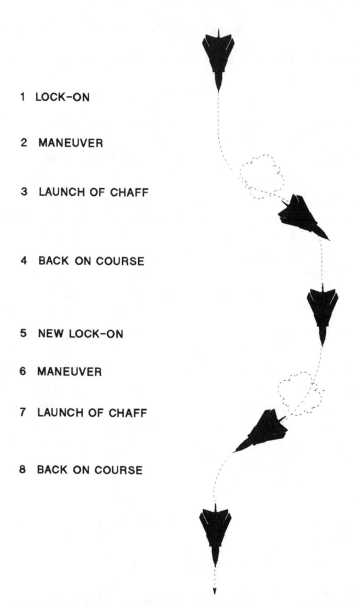

1 LOCK–ON

2 MANEUVER

3 LAUNCH OF CHAFF

4 BACK ON COURSE

5 NEW LOCK–ON

6 MANEUVER

7 LAUNCH OF CHAFF

8 BACK ON COURSE

Figure 8.5 Effectiveness of launching of chaff by command of the RWR.

and the number N_s of platforms, will give the number of plat-forms S saved by the combination RWR + chaff after n missions:

$$S = N'_s - N_s = N(Q'_n - Q_n) = N(0.122 - 0.003) = N \times 0.119$$

The cost/effectiveness ratio will be

$$\frac{C}{E} = \frac{C_{de}}{0.119 C_p}$$

Example No 2: Naval RWR + Chaff System

First of all, it is necessary to determine the reference scenario and the events which contribute to the kill probability of the platform, with their respective probabilities.

In this case the situation consists of an attack by a sea-skimming missile, against which the ship can react with a short-range artillery system, that is, a *close-in weapon system* (CIWS), to which RWR + chaff can be added or not.

Assume that the launching of the missile has taken place from an aircraft which has increased its altitude for an extremely short time with a pop-up maneuver, has detected and identified the target with a few sweeps of its on-board radar, and after the launching, has immediately dropped back to low altitude, without being seen by the equipment on board the ship.

The probabilities of the events which can be hypothesized in this case are the following (Fig. 8.6):

(1) the aircraft detects the ship with its on-board radar, P_d, and launches the missile.
(2) the ship's search radar detects the missile, P_{sd}.
(3) the missile switches on its seeker and acquires the ship, P_a.
(4) the seeker guides the missile onto the ship, P_t.
(5) the CIWS acquires the missile, and fires at it, P_{kc}

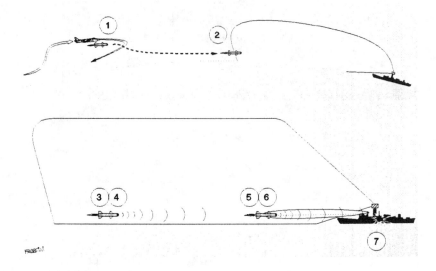

Figure 8.6 Naval scenario

(6) the missile hits the ship P_{km}.

The kill probability of the anti-ship missile, in the case that the ship does not defend itself, is

$$P_k = P_d P_a P_t P_{km}$$

If the ship protects itself only with the ED systems (soft kill), and the effectiveness of the RWR + chaff system is E_t, then

$$P_k' = P_d P_a P_t (1 - E_t) P_{km}$$

If the ship protects itself only with CIWS (hard kill), then

$$P_k'' = P_d P_a P_t (1 - P_{kc}) P_{km}$$

If the ship uses both (hard-kill and soft-kill) systems simultaneously, then

$$P_k''' = P_d P_a P_t (1 - E_t) P_{km} (1 - P_{kc})$$

Here too it is possible to calculate the number of naval platforms surviving after n missions

- with no defense.
- with only the CIWS defense.
- with only the ED system.
- with the simultaneous use of both systems.

Consider again the cost C_{ED} of each ED equipment (life-cycle cost) and the cost C_p of the platform; it is now possible to determine the cost/effectiveness ratio of the equipment

$$\frac{C}{E} = \frac{NC_{de}}{N(Q'_n - Q_n)C_p}$$

where Q'_n is the probability of survival after n missions when the ED equipment is used.

In the same way it is possible to evaluate the effectiveness of ED equipment of the active on-board type. In this case it will be necessary to consider whether the system is also able to affect the detection and acquisition abilities of enemy radars.

8.4 Operational effectiveness

To maintain operational status, one must establish a support program for every system in use with the armed forces. For simple equipment consisting only of hardware which must perform a well-defined function, a logistic support service, that is, a maintenance service, will be enough to ensure continued effectiveness.

For more complex equipment, whose performance depends on many factors, sometimes external to the system and generally comprising hardware and software, the support program cannot be merely of logistic type. It must have a design content as well, since the equipment will frequently require significant modifications.

Consider a command and control center on a ship. The equipment comprises a series of computers, interfaces and displays, but the real heart of the system is the operational software, which requires continuous support and assistance in order to:

- confirm the operating validity of the system.
- continue debugging it, a function which needs to be performed throughout the life cycle of complex equipment.
- modify the system because of changes in the external situation caused by:

(a) new equipment to be integrated.

(b) new combat philosophies.

To maintain the effectiveness of ED equipment it is not enough to begin with good specifications and construction and to ensure good maintenance by a logistic support service. Equipment of this type, whose performance depends largely on knowledge of the characteristics of threats which evolve continually both in number and quality, requires continuous verification of performance with respect to the changing scenario.

It will therefore be necessary to organize an *operational support* (OS) program consisting of ground equipment and software which ensures that the effectiveness of the ED equipment is maintained. The OS will comprise *operational ground support equipment* (OGSE), consisting of a certain number of computers, interfaces, simulators and displays for a continuous testing and simulation activity called *operational ground support* (OGS).

In practice, the OGS will enable performance of the following functions:

- data base management; in general, the armed forces will have centralized data bases (recall what was said in Section 4.5 about ELINT systems).
- study and evaluation of new ED software both for libraries and for operational purposes, including interception and countermeasures, to deal with the continually changing scenario.

- translation of new procedures and new libraries, once developed, into suitable form for the ED equipment and transmission to the bases where the platforms are stationed.
- preparation, at the bases, of data appropriate to the missions and the operational theater, for insertion into equipment memories.
- loading of the new software into the memories of the ED systems by means of the appropriate devices (loaders).
- provision of devices capable of reading and memorizing the contents of messages from missions, so that new data about threats can be transmitted to the data base to keep the ED systems updated and effective.

8.5 Electronic defense and conventional defense

It is now possible to discuss the actual validity of electronic defense by comparing it to conventional defense.

Rather than by a theoretical discussion, this will be done by means of an example: the defense of a naval platform. The results can be applied by analogy to the defense of other platforms or sites.

Consider the defense of a ship from the most probable and most dangerous threat: the anti-ship, sea-skimming missile. Currently only two types of hard-kill systems are used for this purpose: Ultra-fast fire artillery systems (direct impact or fast-fire with fuzed projectiles), or anti-missile missile systems. At present soft defense is only used for jamming of the threat when it is still far from the ship.

However it is possible to exploit one of the radio frequency deceptive jamming systems discussed in chapter 5 to realize an effective system which can:

- shift the radar center of the ship to a false position sufficiently remote, whatever the radar attempting to track it.
- remain immune to all ECCM.
- avoid interference with other on-board activities.

The results of the comparison are summarized in the table. The parameters are self-explanatory, except that the costs are those of the platform and that "limiting target hypotheses" describes the types of targets the system can counter; for ED it has been assumed that 20% of possible missiles can be guided

HARD-KILL/SOFT-KILL COMPARISON			
	HARD KILL		**SOFT KILL**
	Artillery	Missiles	Electronic defense
Mission	Platform defense	Platform defense	Platform defense
Means	Destruction or damage through impact or fragmentation	Destruction or damage through impact or fragmentation	High miss distance because of unavoidable deception jamming
Limiting target hypotheses	Missiles of all types, within geometric, kinematic or numerical limits	Missiles of all types, within geometric, kinematic or numerical limits	RF-Guided missiles, with no additional limit
Probability that the threat is within the limits	0.7	0.9	0.8
Probability of success (one threat)	0.7	0.9	0.9
Probability of success (four simultaneous threats from the same side)	0.2	0.2 0.5*	0.9
Equipment cost per ship	5-10%	10-15%	5%
Munitioning cost per ship	2%	5%	0%
Cost of burst of fire	0.01%	0.5%	0%

* Active-Guidance, Vertical Launcher

by non-RF devices (for example, IR, TV, and so forth) and that the ED system can deal only with RF-guided missiles.

The tabulated results, decidedly in favor of ED systems, are qualitative, and have been displayed provocatively in order to induce serious consideration of the true capabilities of these systems.

To sum up, although traditional weapon systems, based on hard kill, must be granted validity as dissuasive and defensive deterrents, it must be emphasized that ED systems are reaching maturity and are able to perform the mission of true and effective defense in a satisfactory manner. This is very important, especially in a technologically and morally developed world, where the rejection of war as a solution to conflicts is becoming ever more widespread, and where more and more ways of ensuring respect of treaties and of offering proportionate responses aimed at stopping the escalation of conflicts are required.

REFERENCES

[1] O.A. Chembrovskiy, Y.I. Topcheyev, and G.V. Samoylovich, "General Principles of Designing Control Systems," NASA, TT F-782.

[2] S. Carnevale and G. Santi, "Criteri per l'impostazione di un sistema d'arma per la difesa antiaerea di obiettivi fissi," *Rivista Tecnica, Selenia,* Vol. 1, No. 4.

[3] M. Tuccari and R. Vuolo, 'Simulazione digitale del gioco di guerra di un sistema d'arma fisso difensivo di zona contro attacchi a bassa quota',*Rivista Tecnica, Selenia,* Vol. 2, No. 1, 1974.

[4] H.D. Lynch Urban, "Theater Air Defense Engagement Simulation," AGARD-CP-268.

[5] F. Herzmann and H. Sanders, "Design and Simulation of a C^3 System for Surveillance Purposes," AGARD-CP-268.

[6] R. Hutter, "Simulation of Overall Air Defense Command and Control," AGARD-CP-268.

[7] W. Feller, *An Introduction to Probability Theory and Its Applications*, New York: Vols. 1 & 2, John Wiley & Sons, 1968-1971.

[8] R.E. Ball, "The Fundamentals of Aircraft Combat Survivability Analysis and Design," AIAA Education Series, 1985, pp. 188-191.

[9] H.R. Wilhelm, "Simulation of Air Defense Operations and Multiple Air Combat," AGARD-CP-268.

Glossary

α	Absorptance
Δ	Difference channel of a monopulse radar
Δf	Frequency interval or difference
ε	Emissivity; also a general measure of precision
θ_B	Width at half-power (-3 dB) beam, i.e., angle within which the power of the transmitted signal is more than half its maximum level
θ_q	Squint angle
λ	Wavelength: Distance between two adjacent peaks on a sinusoidal wave; for an electromagnetic wave the wavelength is given by c/f where c is the velocity of light and f the frequency of the wave
ρ	Reflectance or reflection coefficient
ρ_r	Reflection coefficient of a rough surface
ρ_s	Specular reflection coefficient
Σ	Sum channel of a monopulse radar
σ	Stefan-Boltzmann constant
σ	Equivalent area
σ	Standard deviation, i.e., square root of the variance of a probability distribution; a quantity whose values have a Gaussian distribution has a :
	67% probability of being within $\pm\,\sigma$ of the mean
	90% probability of being within $\pm\,2\sigma$
	97% probability of being within $\pm\,3\sigma$
σ_p	Standard deviation of a quantity with zero mean
τ	Pulse duration, pulsewidth (PW)
τ_a	Atmospheric transmittance
τ_o	Transmittance of optical system
ϕ	Phase or phase shift angle
Ω	Field of view, measured as a solid angle
ω	Angular velocity or angular frequency of a signal
AAA	Anti-aircraft artillery
AAM	Air-to-air missile

ac	Alternating current
a_e	Radius of the earth (8.5×10^6 m s^{-1})
A_{eff}	Effective area of an antenna
AFC	Automatic frequency control
AFS	Automatic frequency selection
AGC	Automatic gain control
AI	Artificial intelligence
AM	Amplitude modulation
AOA	Angle of arrival
A/R	Amplitude/range: display of signal amplitude as a function of range used with tracking radars
A_r	Attrition rate: rate of loss during military combat
ARGS	Anti-range gate stealing
ARP	Antenna rotation period: *see* ASP
ASM	Air-to-surface missile
ASP	Antenna scan period
B	Generic frequency bandwidth
B_r	Bandwidth used by radar signal
B_{IF}	IF bandwidth
B_j	Bandwidth occupied by jammer
B_{mf}	IF bandwidth
Booster	Rocket for acceleration of a missile or an aircraft
Boresight	Electrical axis of an antenna or a telescope, i.e. the direction in which the radiant intensity is a maximum
Break-lock	Loss of automatic tracking
c	Velocity of light (3×10^8 ms^{-1})
C^3	Command, control, and communications
$C^3 I$	Command, control, communications, and Intelligence
Clutter	Radar echoes, usually unwanted, caused by the operational environment: obstacles, ground, sea, rain, cloud, etc.
CFAR	Constant false alarm rate

chip	Small electrical component comprising microscopic integrated circuit with high performance
CIWS	Close-in weapon system
COM-ECM	Communications electronic countermeasures
COMINT	Communication intelligence
COSRO	Conical-scan on receive only (tracking radar)
CPU	Central processing unit
CW	Continuous wave
D*	Detectivity of an electro-optic sensor (per $cm^{1/2}$ and $Hz^{1/2}$)

dB	Decibel: logarithmic unit of measure, for comparing any two quantities, particularly power levels. If P denotes a power level, and P_r a reference power level then P is Δ decibels above P_r, where $\Delta = 10 = \log_{10}(P/P_r)$ (or $|\Delta|$ decibels below P_r, if Δ is negative). If the reference level is 1 milliwatt, the power is in dBm. For example,
	
	    100 watts is 20 decibels above 1 watt
	    200 watts correspond to 23 decibels above 1 watt
	    1000 watts is 30 decibels above 1 watt
	    1 watt is 20 decibels below 100 watts
	    1000 watts is 30 decibels below 1 megawatt
	    1 watt is 30 dBm
	    1000 watts is 60 dBm

DBF	Digital beam forming: technique used for phased-array radars whereby the phases of signals from the receiving elements are shifted, after conversion to the baseband, by digital methods, which makes possible the synthesis of one or more beams pointing in different directions
dBm	*See* dB
dc	Direct current
deinterleaving	Sorting of pulses received from an ESM system

DF	Direction finding
DINA	Direct noise amplification
DJ	deception jamming
DOA	Direction of arrival (of radiation)
DRFM	Digital radio-frequency memory
duty	For a pulse transmitter, the ratio of effective transmission time to waiting time between pulses
dwell time	Time given to an observation or a measurement
E_a	Effectiveness of a jammer against acquisition
E-d	Against detection
E_t	Against tracking
ECCM	Electronic counter-countermeasures
ECM	Electronic countermeasures
ED	Electronic defense
ELINT	Electronic intelligence
ERP	Effective radiated power = power $\times$ antenna gain
ESM	Electronic support measures
EW	Electronic warfare
F	Noise figure
f	Frequency
F_p	Propagation factor
FEBA	Forward edge of battle area
Feed	Radiating element illuminating an antenna
FFT	Fast Fourier transform: technique for constructing the Fourier transform of a signal in real time
FML	Frequency memory loop
FOV	Field of view
F_R	Pulse repetition frequency (see PRF)
FTC	Fast time constant: device for attenuating signals whose duration is longer than that of a radar pulse
G	Antenna gain
G_t	Antenna gain in transmission
Gr	Antenna gain in reception

G_j	Antenna gain of jammer antenna
G_{SL}	Antenna gain in the direction of sidelobes
gate	Time window, interval
h	Planck's constant
H	Irradiance
h_a, h_R	Height at which a radar is placed
I	Radiant intensity
I, Q	Signals respectively in phase, and in quadrature (90° out of phase) with a reference standard
IFF	Identification of friend or foe: a device for automatic discrimination between friendly and hostile platforms; also called a secondary radar insofar as it supplements a search radar, called primary
IR	Infrared
IRCM	Infrared countermeasures
I/O	Input/output (interface)
J	Radiant intensity (in infrared)
JAFF	Jammer plus chaff
J/S, JSR	Jamming-to-signal ratio (power)
J/N, JNR	Jamming-to-noise ratio (power)
K	Boltzmann's constant
K_a, K_v	Acceleration and velocity constants of a negative feedback servo
K_B	Beaufort's constant, expressing the state of the sea
k_m	Angular gradient of monopulse radar, k_s of conical-scan radar
L	Loss; a subscript denotes the type of loss
lock-on	Automatic tracking by a radar hooked onto a target
loop	A method of increasing the accuracy of an actuator (amplifier, servo) by controling the input through

	the reporting of a portion of the output
LORO	Lobe-switching on receive only (tracking radar)
LPI	Low probability of intercept
LWR	Laser warning receiver
magnetron	Electronic tube capable of generating radiofrequency pulse radiation (up to a few tens of gigahertz) for short periods (up to a few microseconds) and high power (of up to a few megawatts) with very low duty (circa 1/1000)
MAW	Missile approach warner
MIC	Microwave integrated circuit
MIPS	Millions of instructions per second
miss distance	Minimum distance between projectile or missile trajectory and target
mixer	An electronic device able to mix two input signals and output a signal whose amplitude is proportional to the product of the amplitudes of the two input signals and whose frequency is equal to the sum or difference of the two input frequencies; used to shift the frequency of a signal and as a detector of the phase shift between two signals of the same frequency
MLW	Missile launch warner
MMIC	Monolithic microwave integrated circuit
MTBF	Mean time between failures
MTD	Moving target detector
MTI	Moving target indicator: a filter used in radar systems to discriminate moving targets
N	Noise power; radiance
N_i	Number of integrated pulses
NADGE	Nato Air Defense Ground Equipment
NEP	Noise equivalent power
NJ	Noise jamming
Nodding	Oscillations in elevation of the line of sight of

tracking radars because of multipath

p	Power density; probability
P	Power; radiant flux
P_d	Probability of detection
P_{fa}	Probability of false alarm
P_j	Power of a jammer
P_k	Kill probability
P_T	Transmitted power
PLL	Phase-lock loop: a device exploiting phase differences to induce one oscillator to oscillate at the same frequency as another
Plot	Integrated radar signal indicating the presence of a target
Plume	High-temperature gas exhaust from jet or rocket motor
POI	Probability of intercept
PPI	Plan position indicator
PRF	Pulse repetition frequency of radar (*see also* F_R)
PRI	Pulse repetition interval
PW	Pulsewidth
Q	*see* I,Q
r	Rain coefficient (mm/hr)
R	Range
$\mathfrak{R}$	Responsivity
RAF	Range advance factor
RCS	Radar cross section
RF	Radio frequency
RGPI	Range gate pull-in
RGPO	Range gate pull-off
rms	Root-mean-square
Rx, RX	Receiver
RWR	Radar warning receiver

S	Power signal
SAM	Surface-to-air missile
SAR	Synthetic aperture radar
SLAR	Side looking antenna radar
SNR	Signal-to-noise ratio (power)
Sorter	Device incorporated into an ESM system for sorting received pulses and arranging them into homogeneous groups, each of which it may be possible to associate with a single emitter
Spectrum	Distribution of electromagnetic power with respect to frequency, i.e., power density (W/Hz) as a function of frequency
SSM	Surface-to-surface missile
STC	Sensitivity time control: device for control in range of the gain of a radar receiver
t	Time
T	Temperature; period of radar pulses
TEWA	Threat evaluation and weapons assignment
TOA	Time of arrival
T_{ot}	Time on target: time for which a target is illuminated by a scanning antenna
TSS	Tangential signal sensitivity
TWS	Track-while-scan (radar)
TWT	Traveling wave tube: a thermionic tube for the amplification of microwave signals that may be used for weak signals or to generate high power
Tx, TX	Transmitter
v	Voltage signal
VCO	Voltage-controled oscillator
VGPO	Velocity gate pull-off
v_r	Relative velocity
VSWR	Voltage standing-wave ratio

Index

The Artech House Radar Library

David K. Barton, *Series Editor*

Active Radar Electronic Countermeasures by Edward J. Chrzanowski

Airborne Pulsed Doppler Radar by Guy V. Morris

AIRCOVER: Airborne Radar Vertical Coverage Calculation Software and User's Manual by William A. Skillman

Analog Automatic Control Loops in Radar and EW by Richard S. Hughes

Aspects of Modern Radar by Eli Brookner, *et al.*

Aspects of Radar Signal Processing by Bernard Lewis, Frank Kretschmer, and Wesley Shelton

Bistatic Radar by Nicholas J. Willis

Detectability of Spread-Spectrum Signals by Robin A. Dillard and George M. Dillard

Electronic Homing Systems by M.V. Maksimov and G.I. Gorgonov

Electronic Intelligence: The Analysis of Radar Signals by Richard G. Wiley

Electronic Intelligence: The Interception of Radar Signals by Richard G. Wiley

EREPS: Engineer's Refractive Effects Prediction System Software and User's Manual, developed by NOSC

Handbook of Radar Measurement by David K. Barton and Harold R. Ward

High Resolution Radar by Donald R. Wehner

High Resolution Radar Cross-Section Imaging by Dean L. Mensa

Interference Suppression Techniques for Microwave Antennas and Transmitters by Ernest R. Freeman

Introduction to Electronic Defence Systems by F. Neri

Introduction to Electronic Warfare by D. Curtis Schleher

Introduction to Sensor Systems by S.A. Hovanessian

Logarithmic Amplification by Richard Smith Hughes

Modern Radar System Analysis by David K. Barton

Modern Radar System Analysis Software and User's Manual by David K. Barton and William F. Barton

Monopulse Principles and Techniques by Samuel M. Sherman

Monopulse Radar by A.I. Leonov and K.I. Fomichev

MTI and Pulsed Doppler Radar by D. Curtis Schleher

Multifunction Array Radar Design by Dale R. Billetter

Multisensor Data Fusion by Edward L. Waltz and James Llinas

Multiple-Target Tracking with Radar Applications by Samuel S. Blackman

Multitarget-Multisensor Tracking: Advanced Applications, Yaakov Bar-Shalom, ed.

Over-The-Horizon Radar by A.A. Kolosov, et al.

Principles and Applications of Millimeter-Wave Radar, Charles E. Brown and Nicholas C. Currie, eds.

Principles of Modern Radar Systems by Michel H. Carpentier

Pulse Train Analysis Using Personal Computers by Richard G. Wiley and Michael B. Szymanski

Radar and the Atmosphere by Alfred J. Bogush, Jr.

Radar Anti-Jamming Techniques by M.V. Maksimov, *et al.*

Radar Cross Section Analysis and Control by A.K. Bhattacharyya and D.L. Sengupta

Radar Cross Section by Eugene F. Knott, *et al.*

Radar Detection by J.V. DiFranco and W.L. Rubin

Radar Electronic Countermeasures System Design by Richard J. Wiegand

Radar Evaluation Handbook by David K. Barton, *et al.*

Radar Evaluation Software by David K. Barton and William F. Barton

Radar Propagation at Low Altitudes by M.L. Meeks

Radar Range-Performance Analysis by Lamont V. Blake

Radar Reflectivity Measurement: Techniques and Applications, Nicholas C. Currie, ed.

Radar Reflectivity of Land and Sea by Maurice W. Long

Radar System Design and Analysis by S.A. Hovanessian

Radar Technology, Eli Brookner, ed.

Receiving Systems Design by Stephen J. Erst

Radar Vulnerability to Jamming by Robert N. Lothes, Michael B. Szymanski, and Richard G. Wiley

RGCALC: Radar Range Detection Software and User's Manual by John E. Fielding and Gary D. Reynolds

SACALC: Signal Analysis Software and User's Guide by William T. Hardy

Secondary Surveillance Radar by Michael C. Stevens

SIGCLUT: Surface and Volumetric Clutter-to-Noise, Jammer and Target Signal-to-Noise Radar Calculation Software and User's Manual by William A. Skillman

Signal Theory and Random Processes by Harry Urkowitz

Solid-State Radar Transmitters by Edward D. Ostroff, *et al.*

Space-Based Radar Handbook, Leopold J. Cantafio, ed.

Spaceborne Weather Radar by Robert M. Meneghini and Toshiaki Kozu

Statistical Theory of Extended Radar Targets by R.V. Ostrovityanov and F.A. Basalov

The Scattering of Electromagnetic Waves from Rough Surfaces by Peter Beckmann and Andre Spizzichino

VCCALC: Vertical Coverage Plotting Software and User's Manual by John E. Fielding and Gary D. Reynolds